GN00870891

DNA and RNA Profiling in Human Blood

METHODS IN MOLECULAR BIOLOGY™

John M. Walker, Series Editor

502. Bacteriophages: *Methods and Protocols, Volume 2: Molecular and Applied Aspects*, edited by *Martha R. J. Clokie and Andrew M. Kropinski 2009*

501. Bacteriophages: *Methods and Protocols, Volume 1: Isolation*, Characterization, and Interactions, edited by *Martha R. J. Clokie and Andrew M. Kropinski 2009*

496. DNA and RNA Profiling in Human Blood: *Methods and Protocols*, edited by *Peter Bugert, 2009*

493. Auditory and Vestibular Research: *Methods and Protocols*, edited by *Bernd Sokolowski, 2009*

489. Dynamic Brain Imaging: *Methods and Protocols*, edited by *Fahmeed Hyder, 2009*

485. HIV Protocols: *Methods and Protocols*, edited by *Vinayaka R. Prasad and Ganjam V. Kalpana, 2009*

484. Functional Proteomics: *Methods and Protocols*, edited by *Julie D. Thompson, Christine Schaeffer-Reiss, and Marius Ueffing, 2008*

483. Recombinant Proteins From Plants: *Methods and Protocols*, edited by *Loïc Faye and Veronique Gomord, 2008*

482. Stem Cells in Regenerative Medicine: *Methods and Protocols*, edited by *Julie Audet and William L. Stanford, 2008*

481. Hepatocyte Transplantation: *Methods and Protocols*, edited by *Anil Dhawan and Robin D. Hughes, 2008*

480. Macromolecular Drug Delivery: *Methods and Protocols*, edited by *Mattias Belting, 2008*

479. Plant Signal Transduction: *Methods and Protocols*, edited by *Thomas Pfannschmidt, 2008*

478. Transgenic Wheat, Barley and Oats: *Production and Characterization Protocols*, edited by *Huw D. Jones and Peter R. Shewry, 2008*

477. Advanced Protocols in Oxidative Stress I, edited by *Donald Armstrong, 2008*

476. Redox-Mediated Signal Transduction: *Methods and Protocols*, edited by *John T. Hancock, 2008*

475. Cell Fusion: *Overviews and Methods*, edited by *Elizabeth H. Chen, 2008*

474. Nanostructure Design: *Methods and Protocols*, edited by *Ehud Gazit and Ruth Nussinov, 2008*

473. Clinical Epidemiology: *Practice and Methods*, edited by *Patrick Parfrey and Brendon Barrett, 2008*

472. Cancer Epidemiology, Volume 2: *Modifiable Factors*, edited by *Mukesh Verma, 2008*

471. Cancer Epidemiology, Volume 1: *Host Susceptibility Factors*, edited by *Mukesh Verma, 2008*

470. Host-Pathogen Interactions: *Methods and Protocols*, edited by *Steffen Rupp and Kai Sohn, 2008*

469. Wnt Signaling, Volume 2: *Pathway Models*, edited by *Elizabeth Vincan, 2008*

468. Wnt Signaling, Volume 1: *Pathway Methods and Mammalian Models*, edited by *Elizabeth Vincan, 2008*

467. Angiogenesis Protocols: *Second Edition*, edited by *Stewart Martin and Cliff Murray, 2008*

466. Kidney Research: *Experimental Protocols*, edited by *Tim D. Hewitson and Gavin J. Becker, 2008*

465. Mycobacteria, *Second Edition*, edited by *Tanya Parish and Amanda Claire Brown, 2008*

464. The Nucleus, Volume 2: *Physical Properties and Imaging Methods*, edited by *Ronald Hancock, 2008*

463. The Nucleus, Volume 1: *Nuclei and Subnuclear Components*, edited by *Ronald Hancock, 2008*

462. Lipid Signaling Protocols, edited by *Banafshe Larijani, Rudiger Woscholski, and Colin A. Rosser, 2008*

461. Molecular Embryology: *Methods and Protocols, Second Edition*, edited by *Paul Sharpe and Ivor Mason, 2008*

460. Essential Concepts in Toxicogenomics, edited by *Donna L. Mendrick and William B. Mattes, 2008*

459. Prion Protein Protocols, edited by *Andrew F. Hill, 2008*

458. Artificial Neural Networks: *Methods and Applications*, edited by *David S. Livingstone, 2008*

457. Membrane Trafficking, edited by *Ales Vancura, 2008*

456. Adipose Tissue Protocols, Second Edition, edited by *Kaiping Yang, 2008*

455. Osteoporosis, edited by *Jennifer J. Westendorf, 2008*

454. SARS- and Other Coronaviruses: *Laboratory Protocols*, edited by *Dave Cavanagh, 2008*

453. Bioinformatics, Volume 2: *Structure, Function, and Applications*, edited by *Jonathan M. Keith, 2008*

452. Bioinformatics, Volume 1: *Data, Sequence Analysis, and Evolution*, edited by *Jonathan M. Keith, 2008*

451. Plant Virology Protocols: *From Viral Sequence to Protein Function*, edited by *Gary Foster, Elisabeth Johansen, Yiguo Hong, and Peter Nagy, 2008*

450. Germline Stem Cells, edited by *Steven X. Hou and Shree Ram Singh, 2008*

449. Mesenchymal Stem Cells: *Methods and Protocols*, edited by *Darwin J. Prockop, Douglas G. Phinney, and Bruce A. Brunnell, 2008*

448. Pharmacogenomics in Drug Discovery and Development, edited by *Qing Yan, 2008*

447. Alcohol: *Methods and Protocols*, edited by *Laura E. Nagy, 2008*

446. Post-translational Modifications of Proteins: *Tools for Functional Proteomics, Second Edition*, edited by *Christoph Kannicht, 2008*

445. Autophagosome and Phagosome, edited by *Vojo Deretic, 2008*

444. Prenatal Diagnosis, edited by *Sinhue Hahn and Laird G. Jackson, 2008*

443. Molecular Modeling of Proteins, edited by *Andreas Kukol, 2008*

METHODS IN MOLECULAR BIOLOGY ™

DNA and RNA Profiling in Human Blood

Methods and Protocols

Edited by

Peter Bugert, PhD

Institute of Transfusion Medicine and Immunology, Mannheim, Germany

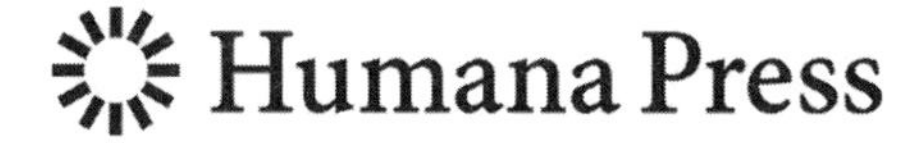

Editor
Peter Bugert
Institute of Transfusion Medicine and Immunology
Heidelberg University, Medical Faculty of Mannheim
Mannheim, Germany
p.bugert@iti-ma.blutspende.de

Series Editor
John M. Walker
University of Hertfordshire
Hatfield, Herts
UK

ISSN: 1064-3745 e-ISSN: 1940-6029
ISBN: 978-1-934115-93-0 e-ISBN: 978-1-59745-553-4
DOI 10.1007/978-1-59745-553-4

Library of Congress Control Number: 2008937584

Printed on acid-free paper

springer.com

Preface

Blood samples are widely used as biological specimens for diagnostic or research purposes. There are many examples of important disease markers which can be investigated using peripheral blood. In the context of blood transfusion and immunohematology, blood itself is the target of investigation with regard to the determination of blood groups and the screening for antibodies. Furthermore, blood samples represent the main source of genetic material, i.e., DNA and RNA, for analyzing gene mutations, polymorphisms, and expression at the molecular level.

The development of novel bioanalytical technologies for complex and quantitative molecular analysis of DNA, RNA, proteins, and cell functions led to the introduction of '-omics' terms such as genomics, transcriptomics or proteomics. These molecular profiling approaches have opened up a broad field of research and may help to identify further disease markers in blood. Recently developed techniques, such as microarrays or bead arrays, represent the basis of high-throughput multiplex DNA typing and RNA profiling. Such methods have been applied already to blood cells or plasma and many of them were adapted to the special characteristics of blood cells. Thus, protocols have been developed to achieve diagnostic systems in the fields of genotyping for blood cell antigens including Human Leukocyte Antigens (HLA) and blood groups. The diagnostic systems are of great importance in blood transfusion and organ transplantation. Furthermore, special protocols were adapted to particularities of certain blood cell types such as platelets or reticulocytes in order to address questions in the field of gene expression analysis.

The aim of DNA AND RNA PROFILING IN HUMAN BLOOD is to bring together established, standardized, and recently developed protocols for complex and/or high-throughput DNA and RNA profiling. This book consists of two sections, Part I: *DNA Profiling for Blood Cell Antigens*, and Part II: *RNA Profiling in Blood Cells*. In Part I, a number of methods and protocols describe high-throughput multiplex approaches for genotyping of various blood cell antigens (*see* **Chapters 1–5, 8,** and **9**). Blood grouping by DNA typing also includes a step-by-step protocol for prenatal RhD determination using of maternal plasma (*see* **Chapter 11**). Other DNA protocols describe modern techniques for SNP typing other than blood cell antigen SNPs (*see* **Chapters 6, 7, 10** and **12**) that may serve as examples to establish protocols for the own purposes.

Part II is focused on RNA profiling methods and protocols that have been adapted to the special characteristics of certain blood cell types such as platelets (*see* **Chapters 16–18**), reticulocytes (*see* **Chapter 20**) or megakaryocytes (*see* **Chapter 19**). Furthermore, methods and protocols are included to describe recently developed techniques which have been applied to blood samples (see **Chapters 13, 14, 21,** and **22**) or which may be applied to RNA samples of any type of biological source (see **Chapters 12** and **15**).

This book summarizes contributions from leading international experts in the fields of DNA and RNA profiling. As editor of this volume, I am very grateful indeed to themfor their willingness to provide an insight into their knowledge and to provide the detailed step-by-step protocols. I also wish to thank Steffanie Bickelhaupt and Daniela Griffiths for considerable editorial and secretarial assistance.

Peter Bugert

Contents

Contributors

PARVIZ AHMAD-NEJAD, MD • *Institute for Clinical Chemistry, University of Heidelberg, Medical Faculty Mannheim, Mannheim, Germany*
FERNANDO ARAUJO, MD, PHD • *Molecular Biology Center, Department of Transfusion Medicine and Blood Center, Medicine Faculty of Oporto University and S. João Hospital, Portugal*
WADIE F. BAHOU, MD • *Division of Hematology/Oncology, Department of Medicine, State University of New York at Stony Brook, Stony Brook, NY, USA*
JULIAN BOEHM • *Division of Immunogenetics, Department of Pediatrics, Rangos Research Center Children's Hospital of Pittsburgh, University of Pittsburgh School of Medicine, Pittsburgh, PA, USA*
BÉATRICE BONAFOUX, MD, PHD • *Laboratoire de Biologie, Centre Hospitalier, Ales, France*
SUSAN BORTOLIN, PHD • *Luminex Molecular Diagnostics, Toronto, ON, Canada*
COLIN J. H. BRENAN, PHD • *BioTrove Inc., Woburn, MA, USA*
THÉRÈSE COMMES, PHD • *Institut de Génétique Humaine, Montpellier, France*
CLARISSA CONSOLANDI, PHD • *Institute for Biomedical Technologies, National Research Council, Milano, Italy*
GEOFF DANIELS, PHD, F.R.C. PATH. • *International Blood Group Reference Laboratory, NHS Blood and Transplant, Bristol, UK*
SVENJA DEBEY-PASCHER • *Department for Genomics, Life and Medical Sciences, University of Bonn, Bonn, Germany*
GREGORY A. DENOMME, PHD • *Research & Development, Canadian Blood Service, Toronto, Canada*
JOHN J. DUNN, PHD • *Biology Department, Brookhaven National Laboratory, Upton, NY, USA*
EIMEAR DUNNE • *Molecular and Cellular Therapeutics, Royal College of Surgeons in Ireland, Dublin, Ireland*
LOUISE EDVARDSSON • *Division of Hematology and Transfusion Medicine, Department of Laboratory Medicine, Biomedical Center, Lund University, Lund, Sweden*
DANIELA EGGLE • *Department for Genomics, Life and Medical Sciences, University of Bonn, Bonn, Germany*
GIANMARCO FERRI, PHD • *Department of Diagnostic and Laboratory Service and Legal Medicine, Section of Legal Medicine, University of Modena Reggio Emilia, Italy*
KIRSTIN FINNING, PHD • *International Blood Group Reference Laboratory, NHS Blood and Transplant, Bristol, UK*
RAMIRO GARZON, MD • *Department of Medicine, Division of Hematology and Oncology, Comprehensive Cancer Center, The Ohio State University, USA*
DMITRI V. GNATENKO, PHD • *Division of Hematology/Oncology, Department of Medicine, State University of New York at Stony Brook, Stony Brook, NY, USA*
ANDREW HILLMANN, PHD • *Regenerative Medicine Institute (REMEDI), National University of Ireland, Galway, Ireland*

JAMES HURLEY • *BioTrove Inc., Woburn, MA, USA*
DERMOT KENNY, MD • *Molecular and Cellular Therapeutics, Royal College of Surgeons in Ireland, Dublin, Ireland*
CHARLES H. LIN • *Illumina, Inc., San Diego, CA, USA*
YING LU • *Division of Immunogenetics, Department of Pediatrics, Rangos Research Center Children's Hospital of Pittsburgh, University of Pittsburgh School of Medicine, Pittsburgh, PA, USA*
PETE MARTIN • *International Blood Group Reference Laboratory, NHS Blood and Transplant, Bristol, UK*
SIMON E. MCBRIDE • *Department of Platelet Immunology, NHS Blood and Transplant, Cambridge, UK*
TIMOTHY K. MCDANIEL • *Illumina, Inc., San Diego, CA, USA*
JENS MÜLLER, MSc • *Institute of Experimental Haematology and Transfusion Medicine, University of Bonn, Germany*
LYNN NICHOL • *Division of Immunogenetics, Department of Pediatrics, Rangos Research Center Children's Hospital of Pittsburgh, University of Pittsburgh School of Medicine, Pittsburgh, PA, USA*
TOR OLOFSSON • *Division of Hematology and Transfusion Medicine, Department of Laboratory Medicine, Biomedical Center, Lund University, Lund, Sweden*
SUSI PELOTTI, MD • *Department of Medicine and Public Health, Section of Legal Medicine, University of Bologna, Italy*
BERND PÖTZSCH, MD • *Institute of Experimental Haematology and Transfusion Medicine, University of Bonn, Germany*
MANGALATHU S. RAJEEVAN, PHD • *National Center for Zoonotic, Vector-Borne, and Enteric Diseases (NCZVED), Centers for Disease Control and Prevention (CDC), Atlanta, GA, USA*
STEVEN RINGQUIST • *Division of Immunogenetics, Department of Pediatrics, Rangos Research Center Children's Hospital of Pittsburgh, University of Pittsburgh School of Medicine, Pittsburgh, PA, USA*
DOUGLAS ROBERTS • *BioTrove Inc., Woburn, MA, USA*
NINA ROLF, MD • *Children's Hospital, University Hospital Carl Gustav Carus, Technical University Dresden, Dresden, Germany*
JUTTA M. ROX, MD • *Institute for Transplantation Diagnostics and Cellular Therapeutics, University of Düsseldorf, Germany*
ANGELIKA SCHEDEL, MSc • *Institute of Transfusion Medicine and Immunology, University of Heidelberg, Medical Faculty of Mannheim, Mannheim, Germany*
JOACHIM L. SCHULTZE • *Department of Genomics, Life and Medical Sciences, University of Bonn, Bonn, Germany*
JOHN SCHWEDES, MS • *University DNA Microarray Facility, Office of Scientific Affairs, School of Medicine, State University of New York at Stony Brook, Stony Brook, NY, USA*
RICHARD SHEN • *Illumina, Inc., San Diego, CA, USA*
MARTIN STEINAU, PHD • *National Center for Zoonotic, Vector-Borne, and Enteric Diseases (NCZVED), Centers for Disease Control and Prevention (CDC), Atlanta, GA, USA*
MARYSE ST-LOUIS, PHD • *Research & Development, Héma-Québec, Québec, Canada*
PATRICK STORDEUR, PHD • *Départment d'Immunobiologie-Hémobiologie- Transfusion, Hôpital Erasme, Université Libre de Bruxelles, Belgium*

MASSIMO TRUCCO • *Division of Immunogenetics, Department of Pediatrics, Rangos Research Center Children's Hospital of Pittsburgh, University of Pittsburgh School of Medicine, Pittsburgh, PA, USA*
JOANNE M. YEAKLEY • *Illumina, Inc., San Diego, CA, USA*

Part I

DNA Profiling for Blood Cell Antigens

Chapter 1

PCR–ELISA for High-Throughput Blood Group Genotyping

Maryse St-Louis

Abstract

During the last decade, blood bank specialists have shown an increased interest in molecular analyses to complement serology work in determining blood group antigens. To efficiently respond to the numerous demands made for hemolytic disease of the newborn cases and polytransfused patients, we designed an inexpensive colorimetric high-throughput method to genotype several blood group antigens rapidly. Three simple steps are required to perform this technique: genomic DNA extraction, PCR amplification, and amplicon detection by a microplate ELISA. The 96-well plate format facilitates the manipulations and enables the analysis of multiple samples at once or the analysis of multiple antigens for fewer samples.

The most common and clinically relevant minor blood group antigens were adapted to this method and are described in this work: Rh (D, C, c, E, e), Kell (K, k), Duffy (Fy^a, Fy^b), and Kidd (Jk^a, Jk^b). Other blood group antigens could be easily tested this way as long as their molecular basis is well established.

Key words: Blood groups, genotyping, PCR–ELISA, microplate, biotin, DIG.

1. Introduction

Red blood cell antigens identification by agglutination-based assays remains the gold standard in the field of blood banks. However, molecular biology analyses have taken up more space over the years to complement the serology work especially for complex cases. A constant increase for DNA-based analyses has been observed since. In order to efficiently respond to these demands, we developed an inexpensive colorimetric high-throughput method in a microtiter plate format to detect DIG-labeled PCR products by specific hybridization to a capture oligonucleotide bound to a microplate well *(1)*.

This PCR–ELISA technique is done in three easy steps: genomic DNA extraction, PCR amplification, and detection by an

Peter Bugert (ed.), *DNA and RNA Profiling in Human Blood: Methods and Protocols, vol. 496*

DOI 10.1007/978-1-59745-553-4_1 Springerprotocols.com

ELISA. The DNA extraction is performed by using commercially available reagents. During the PCR amplification, a small portion of the regular nucleotides mix is replaced by DIG-labeled dUTP, resulting in DIG-labeled amplicons. The detection is done by ELISA in a 96-well plate. The denatured PCR product is added to the well already coated with streptavidin to which bind the 5′-biotinylated capture oligonucleotide specific to our target DNA. An hybridization will result between the PCR product and the capture oligonucleotide. After removal of non-hybridized PCR products, the specifically bound products are detected by an anti-DIG antibody coupled to a peroxidase molecule which is then revealed by the enzyme substrate. The resulting coloration is read on a regular plate reader.

This method presents several advantages: high specificity (specific-capture oligonucleotide), high sensitivity (up to 100 times more sensitive than ethidium bromide stained gels), and flexibility (capture oligonucleotide can be changed to detect other amplicons). This technique and variations of it have been used by many groups to detect genes in different types of organisms such as parvovirus B19 *(2)*, hepatitis B virus *(3)*, MRSA *(4)*, *Plasmodium falciparum (5)*, CMV *(6)*, HIV *(7)*, and *Aspergillus fumigatus (8)*. In 2003, this methodology was adapted by our group to detect the presence of West Nile virus in blood donor plasma before any commercial assays were made available *(9)*.

2. Materials

2.1. Genomic DNA Extraction

1. Peripheral blood sample of about 7 mL is collected on EDTA and stored in aliquots at –20°C until processed (see **Note 1**).
2. QIAamp DNA blood mini kit (Qiagen, Mississauga, ON, Canada) (see **Note 2**).

2.2. PCR Amplification

1. All primers for the PCR amplification were synthesized by Invitrogen (Burlington, ON, Canada) or Sigma-Genosys (Oakville, ON, Canada). **Table 1.1** lists the primers used in each assay, including primers for internal controls along with their concentration. Internal controls are essential in this type of assay to monitor the amplification success. Primer pools are prepared and stored in aliquots at –35°C before use (*see* **Note 3**). The volumes are calculated to have the required concentration in 5 μL of primer pools for a 25 μL reaction.
2. AmpliTaq Gold (Applied Biosystems, Foster City, CA) is used (*see* **Note 4**) along with its 1x GeneAmp PCR Buffer II and 1.5–2.5 mM $MgCl_2$. All reagents are stored at –20°C, except $MgCl_2$, which is kept at 4°C to avoid ice crystals formation.

Table 1.1
PCR amplification primers (reproduced in part from ref. 1 with permission from Blackwell publishing)

Antigen	Primer	Sequence (5′ to 3′)	Quantity (ng)	Product	Size (bp)
D	53 F	GTG GAT GTT CTG GCC AAG TT	50	Exon 5	157
	54 R	cac CTT GCT GAT CTT ACC	50		
	58 F	aa cag GTT TGC TCC TAA ATA TT	50	Exon 9	71
	59 R	AAA CTT GGT CAT CAA AAT ATT TAA CCT	50		
	60 F	CCT TCC TGG GCA TGG AGT CCT G	5	β-Actin	200
	61 R	GGA GCA ATG ATC TTG ATC TTC	5		
C/c	63 F	ca ggg cca cca ttt gaa	50	Intron 2	360
	64 R	gaa cat gcc act tca ctc cag	50		
	65 F	TCG GCC AAG ATC TGA CCG	15	Exon 2	177
	66 R	TGA TGA CCA CCT TCC CAG G	15		
	90 F	TGC CTT CCC AAC CAT TCC CTT A	15	HGH	432
	91 R	CCA CTC ACG GAT TTC TGT TGT GTT TC	15		
E	67 F	CCA AGT GTC AAC TCT **C**	37.5	Exon 5	141
	72 R	CAT GCT GAT CTT CCT	37.5		
	69 F	ACA GAC TAC CAC ATG AAC	15	*RHD/RHCE* exon 4	94
	70 R	GCT TTG GCA GGC ACC AGG CCA C	15		
e	71 F	CCA AGT GTC AAC TCT **G**	37.5	Exon 5	141
	72 R	CAT GCT GAT CTT CCT	37.5		
	69 F	ACA GAC TAC CAC ATG AAC	15	*RHD/RHCE* exon 4	94
	70 R	GCT TTG GCA GGC ACC AGG CCA C	15		
K	112 F	t ctc tct cct tta aag CTT GGA	100	Exon 6	83
	93 R	TCA GAA GTC TCA GC**A**	100		
	60 F	CCT TCC TGG GCA TGG AGT CCT G	10	β-Actin	200
	61 R	GGA GCA ATG ATC TTG ATC TTC	10		

(continued)

Table 1.1 (continued)

Antigen	Primer	Sequence (5′ to 3′)	Quantity (ng)	Product	Size (bp)
k	112 F	t ctc tct cct tta aag CTT GGA	100	Exon 6	83
	92 R	TCA GAA GTC TCA GC**G**	100		
	60 F	CCT TCC TGG GCA TGG AGT CCT G	10	β-Actin	200
	61 R	GGA GCA ATG ATC TTG ATC TTC	10		
Fy[a]	158 F	ctc att agt cct tgg ctc tta t	132	Exon 1	711
	159 R	AGC TGC TTC CAG GTT GGC A**C**	122		
	90 F	TGC CTT CCC AAC CAT TCC CTT A	20	HGH	432
	91 R	CCA CTC ACG GAT TTC TGT TGT GTT TC	20		
Fy[b]	158 F	ctc att agt cct tgg ctc tta t	132	Exon 1	711
	160 R	AGC TGC TTC CAG GTT GGC A**T**	122		
	90 F	TGC CTT CCC AAC CAT TCC CTT A	20	HGH	432
	91 R	CCA CTC ACG GAT TTC TGT TGT GTT TC	20		
Jk[a]	162 F	GTC TTT CAG CCC CAT TTG CG**G**	63	Exon 4	131
	161 R	GCA CAG CCA AGA GCC AGG AGG	63		
	90 F	TGC CTT CCC AAC CAT TCC CTT A	15	HGH	432
	91 R	CCA CTC ACG GAT TTC TGT TGT GTT TC	15		
Jk[b]	163 F	GTC TTT CAG CCC CAT TTG CGA	63	Exon 4	131
	161 R	GCA CAG CCA AGA GCC AGG AGG	63		
	90 F	TGC CTT CCC AAC CAT TCC CTT A	15	HGH	432
	91 R	CCA CTC ACG GAT TTC TGT TGT GTT TC	15		

Primer orientation: F = forward; R = reverse; lowercase letters: intronic sequences; bold letters: polymorphisms; quantity for a 25 μL reaction; HGH = human growth hormone.

3. Digoxigenin-dUTP (1 mM) (Roche Diagnostics, Laval, QC, Canada) stored in aliquots at –20°C from which 5 μL is added to the 100 μL dNTPs solution described below.
4. dNTPs (1 mM) are prepared from a 10 mM stock (Invitrogen) and stored in aliquots at –20°C. The dTTP concentration is reduced to 0.95 mM to be replaced by an equivalent concentration of digoxigenin-dUTP (0.05 mM) (ratio of dUTP/dTTP is 1:19).

2.3. PCR–ELISA

1. Flat-bottom 96-well Immulon II plates (Dynatech Laboratories, Chantilly, VA) (*see* **Note 5**).
2. Solution of 1x PBS: 10 mM phosphate buffer, pH 7.4; 136 mM NaCl; Dubelcco's PBS), stored at 4°C (Invitrogen).
3. Solution of 1x PBS–0.01% thimerosal (Sigma-Aldrich, St. Louis, MO, USA), stored at 4°C light protected.
4. Solution of 1x PBS–0.25% casein–0.01% thimerosal (casein: EM Science, Gibbstown, NJ, USA), stored at 4°C.
5. Solution of 250 ng streptavidin (Jackson ImmunoResearch, West Grove, PA, USA) is prepared from a stock solution of 1 mg/mL diluted in 1x PBS–0.01% thimerosal. Stored in aliquots at 4°C.
6. Carbonate buffer (115 mM), pH 9.6: 31 mM Na_2CO_3; 84 mM $NaHCO_3$; 0.02% NaN_3 (Sigma-Aldrich). The pH is adjusted with concentrated NaOH and stored at 4°C.
7. All biotinylated capture oligonucleotides were synthesized by Invitrogen or Sigma-Genosys (*see* **Table 1.2**). They hybridize to the middle part of the amplicon (*see* **Note 6**). A stock solution is prepared at 50 μM and stored in aliquots at –20°C.
8. NaOH (1 N)-0.5 mg/mL thymol blue is prepared from 10 N NaOH and 10 mg/mL thymol blue (Sigma-Aldrich) (*see* **Note 7**) and stored at room temperature.
9. NaH_2PO_4 (0.5 M) to neutralize NaOH (stored at room temperature).
10. Solution of 150 U/mL anti-DIG-POD Fab fragment (Roche Diagnostics) is prepared in H_2O and stored in aliquots at 4°C away from light (*see* **Note 8**).
11. Ready to use peroxidase substrate TMB (ScyTek Laboratories, Logan, UT) Stored at 4°C away from light.
12. H_2SO_4 (1 N) (stop solution) prepared from 10 N H_2SO_4 and stored at 4°C.

3. Methods

This technique is based on PCR amplification of specific single nucleotide polymorphisms (SNPs). Special care should be given to the preparation of the PCR mix to avoid contamination. We

Table 1.2
PCR–ELISA capture oligonucleotides (reproduced from ref. 1 with permission from Blackwell publishing)

Antigen	Primer	Sequence (5′ to 3′)	Specificity
D	48	Biotin-TGT TCA ACA CCT ACT ATG CT	Exon 5
	125	Biotin-AAT ATG GAA AGC ACC TCA TG	Exon 9
	126	Biotin-AAG ACC TGT ACG CCA ACA CA	β-Actin
C/c	171	Biotin-$(T)_{12}$ TTC AGG GTG CCC TTT GTC ACT T	Intron 2 (C)
	170	Biotin-$(T)_{12}$ TTC ACC TCA AAT TTC CGG AGA C	Exon 2 (c)
	173	Biotin-$(T)_{12}$ CTC AGA GTC TAT TCC GAC ACC	HGH
E/e	48	Biotin-TGT TCA ACA CCT ACT ATG CT	Exon 5
	127	Biotin-TCT ACG TGT TCG CAG CCT AT	*RHD/RHCE* exon 4
K/k	180	Biotin-$(T)_{12}$ CGC ATC TCT GGT AAA TGG AC	Exon 6
	126	Biotin-AAG ACC TGT ACG CCA ACA CA	β-Actin
Fy^a/Fy^b	189	Biotin-$(T)_{12}$ GAT TCC TTC CCA GAT GGA GA	Exon 1
	173	Biotin-$(T)_{12}$ CTC AGA GTC TAT TCC GAC ACC	HGH
Jk^a/Jk^b	190	Biotin-$(T)_{12}$ AAT GTT CAT GGC GCT CAC CT	Exon 4
	173	Biotin-$(T)_{12}$ CTC AGA GTC TAT TCC GAC ACC	HGH

HGH = human growth hormone.

usually work in separate rooms for the pre-PCR, PCR, and post-PCR steps. Each room contains dedicated equipment and materials to reduce even more the risk of contaminations. No DNA is allowed in the pre-PCR room.

3.1. Genomic DNA Extraction

1. Extract DNA from whole blood according to the manufacturer‘s specifications.
2. Measure the DNA concentration by spectrophotometry or other means.
3. Store purified DNA at –35°C.

3.2. PCR Amplification

1. Amplify DNA following a classic PCR procedure, except that nucleotides include digoxigenin-dUTP to label the amplicons. **Table 1.1** lists the oligonucleotides and their concentration, **Table 1.3** describes the reaction mix and **Table 1.4**, the thermocycler program profile for each assay (*see* **Note 9**). Control DNA with known phenotypes are essential to set up the assay.

3.3. PCR–ELISA

1. Coat Immulon II plate by adding 100 μL of 2.5 μg/mL streptavidin diluted in 115 mM carbonate buffer at pH 9.6 to each well. Cover plates with adhesive lid and incubate at 4°C overnight or until later use (*see* **Note 10**).

Table 1.3
PCR reaction mix for a 25 μL reaction

Materials	D	C/c	E/e	K/k	Fy^a/Fy^b	Jk^a/Jk^b
H_2O	Varies	Varies	Varies	Varies	Varies	Varies
10x PCR Buffer II (μL)			2.50			
25 mM $MgCl_2$ (μL)	2.50	2.50	1.50	3.00	1.50	1.50
1 mM dNTPs-DIG (μL)			1.25			
Primer pool[a] (μL)			5			
AmpliTaq Gold (μL)	0.50	0.25	0.50	0.50	0.25	0.25
DNA (~50 ng) or H_2O	Varies	Varies	Varies	Varies	Varies	Varies
Final volume (μL)			25			

[a]Primer pools are prepared to genotype RhE, Rhe, K, k, Fy^a, Fy^b, Jk^a, and Jk^b separately as presented in **Table 1.1**.

Table 1.4
Thermocycler program profile for each assay

Program profile	D	C/c	E/e	K/k	Fy^a/Fy^b	Jk^a/Jk^b
Initial denaturation[a]			94°C/9 min			
32 cycles						
Denaturation			94°C/30 s			
Annealing	55°C/30 s	65°C/30 s	49°C/30 s	58°C/30 s	64°C/30 s	61°C/30 s
Extension			72°C/30 s			
Final extension			72°C/5 min			
Holding			4°C			

[a]Recommended to activate AmpliTaq Gold. It may be different for other enzymes. It is possible to use a gradient thermocycler to amplify multiple alleles in one run (*see* **Note 9**).

2. Wash wells six times with 1x PBS manually or with an automated plate washer. Tap the plate on a dry, lint-free, absorbent cloth.
3. Block wells by adding 300 μL of 1x PBS–0.25% casein–0.01% thimerosal and incubate for at least 30 min at 37°C (*see* **Note 11**).
4. Empty the wells by inverting the plate.
5. Add appropriate capture oligonucleotide to each well (20 ng/100 μL 1X PBS–0.25% casein–0.01% thimerosal) and incubate for at least 1 h at 37°C (*see* **Table 1.2** for the list of capture oligonucleotides).
6. Repeat washes as described at step 2.

7. To each well, add 48 μL 1× PBS–0.25% casein–0.01% thimerosal.
8. Add 2 μL of PCR products to each specific well (see **Note 12**).
9. Add 10 μL of 1 N NaOH–0.5 mg/ml thymol blue to each well to denature PCR products. Mix well until a uniform blue coloration is obtained. Incubate for 10 min at room temperature.
10. Add 30 μL of 0.5 M NaH_2PO_4 to each well to neutralize NaOH. Mix well until a uniform yellow coloration is obtained. Incubate for at least 1 h at 37°C (RhD, RhE, Rhe, K, k) or 55°C (RhC, Rhc, Fy^a, Fy^b, Jk^a, Jk^b) (*see* **Note 13**).
11. Repeat washes as described at step 2.
12. Add to each well 100 μL of anti-DIG-POD diluted 1:1000 (0.15 U/mL) to 1:5000 (0.03 U/mL) in 1× PBS–0.25% casein–0.01% thimerosal. Incubate at least 1 h at 37°C away from light.
13. Repeat washes as described at step 2.
14. Add to each well 100 μL of TMB to reveal the results. Incubate 30 min at room temperature away from light. Positive wells should show a yellow coloration.
15. Add to each well 100 μL of 1 N H_2SO_4 to stop the reaction.
16. Read the plate on a plate reader at 450 nm with a reference at 630 nm.

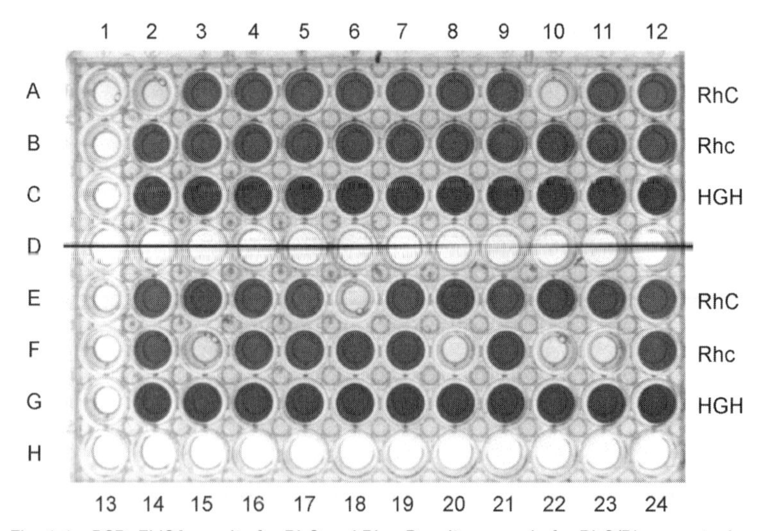

Fig. 1.1. PCR–ELISA results for RhC and Rhc. Results example for RhC/Rhc genotyping. Rows A and E: RhC; rows B and F: Rhc; and rows C and G: HGH (internal control). Rows D and H were empty. Samples 1 and 13: water control; samples 2, 10, and 18: Rhc/Rhc, samples 3–9, 11–12, 14, 16–17, 19, 21, and 24: RhC/Rhc; samples 15, 20, 22–23: RhC/RhC. Internal control is positive for every DNA sample.

Table 1.5
PCR–ELISA mean OD values (reproduced in part from ref. 1 with permission from Blackwell publishing)

Antigen	Water control	Antigen-negative samples	Antigen-positive samples	Ratio (positive/negative)
D	5 (0.050 ± 0.040)	15 (0.187 ± 0.167)	85 (3.252 ± 0.340)	17.4
C	10 (0.086 ± 0.065)	30 (0.183 ± 0.097)	70 (3.478 ± 0.102)	19.0
c	7 (0.115 ± 0.016)	20 (0.239 ± 0.098)	80 (3.500 ± 0.000)	14.6
E	10 (0.050 ± 0.036)	70 (0.047 ± 0.027)	30 (3.186 ± 0.448)	67.2
e	10 (0.078 ± 0.023)	4 (0.082 ± 0.027)	96 (3.496 ± 0.039)	42.6
K	10 (0.075 ± 0.047)	95 (0.093 ± 0.045)	5 (2.837 ± 0.615)	30.5
k	9 (0.076 ± 0.033)	0 (no samples)	100 (3.494 ± 0.052)	–
Fy^a	10 (0.084 ± 0.045)	40 (0.223 ± 0.089)	60 (3.420 ± 0.229)	15.3
Fy^b	10 (0.078 ± 0.046)	22 (0.181 ± 0.084)	78 (3.302 ± 0.418)	18.2
Jk^a	10 (0.057 ± 0.032)	30 (0.280 ± 0.104)	70 (3.484 ± 0.044)	12.4
Jk^b	10 (0.093 ± 0.043)	24 (0.598 ± 0.246)	76 (3.482 ± 0.047)	5.8

Data presented as number of samples (mean OD ± SD).

3.4. Results Interpretation

The optical density (OD) results obtained for the water controls should be low (<0.125, *see* **Note 14**) and the OD for the internal control should be strongly positive for each DNA sample (often "over"). To consider a sample positive, its OD should be high (>2.5). A negative sample should have a low OD. **Figure 1.1** illustrates a result example for RhC and Rhc genotyping. Positive wells are dark while negative wells are light or clear. **Table 1.5** shows the mean OD values obtained for each genotyping assay. The ratio (positive/negative) is also indicated. These ratios varied from 5.8 (Jk^b) to 67.2 (RhE).

4. Notes

1. EDTA tubes work fine for DNA preparation as well as sodium citrate. Heparin tubes are to be avoided, since heparin inhibits the polymerase.
2. Qiagen was used for DNA preparation, but any other renowned manufacturer should be appropriate.
3. We have seen occasionally problems when using primer pools freshly prepared. We now store every pool at –35°C at least overnight before use.
4. All our PCR assays were developed with AmpliTaq Gold. No other enzymes were used. The AmpliTaq Gold has been

modified chemically and is stable at room temperature. When preparing the PCR mix, all reagents were kept at room temperature. Other conditions could be necessary if you use a different DNA polymerase. Read carefully the packing insert provided with the enzyme. The specific SNP amplification could not work properly if using a more basic type of polymerase. During development, make sure the amplification is valid (e.g., agarose gel) before continuing to the ELISA step.

5. Costar and Crosslink plates were compared to Immulon II. Costar did not bind any capture oligonucleotides. Crosslink were great at binding, but the cost was considered to high for this type of technique.
6. A T-linker had to be added in 5′ of the capture oligonucleotide for some assays to increase the signal to noise ratio.
7. Thymol blue is a pH indicator. It is optional.
8. Do not freeze this antibody and keep away from light.
9. To accelerate the procedure, a gradient thermocycler should be used. For example, to genotype RhE, Rhe, RhD, K, and k in one PCR run, set the annealing temperature at 49°C with a gradient of 9°C. The temperature should be: row 1, 49.0°C (RhE); row 2, 49.3°C (Rhe); row 8, 55.2°C (RhD); row 11, 57.8°C (K); and row 12, 58.1°C (k).
10. Coated plates were stored at 4°C for as long as 3 months with same quality of results.
11. A dark dry incubator was used throughout the PCR–ELISA.
12. A schematic plate or work sheet should be used to follow the samples within the PCR–ELISA plate.
13. The hybridization temperature could vary depending on the length and the T_m of the capture oligonucleotide used.
14. The optical density (OD) results may vary depending on the plate reader used. The one we used states "over" when the OD is above 3.5 (software: Revelation 4.02 from Dynatech Laboratories and plate reader: MRX from Dynex Technologies).

Acknowledgements

The author would like to thank Josée Perreault for her technical support and for reviewing the manuscript.

References

1. St-Louis, M., Perreault, J., Lemieux, R. (2003) Extended blood grouping of blood donors with automatable PCR–ELISA genotyping. *Transfusion* 43, 1126–1132.
2. Bonvicini, F., Gallinella, G., Cricca, M., Venturoli, S., Musiani, M., Zerbini, M. (2004) A new primer set improves the efficiency of competitive PCR–ELISA for the detection of B19 DNA. *J Clin Virol* 30, 134–136.
3. Kim, J. -W., Shim, J. -H., Park, J. -W., Jang, W. -C., Chang, H. K., Song, I. H., Baek, S. -Y., Lee, S. -H., Yoon, D. -Y., Park, S. -N.

(2005) Development of PCR–ELISA for the detection of hepatitis B virus x gene expression and clinical application. *J Clin Lab Anal* 19, 139–145.
4. Daeschlein, G., Assadian, O., Daxboeck, F., Kramer, A. (2006) Multiplex PCR–ELISA for direct detection of MRSA in nasal swabs advantageous for rapid identification of non-MRSA carriers. *Eur J Clin Microbiol Infect Dis* 25, 328–330.
5. Heidari, A., Dittrich, S., Jelinek, T., Kheirandish, A., Banihashemi, K., Keshavarz, H. (2007) Genotypes and in vivo resistance of *Plasmodium falciparum* isolates in an endemic region of Iran. *Parasitol Res* 100, 589–592.
6. Mas, V., Alvarellos, T., Albano, S., de Boccardo, G., Giraudo, C., Garrett, C. T., Ferreira-Gonzalez, A. (1999) Utility of cytomegalovirus viral load in renal transplant patients in Argentina. *Transplantation* 67, 1050–1055.
7. Rapier, J. M., Villamarzo, Y., Schochetman, G., Ou, C. -Y., Brakel, C. L., Donegan, J., Maltzman, W., Lee, S., Kirtikar, D., Gatica, D. (1993) Nonradioactive, colorimetric microplate hybridization assay for detecting amplified human immunodeficiency virus DNA. *Clin Chem* 39, 244–247.
8. Fletcher, H. A., Barton, R. C., Verweij, P. E., Evans, E. G. V. (1998) Detection of *Aspergillus fumigatus* PCR products by a microtitre plate based DNA hybridisation assay. *J Clin Pathol* 51, 617–620.
9. St-Louis, M., Thibault, L., Chevrier, M.-C., Perreault, J., Richard, M., de Grandmont, M. J., Beauséjour, A., Nolin, M., Guérin, M., Vachon, A., Lemieux, R. (2003) In-house development and production of a WNV NAT assay for possible contingency testing of blood donors in June 2003. *Transfusion* 43, 128A (meeting abstract).

Chapter 2

Single Base Extension in Multiplex Blood Group Genotyping

Gregory A. Denomme

Abstract

Transfusion recipients who become alloimmunized to blood group antigens require antigen-negative blood to limit adverse transfusion reactions. An alternative strategy to phenotyping blood is to assay genomic DNA for the associated single nucleotide polymorphisms (SNPs). A multiplex PCR coupled with a single base oligonucleotide extension assay using genomic DNA can identify SNPs related to D, C/c, E/e, S/s, K/k, $Kp^{a/b}$, $Fy^{a/b}$, Fy0 (−33 promoter silencing polymorphism), $Jk^{a/b}$, $Di^{a/b}$, and HPA-1a/b. Using this technology, individual SNP call rates vary from 98 to 100%. The platform has the capacity to genotype thousands of samples per day. The suite of SNPs provides rapid data for both blood donors and transfusion recipients and is poised to change whose blood is provided for potential transfusion recipients. The potential to dramatically lower the incidence of alloimmunization and to avoid serious hemolytic complications of transfusions can be realized with the implementation of this technology.

Key words: Blood group genotyping, multiplex PCR, single base extension, SNP analysis.

1. Introduction

With over 100 years of research and discovery in the expression of and alloimmune response to red cell antigens, the field of immunohematology has characterized over 300 blood group antigens *(1)*. Transfusion recipients who become alloimmunized require antigen-matched red cells. The frequency of alloimmunization increases with the frequency of transfusion *(1, 2)* and blood banks invest considerable resources in the investigation and acquisition of antigen-compatible blood. The serological analyses of blood group antigen phenotypes and compatibility testing have remained largely unchanged for more than 60 years. The direct hemagglutination and indirect Coomb's tests are the

Peter Bugert (ed.), *DNA and RNA Profiling in Human Blood: Methods and Protocols, vol. 496*

DOI 10.1007/978-1-59745-553-4_2 Springerprotocols.com

mainstay of blood group phenotyping. The same clinical necessity applies to alloimmunized platelet transfusion recipients. Although a smaller group of platelet antigen systems have been identified, their clinical importance is nonetheless important; alloimmune thrombocytopenic purpura, platelet alloimmune refractoriness, and post-transfusion purpura require antigen-matched platelets for an effective platelet incremental response. The characterization of human neutrophil antigens also has a number of defined polymorphic systems *(4)*. So, it is not a surprise that nearly all of the defined blood group and platelet antigen systems are ipso facto polymorphic. It is the diversity among humans and the development of allogeneic transfusion that created such a repertoire of alloantibodies due to the vast immune response to the proteins, glycoproteins, and carbohydrate moieties expressed on red cells, platelets, and most likely neutrophils.

Starting in 1987, the genes responsible for the expression of blood group antigens were cloned and their allelic variants defined (reviewed in *(5)*). The molecular mechanisms responsible for the diversity of blood group and platelet antigens are diverse; missense, nonsense, deletions, insertions, and other alterations modify the expressed moieties. However, a vast majority of the most common clinically important blood group and platelet antigens are defined by single nucleotide polymorphisms (SNPs, pronounced SNiPs). Moreover, the knowledge of the molecular basis for blood group antigen expression has resulted in a paradigm shift in the goal to obtain antigen-matched red cells and platelets. Blood group genotyping was developed for several reasons *(6)*. First, it is advantageous to have an inexpensive and redundant screening assay to obtain large amounts of data for blood groups and platelet antigens. Blood group antigen screening using antibodies is expensive and time consuming. Second, DNA testing facilitates the identification of rare antigen-negative blood and blood with rare combinations of antigens, which allows screening outside family members of recipients with rare blood types *(7)*. Third, DNA testing can be performed in the absence of available antisera. Fourth, the accuracy of antigen typing (phenotype plus genotype) can be improved if both are performed and vast databases of blood group SNPs can be 'mined' for unusual alleles and their gene products. In other words, knowledge of a blood group phenotype and the underlying genotype can identify phenotyping deficiencies (i.e., weak or partial expression of a blood group antigen), and phenotype/genotype conundrums (pseudogenes, null alleles, and alterations that affect epitope expression) that can be investigated to add to our overall knowledge of the molecular basis of blood group and platelet antigen expression.

Presently, blood group genotyping can be performed using a few distinct platforms that differ in their nucleic acid chemistries.

To date, all platforms are designed to identify SNPs and a few applications are amiable to multiplex and high-throughput capacity, thereby facilitating their use as a screening tool to test large numbers of samples for multiple antigens. Single base extension technology is particularly suited as it uses a recombinant DNA polymerase and synthetic oligonucleotides with proprietary hybridization tags in a microchip array format *(8)*, which can be modified to suit the needs of the desired SNPs to 'interrogate'; a term used to indicate the SNP being evaluated. The most recent platforms are designed to identify nearly 50 SNPs simultaneously and in a high-throughput format. Future platforms, designed using single base extension technology and detection instrumentation, will identify more than 1000 SNPs simultaneously. Those instrumentations will redefine the ability to meet the need to genotype red cell, platelet, neutrophil, and HLA markers along with infectious marker detection, all in a single platform.

The purpose of this chapter is to provide the summary methodology of the Beckman SNPstream suite of instrumentation and software to use single base extension technology for the purpose of screening blood donors for clinically important blood group antigens. The SNP design identifies low frequency antigen-negative donors and combinations of antigen-negative donors. The technology has a high degree of fidelity and, if desired, redundancy can be incorporated to improve output beyond 98% reliability. SNP analysis of both the sense and antisense strand can reduce genotyping failures and minimize the number of samples requiring repeat analysis.

Single base extension is also known as mini-sequencing or template-directed nucleotide incorporation. The advantage of single base extension technology is that the detection phase can be simplified with the use of a combined oligonucleotide address tag and complementary probe for the detection of gene-specific base incorporation. Genomic DNA is the source material and the PCR phase is designed to amplify the flanking region surrounding a SNP of interest. A complementary oligonucleotide 'capture' primer is designed to lie proximal to the nucleotide of interest and DNA polymerase incorporates the complementary base (a fluorescent-conjugated dideoxy nucleotide) into the primer (**Fig. 2.1A**). The primer is a hybrid oligomer that includes an 'address' nucleotide that will hybridize to a complementary 'tag' sequence fixed to a glass chip using one of several microarray chemistries. Laser excitation of the fluorescent-conjugated oligonucleotide hybridized to the microarray tags can resolve the corresponding SNPs in a bi-fluorometric system (**Fig. 2.1B**). The methodology relies on proprietary instrumentation and software for accurate imaging, interpretation, and review of fluorescent output.

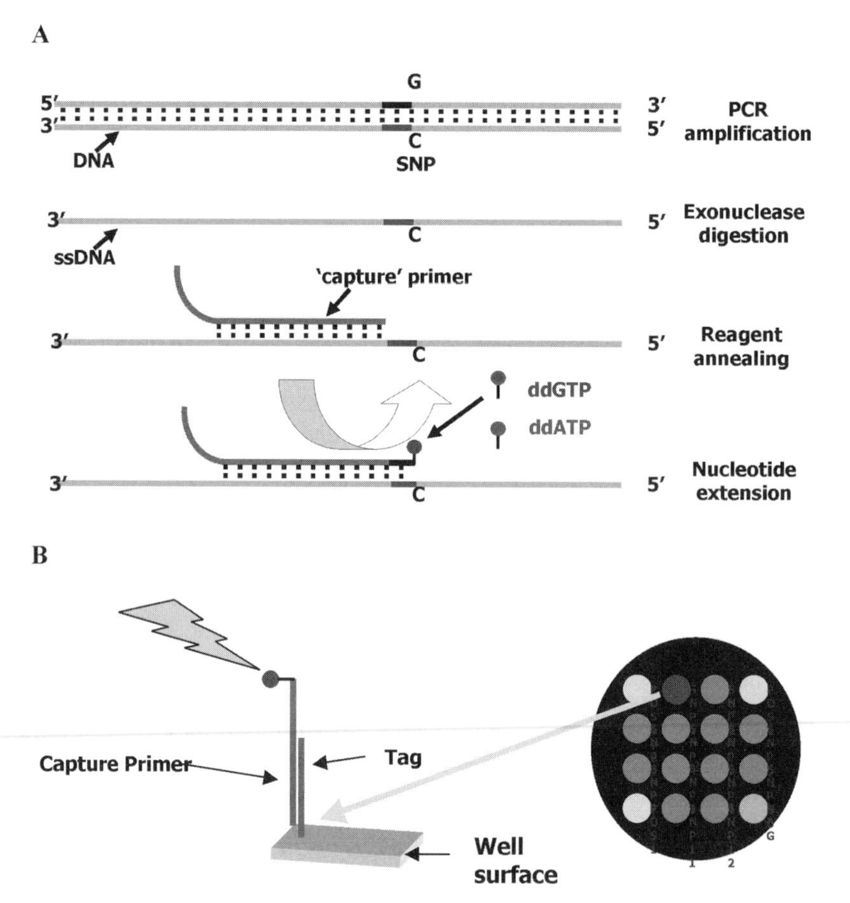

Fig. 2.1. Single base nucleotide extension in blood group genotyping. **A**: The major steps of single base extension technology. The SNP of interest is PCR amplified with flanking DNA (PCR amplification). One strand of the PCR product is digested leaving a single-stranded fragment (exonuclease digestion – optional). A capture primer is annealed proximal to the SNP of interest (reagent annealing). The capture primer has a region that is complementary to a tag oligonucleotide spotted on a chip in a 12 (shown) or 48 microarray format (nucleotide extension). **B**: Laser excitation (*jagged arrow*) of the bi-fluorescent nucleotides results in color identification of the interrogated SNP. A single reaction is shown.

2. Materials

2.1. Genomic DNA Extraction

1. QiaAmp DNA blood kit (Qiagen, Hilden, Germany), 24 or 96 sample preparations.
2. UV spectrophotometer.

2.2. Multiplex PCR

1. PCR buffer (provided with the *Taq* polymerase).
2. *Taq* polymerase (AmpliTaq Gold: Applied Biosystems, Foster City, CA, USA).
3. Deoxynucleotides (dNTPs: dATP, dCTP, dGTP, dTTP).
4. $MgCl_2$.

5. Primer oligonucleotides with sequences as published previously (10, 11).
6. PCR cycler (MJ Research Inc., Waltham, MA, USA).

2.3. Post-PCR Clean-Up

1. Exonuclease I (USB Corporation, Cleveland, OH, USA).
2. Shrimp alkaline phosphatase (SAP) and SAP buffer (SAP; USB Corporation).
3. Nuclease free water.
4. UHT high-throughput salt solution (Beckman Coulter Inc., Fullerton, CA, USA).

2.4. Single Base Extension

1. Tag primers with sequences as published previously (10, 11).
2. Fluorescent ddCTP and ddTTP nucleotides (TAMRA and BODIPY labeled).
3. Extension mix diluent (Beckman Coulter).
4. DNA polymerase (Beckman Coulter).

2.5. Post-Extension and Transfer to the Microarray Plate

1. Hybridization additive solution (Beckman Coulter).
2. UHT wash buffer (Beckman Coulter).
3. SNPware® microarray plate, 384 well (Beckman Coulter).
4. Humidified 42°C incubator.
5. SNPScope plate reader (Beckman Coulter).

3. Methods

The methodology for single base extension is based on the Beckman SNPstream platform and accompanying product monograph documentation. Full disclosure of the methodology is available from Beckman Coulter Inc.

3.1. Genomic DNA Extraction

Consult molecular methods manual for protocols and discussion on genomic DNA extraction (9). The goal is to obtain approximately 100–200 μg of DNA with a 260/280 ratio of 1.8 ± 0.2 (*see* **Note 1**).

3.2. Multiplex PCR

PCR optimization may be necessary. Protocols for the optimum annealing temperature and Mg concentration are likely necessary due to the minor variations in instruments and synthetic reagents. Helpful protocols for PCR optimization are available (9).

1. Stock primers are diluted to 12 μM each and mixed as a pool in nuclease free water (*see* **Note 2**).
2. Prepare a 10 μL multiplex PCR master mix containing 50 nM of primer pool, 75 μM each dNTP, 5 mM $MgCl_2$, 1 × PCR buffer, and 0.5 U/μL *Taq* polymerase.

3. Transfer 2 μL of genomic DNA to each well of 96-well PCR microplates.
4. Transfer 8 μL of the PCR master mix to the DNA samples. Centrifuge the microplate at 340 × *g* if the contents require mixing using a table-top centrifuge with a head design for microplates.
5. Place in a PCR cycler and cycle under the following conditions:

Initial denaturation:	94°C, 8 min
35 cycles of:	94°C, 30 s; 55°C, 30 s; 72°C, 1 min
Hold temperature:	4°C, until ready to proceed

3.3. Post-PCR Clean-Up

PCR clean-up can be performed if one of the amplimers is 5'-capped with phosphorothioate. The capping protects the strand from exonuclease degradation.

1. Prepare a master mix containing 0.13 U/μL exonuclease I and 0.33 U/μL SAP in 1 ? SAP buffer diluted in nuclease free water.
2. Combine the Exo/SAP pool mix, UHT high-throughput salt solution and the PCR reaction to the 96-well microplates and centrifuge to mix the contents.
3. Place in MJ thermalcycler and run the following program: 37°C, 30 min; 100°C, 10 min; 4°C, until ready to proceed.

3.4. Single Base Extension

1. Prepare the extension mix containing oligonucleotide tag primers, fluorescent ddCTP and ddTTP nucleotides (TAMRA and BODIPY labeled), extension mix diluent, and DNA polymerase as given in **Table 2.1**.
2. Transfer 7 μL of the extension mix to the post-PCR-cleaned products and centrifuge at 340 × *g* to mix the contents.
3. Run the following extension primer program:

Initial denaturation:	96°C, 3 min
45 cycles of:	94°C, 20 s; 40°C, 11 s
Hold temperature:	4°C, until ready to proceed

3.5. Post-Extension and Microarray Plate Reading

1. Preheat the incubator to 42°C.
2. Prepare 3.7 mL of a hybridization master mix consisting of 3.7 mL and 200 μL of hybridization additive solution.
3. Wash the 96-well microplate containing the extension reactions with 3 × UHT wash buffer.
4. Add 8.0 μL of hybridization solution master mix to each SNP extension reaction and subsequently transfer 8.0 μL of this mixture to the prepared 384-well microarray plate.
5. Place the microarray plate in humidified 42°C incubator for 2 h.
6. Wash the microarray plate with 3 × high stringency wash solution.

Table 2.1
Components of the extension mix

Component	Volume per well(μL)
Tag oligonucleotide primer pool	3.2
C/T ddNTP extension mix	21.4
Extension mix diluent	402
DNA polymerase	2.2
Nuclease $dddH_2O$	318

7. Completely dry the 384-well microarray plate using vacuum suction and a micropipette tip.
8. Read the microarray plate in a SNPScope plate reader (*see* **Note 3**).

3.6. Data Analysis

An important step in SNP analysis is the evaluation of the fluorescence emissions. The software suite analyses both the single or combined emission spectra (scatter plots) and the intensity of the emission. Therefore, weak fluorescence can be excluded on the basis that it does not achieve a minimum threshold or cannot be distinguished from background fluorescence. This threshold is particularly useful where alternate genes are co-amplified in the PCR stage.

1. *RHCE*C* allele has similar DNA sequence in exon 2 and it is difficult to obtain short PCR amplimers in the presence of *RHD*. However, if *RHCE* is preferentially amplified, the analysis can exclude the background reading of *RHD* amplification from that produced by *RHCE*. Some background fluorescence exists, but the resultant analysis provides good positive predictive values to identify the R_1R_1 phenotype *(10)*. An intron 2 surrogate marker can be used but does not correlate 100% with the *RHCE*C* allele *(12)*. *RHCE*E/e* SNP analysis does not suffer from this problem because a seperate PCR amplification stage can be designed to specifically target the *RHCE* gene.
2. The emission scatter plots can be grouped into three categories: as independent values and combined fluorescence emissions. **Figure 2.2** is a representative example of fluorescent output data (*see* **Note 4**).
3. When testing includes several hundred samples, acceptance criteria for the analysis of blood group SNPs should include a calculation of the allele frequencies for the population under evaluation for comparison to known frequencies (*see* **Note 5**).

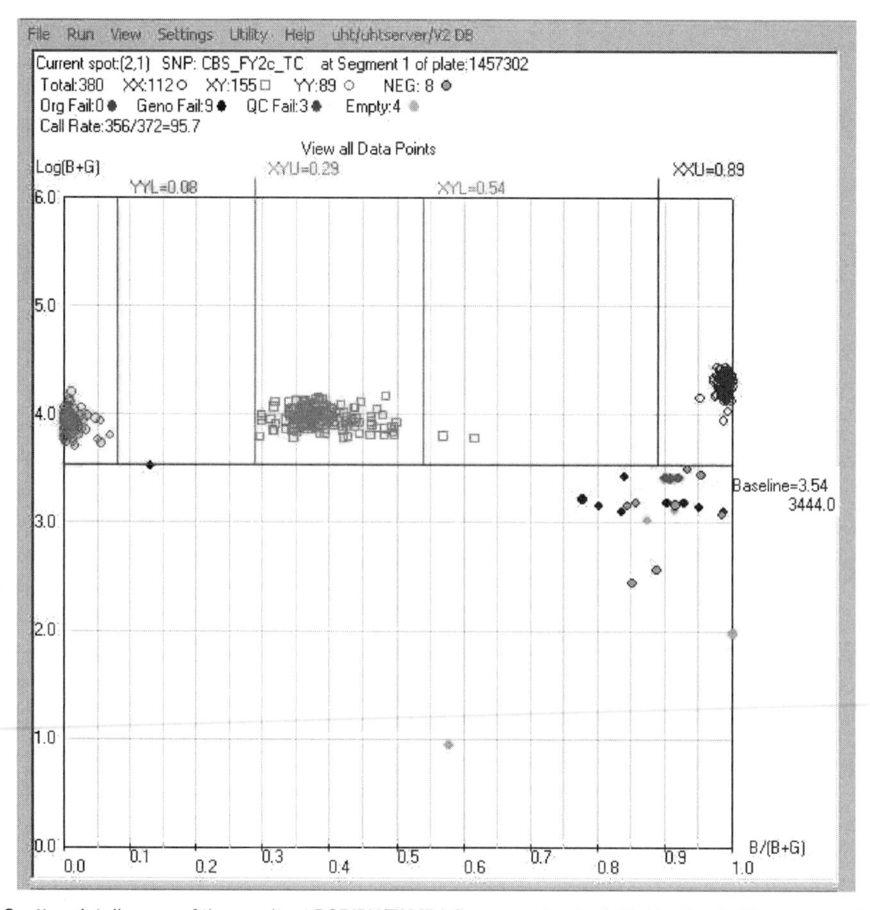

Fig. 2.2. Scatter plot diagram of the resultant BODIDY/TAMRA fluorescent output. The horizontal line represents baseline output (<3.54 units) and identifies genotype, QC, or origin failures and empty wells. The two vertical lines identify the genotypic groupings (XX, XY, and YY) and any outliers. Two outliers are identified in Fig. 2.2 and these samples can be selected for repeat analysis. (*See* Color Insert)

Blood group, platelet, and neutrophil SNPing is poised to improve the database of blood donors with known antigen profiles and allow studies to analyze the benefits of blood product matching on the basis of genotypes alone to minimize adverse immune transfusion outcomes and reactions in both selected patient populations *(13–15)* and possibly, in the near future, the general transfusion recipient population including those who are at risk of alloimmunization to high frequency antigens *(16)*.

4. Notes

1. A number of protocols are suitable for the extraction of genomic DNA. The QiaAmp kits or MagNA Pure DNA isolation platform (F. Hoffmann-La Roche Ltd, Switzerland) provides adequate genomic DNA. The DNA should show a signal

band on agarose gel electrophoresis indicating little DNA degradation or shearing.

2. Oligonucleotides for single base extension are designed to minimize inter-primer hybridization, intra-primer hairpin formation, and with similar annealing temperatures. Primer sequences for the detection of D, C/c, E/e, M, N, S, s, K, k, Fy(a/b), Fy0 (GATA-1 silencing), Jk(a/b), and HPA-1a/b were published previously *(10, 11)*.
3. The SNPScope plate reader will excite and capture emission images of BODIPY- and TAMRA-labeled ddNTPs. All genotype calls are automatically generated using the SNPStream Software Suite consisting of MegaImage, UHTGetGenos, and QCReview.
4. Note that three independent fluorescent groups can be identified, i.e., TAMRA or BODOPY fluorescence only, and combined TAMRA and BODOPY fluorescence. The groupings are distinct and there is no overlap fluorescence between the groups. Samples falling outside defined groups represent an inability to assign the nucleotide(s) occupying the SNP position. These failures are referred to as "call" failures. PCR optimization should be continued until call failures are less than 1.0%. DNA amplification failures should be rare but with appropriate sample process controls in place, samples can be re-analyzed in a subsequent batch assay.
5. For allele frequencies that do not fall within the expected population frequencies, a subset of the test population sho^uld be evaluated by manual molecular tests for validation. For populations with unknown allele frequencies, serological analysis on a subset can be performed to confirm the genotype/phenotype association.

References

1. Daniels, G., Flegel, W. A., Fletcher, A., Garratty, G., Levene, C., Lomas-Francis, C., Moulds, J. M., Moulds, J. J., Olsson, M. L., Overbeeke, M. A., Poole, J., Reid, M. E., Rouger, P., van der Schoot, C. E., Scott, M., Sistonen, P., Smart, E., Storry, J. R., Tani, Y., Yu, L. C., Wendel, S., Westhoff, C. M., Zelinski, T. (2007) International society of blood transfusion committee on terminology for red cell surface antigens: cape town report. *Vox Sang* 92, 250–253.
2. Coles, S. M., Klein, H. G., Holland, P. V. (1981) Alloimmunization in two multitransfused patient populations. *Transfusion* 21, 462–466.
3. Fluit, C. R., Kunst, V. A., Drenthe-Schonk A. M (1990) Incidence of red cell antibodies after multiple blood transfusion. *Transfusion* 30, 532–535.
4. *Stroncek*, D. (2002) Neutrophil alloantigens. *Transfus Med Rev* 16, 67–75.
5. Logdberg, L., Reid, M. E., Lamont, R. E., Zelinski, T. (2005) Human blood group genes 2004: chromosomal locations and cloning strategies. *Transfus Med Rev* 19, 45–57.
6. Denomme, G. A., Lomas-Francis, C., Reid, M. E., Storry, J. R. (2004) Blood group molecular genotyping and its applications, in (Stowell, C., Dzik, S., eds.), *Emerging Diagnostic and Therapeutic Technologies in Transfusion Medicine*. American Association of Blood Banks, Bethesda.
7. Kanter, M. H., Hodge, S. E. (1990) The probability of obtaining compatible blood

from related directed donors. *Arch Pathol Lab Med* 114, 1013–1016.

8. Bell, P. A., Chaturvedi, S., Gelfand, C. A., Huang, C. Y., Kochersperger, M., Kopla, R., Modica, F., Pohl, M., Varde, S., Zhao, R., Zhao, X., Boyce-Jacino, M. T., Yassen, A. (2002) SNPstream UHT: ultra-high throughput SNP genotyping for pharmacogenomics and drug discovery. *Biotechniques* Suppl, 70–77.
9. Denomme, G. A., Rios, M., Reid, M. E. *Molecular Protocols in Transfusion Medicine.* Academic Press, New York.
10. Denomme, G. A., van Oene, M. (2005) High-throughput multiplex single-nucleotide polymorphism analysis for red cell and platelet antigen genotypes. *Transfusion* 45, 660–666.
11. Montpetit, A., Phillips, M. S., Mongrain, I., Lemieux, R., St-Louis, M. (2006) High-throughput molecular profiling of blood donors for minor red blood cell and platelet antigens. *Transfusion* 46, 841–848.
12. Tax, M. G., van der Schoot, C. E., van Doorn, R., Douglas-Berger, L., van Rhenen, D. J., Maaskant-vanWijk, P. A. (2002) RHC and RHc genotyping in different ethnic groups. *Transfusion* 42, 634–644.
13. Aygun, B., Padmanabhan, S., Paley, C., Chandrasekaran, V. (2002) Clinical significance of RBC alloantibodies and autoantibodies in sickle cell patients who received transfusions. *Transfusion* 42, 37–43.
14. Schonewille, H., Haak, H. L., van Zijl, A. M. (1999) Alloimmunization after blood transfusion in patients with hematologic and oncologic diseases. *Transfusion* 39, 763–771.
15. Sirchia, G., Zanella, A., Parravicini, A., Morelati, F., Rebulla, P., Masera, G. (1985) Red cell alloantibodies in thalassemia major. Results of an Italian cooperative study. *Transfusion* 25, 110–112.
16. Seltsam, A., Wagner, F. F., Salama, A., Flegel, W. A. (2003) Antibodies to high-frequency antigens may decrease the quality of transfusion support: an observational study. *Transfusion* 43, 1563–1566.

Chapter 3

Real-Time PCR Assays for High-Throughput Blood Group Genotyping

Fernando Araujo

Abstract

There are multiple situations in the context of transfusion medicine where the classic serologic methods are unable to provide an adequate response, for example, recently polytransfused patients, patients with positive direct human antiglobulin tests, and hemolytic disease of the newborn. The traditional polymerase chain reaction techniques are slow and sometimes difficult to carry out and interpret. Thus there is a need for the development and validation of rapid and effective molecular methods. The genetic basis of the main alleles of the most important blood groups are known, but the frequencies vary in the different populations, thus for the genetic techniques to be efficient it is important to evaluate them, in order to adapt the molecular approaches.

Key words: Blood groups, *RHD, RHCE, KEL, JK, FY, DO*, hemolytic disease of newborn, real-time PCR.

1. Introduction

The study of the human blood group antigens has provided useful markers of many proteins, glycoproteins, and glycolipids that are found on the surface of the red blood cells. The initial biochemical analysis and the recent molecular characterization allowed the study of these molecules in healthy and pathological conditions *(1)*. Many of them are found distributed in various tissues in the human organism, whereas others seem to be specific to the red blood cells.

The blood groups have been found to participate in almost all human physiological processes, including functions such as enzymes, adhesion molecules, transporters or gas and ions channels, as well as an important role in immunological defence

Peter Bugert (ed.), *DNA and RNA Profiling in Human Blood: Methods and Protocols, vol. 496*

DOI 10.1007/978-1-59745-553-4_3 Springerprotocols.com

systems like the complement pathway. Certain blood group phenotypes have demonstrated the possibility of being used as evidence of disease and even details, such as the strength of the antigen expression, may be correlated to the progression of some tumors. The blood groups in the red cells provide a "window" to look at body physiology, both in health and in illness.

The classical methodology based on the serological determination of the antigens present on the surface of the red cells and in the detection of the antibodies, is extremely simple to carry out, economical and, when correctly executed, has a sensitivity and specificity that are adequate for the resolution of the greater part of the cases. However, hemagglutination techniques have limitations, and therefore molecular genetics has an extremely valid and necessary role in this area.

The majority of the genes encoding blood groups have already been sequenced, making it thus possible to understand their many functions and formulate methods for their identification. There are various molecular changes which give rise to a variety of antigens and phenotypes, namely recombination or gene conversion, exon duplication, deletion, insertion or nucleotide substitution. On the other hand, the analysis of null phenotypes has demonstrated that different genetic events may give rise to the same phenotype.

Several techniques are described in order to detect these variations, for example, polymerase chain reaction-restriction fragment length polymorphisms (PCR-RFLP), polymerase chain reaction-sequence-specific primers (PCR-SSP), real-time PCR, pyrosequencing and in far reaching development, "chip" technology (microarrays). The clinical applications in the context of transfusion medicine may be multiple, but at present the most relevant seem to be the following:

1. In recently polytransfused patients: the presence of circulating red cells of blood donors, in relevant quantities, makes phenotyping by hemagglutination techniques complex, slow, and potentially incorrect *(2, 3)*
2. In patients with positive direct antiglobulin tests: in patients with auto-immune hemolytic anemia, in which the red cells are covered with immunoglobulins, molecular biology is important when the immunological tests are not efficient in removing the linked immunoglobulins or may destroy clinically relevant antigens.
3. In blood donors: laboratory tests based on the study of the deoxyribonucleic acid (DNA) may be used to type blood donor antigens for transfusion and to constitute reagent panels for the identification of antibodies. This is particularly useful in cases where the antibodies are weakly reactive or commercially unavailable. Recent technologies have also the potential to carry out DNA determinations in large quantities of samples,

with relevance in rare blood groups. Because the new automatic procedures obtain results in a quicker, simple, and eventually more economical manner, the typing of blood donors using tests based on DNA may have great potential.

4. In the pre-natal setting: the hemagglutination techniques only provide indirect indications on the risk and severity of the hemolytic disease of the newborn (HDN). Thus, the typing of antigens by techniques based on DNA has an important value in the pre-natal setting, namely in the identification of fetuses that are not under the risk of HDN, preventing unnecessarily aggressive monitoring of the mother. Fetal DNA may be isolated from cells obtained from invasive procedures such as amniocentesis or chorionic villi samples, and by non-invasive procedures from the trophoblasts collected from transcervical samples and from fetal erythroblasts in maternal circulation, but with many limitations. Free fetal DNA may also be successfully extracted from the maternal plasma and makes typing of the blood groups possible. In this simple manner, it is possible to predict the phenotype of the fetus and correctly orientate the mother's evaluation, without risks to both of them.

The classic genotyping methods pose several problems, by using separate tubes for specifically amplifying each allele (PCR-SSP needs amplification of internal control fragments) or by using restriction enzymes (with the risk of incomplete digestion of the amplicons or generating false-positive results if silent mutations occurred near the mismatch not allowing the enzyme to cut). The potential benefits of homogeneous detection systems, in relation to the classic methods, have long been recognized: simple processing and rapid analysis, no post-amplification processing, eliminating sampling tracking errors, and end-product contamination. The methods described are robust and reliable, with few manuals processing steps: the "hands-on" time for setting up the assays is shorter and no manual intervention is required after loading in the instrument. They allow high throughput and rapid results in a very economical manner, using small volumes lowers the reagent costs, resulting in a highly competitive technology for a routine laboratory.

The development of new molecular tools for diagnostic use, complementing the classical methods applied in transfusion services, with results available in a timely manner to allow their use on a routine basis was the aim of these methods. In order to achieve that, we used the LightCycler® (Roche Molecular Biochemicals) PCR instrument: a microvolume fluorometer integrated with a thermal cycler that combines rapid cycle PCR with real-time fluorescence monitoring. This assay is based on the fluorescence resonance energy transfer (FRET) principle. Incorporation of labeled hybridization probes with the initial reaction mix and target nucleic acid allows detection and analysis of PCR products with the LightCycler system in a closed reaction vessel. In our protocols we used the two fluorescence-based methods for the detection of amplification products: DNA binding dye SYBR Green I and hybridization probes.

With these assays, a short fragment harboring the particular polymorphic site is amplified. In the hybridization probes protocol, the amplicons are detected by fluorescence using a 3′-fluorescein-labeled probe and a 5′-LightCycler Red 640-labeled probe that are in FRET when hybridized to the same strand internal to the unlabeled PCR-primers. Homogeneous genotyping is achieved by positioning one of the probes over the polymorphic nucleotide. When fluorescence is monitored as the temperature increases through the melting point (T_m) of the probe/product duplex, a characteristic melting profile is obtained, depending on the presence or absence of a base pair mismatch in the heteroduplex. The fluorescence signal is then plotted in real time against temperature to produce melting curves for each sample, and then converted to derivative melting curves by plotting the negative derivative of the fluorescence with respect to temperature against temperature.

2. Materials

1. Anticoagulated blood samples (*see* **Note 1**).
2. Gel cards (DiaMed AG; Cressier sur Morat, Switzerland)
3. Antisera obtained from Gamma Biologicals Inc. (Houston, TX, USA) and Diagast (Immucor Medizinische Diagnostik GmbH, Rodermark, Germany)
4. Magna Pure LC® and the High Pure PCR-Template-Preparation® kit (Roche, Mannheim, Germany) or QIAamp® (Qiagen GmbH, Hilden, Germany).
5. LightCycler (Roche)
6. Oligonucleotides and probes as given in **Table 3.1**.
7. PCR reagents for SYBR Green I or hybridization probes protocols (Roche).

3. Methods

The red cells could be phenotyped by routine hemagglutination with gel cards and antisera according to the instructions of the manufacturers.

3.1. Real-Time PCR Assays

1. DNA is extracted using commercial kits according to the instructions of the manufacturers (*see* **Note 2**).

Table 3.1
Sequences of the primers and probes used to genotype blood group alleles and the length of the PCR products

Primers and probes	Sequences	PCR product (bp)
KEL		
Anchor	5′-GCTTGGAGGCTGGCGCAT-3′	148
	5′-CTGGATGACTGGTGTGTGTGGA-3′	
Sensor	AGTCAGTATGGCCATTTCCCTTTCTTCATTAACCG	
	AACGCTGAGACTTCTGA	
JK		
Anchor	5′- ATCCCACCCTCAGTTTCCT-3′	165
	5′- ATGAACATTCCTCCCATTGC-3′	
Sensor	ACTCTGGGGTTTCAACAGCTCTCTGCCCCATTTGA	
	GAACATCTACTTTG	
DO		
Anchor	5′-GAGTTTGGGAACCAGACACTA-3′	196
	5′-TTTAGCAGCTGACAGTTATATGT-3′	
Sensor	CTTGGGTGGTAGCTCATATTTATAACTTTAAACCT	
	CAACTGCAACCAGTCTCC	
FY		
125G>A		213
Anchor	5′-CAGCTGGACTTCGAAGATGTA-3′	
	5′-GCGGAAGAGAGGTCTGAAA-3′	
Sensor	TGGGAAGGAATCATTCACACCATAGGAAGAA	
	GTTGGCACCATAGTCTCCAT	
-33T>C		
Anchor	5′-AGGGGCATAGGGATAAGGGACT-3′	
	5′-AGGGGCATAGGGATAAGGGACT-3′	287
Sensor	CGCTGACAGCCGTCCCAGCCCTTCTTGGCTCTTA	
	CCTTGGAAGCACA	
RHC	5′-GATGCCTGGTGCTGGTGGAAC-3′	112
	5′-GCTGCTTCCAGTGTTAGGGCG-3′	
RHc	5′-TCGGCCAAGATCTGACCG-3′	177
	5′-TGATGACCACCTTCCCAGG-3′	

(continued)

Table 3.1 (continued)

Primers and probes	Sequences	PCR product (bp)
RHE/RHe		
Anchor	5′-GCAACAGAGCAAGAGTCCATC-3′	392
	5′-GAACATGGCATTCTTCCTTTG-3′	
Sensor	CGCCCTCTTCTTGTGGATGTTCTGCCAAGTGTCAA	
	CTCTGCTCTGCT	
RHD (intron 4)	5′-TGACCCTGAGATGGCTGT-3′	600
	5′-ACGATACCCAGTTTGTCT-3′	
RHD (exon 7)	5′-AGCTCCATCATGGGCTACAA-3′	96
	5′-ATTGCCGGCTCCGACGGTATC-3′	
RHD (exon 4)	5′-GCCGACACTCACTGCTCTTAC-3′	381/418
	5′-TCCTGAACCTGCTCTGTGAAGTGC-3′	
RHD zygosity	5′-GGCCAACAAAACCATTTTTTCCTGATAC-3′	943
	5′- CTCTGTCTCAAAAAAAAAAAAAAAACAAGTG-3′	

2. Prepare PCR reaction mixes as described in **Table 3.2**, using the primers of the **Table 3.1** and according the amplification conditions edited in **Table 3.3** (*see* **Note 3**).
3. For *KEL1/KEL2*, *JK1/JK2*, *FY1/FY2/FY*, *DO1/DO2*, and *RHE/RHe* (exon 5) we utilized the hybridization probes protocol, according to the instructions of the manufacturers (*see* **Note 4**).
4. For *RHC/RHc* (the analysis was done in sequences of the exon 1 and intron 2 –*RHC*- and exon 2 –*RHc*), *RHD* (sequences of intron 4, exon 7, and exon 4 – pseudogene) and *RHD* zygosity (sequences from the identity region of the hybrid *Rhesus box*), the SYBR Green I protocol is used, according to the instructions of the manufacturers.

3.2. Result Evaluation

1. *KEL1/KEL2*: The melting point of *KEL1* is at 66°C and the *KEL2* at 59°C. In different runs, the positions and distances of the melting peaks are identical and differed by less than 1°C for the same allele (**Figs. 3.1A and 3.2**).
2. *JK1/JK2*: The melting point of *JK1* is at 56°C and the *JK2* at 62°C. In different runs, the positions and distances of the melting peaks are identical and differed by less than 1°C for the same allele (**Fig. 3.1B**).
3. *FY1/FY2/FY*: The melting point of the *FY1* is at 63°C and the *FY2* at 55°C (125 G>A); while the allele without mutation at the promoter region has a melting point at 64°C and the *FY2*

Table 3.2
PCR reaction mixes used in the SYBR Green I and hybridization probe protocols of the LightCycler

	H_2O (μL)	$MgCl_2$ (μL (mM))	Primer (μL (μM))	Probe (μL (μM))	Master mix (μL)	DNA (μL (ng))
KEL1/KEL2	10.8	1.6 (3)	1 (10)	0.8 (5)	2	2 (100)
JK1/JK2	10.8	1.6 (3)	1 (10)	0.8 (5)	2	2 (100)
FY1/FY2	10.8	1.6 (3)	1 (10)	0.8 (5)	2	2 (100)
DO1/DO2	10.8	1.6 (3)	1 (10)	0.8 (5)	2	2 (100)
RHE/Rhe	10.8	1.6 (3)	1 (10)	0.8 (5)	2	2 (100)
FY	11.6	0.8 (2)	1 (10)	0.8 (5)	2	2 (100)
RHC/RHc	13.2	0.8 (2)	1 (10)	0.8 (5)	2	2 (100)
RHD (intron 4)	12.8	1.2 (2.5)	1 (10)	0.8 (5)	2	2 (100)
RHD (exon 7)	12.4	1.6 (3)	1 (10)	–	2	2 (100)
RHD (exon 4)	12.4	1.6 (3)	1 (10)	–	2	2 (100)
RHD zygosity	14[a]	2 (3.5)	0.5 (5)	0.8 (5)	2	1 (50)

[a]1% v/v DMSO instead of H_2O.

Table 3.3
Conditions of amplification of SYBR Green I and hybridization probes protocols used in the LightCycler

	Initial denaturation	Amplification cycles				Melting curve
		Denaturation	Annealing	Extension	Cycles (n)	
KEL1/KEL2	95°C; 30 s	95°C; 0 s	62°C; 10 s	72°C; 10 s	45	50–75°C
JK1/JK2	95°C; 30 s	95°C; 0 s	55°C; 10 s	72°C; 10 s	40	45–75°C
FY1/FY2/FY	95°C; 30 s	95°C; 0 s	60°C; 10 s	72°C; 28 s	45	45–75°C
RHE/RHe	95°C; 30 s	95°C; 0 s	56°C; 60 s	72°C; 16 s	40	53–90°C
DO1/DO2	95°C; 30 s	95°C; 0 s	65°C; 10 s	72°C; 10 s	45	45–75°C
RHC	95°C; 600 s	95°C; 0 s	72°C; 3 s	72°C; 6 s	40	86–95°C
RHc	95°C; 600 s	95°C; 0 s	63°C; 4 s	72°C; 8 s	40	78–95°C
RHD (intron 4)	95°C; 600 s	95°C; 0 s	70°C; 3 s	72°C; 72 s	45	84–99°C
RHD (exon 7)	95°C; 600 s	95°C; 0 s	71°C; 1 s	72°C; 10 s	40	80–95°C
RHD (exon 4)	95°C; 600 s	95°C; 0 s	66°C; 5 s	72°C; 17 s	45	80–95°C
RHD zygosity	95°C; 600 s	95°C; 0 s	69°C; 4 s	72°C; 38 s	55	80–95°C

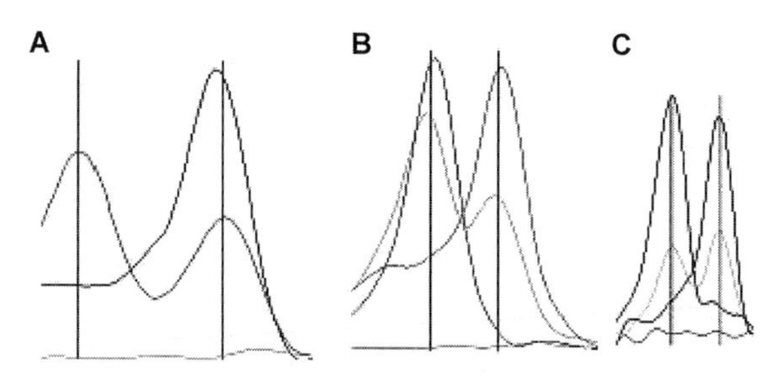

Fig. 3.1. Genotyping by real-time PCR, using the hybridization probes protocols. (A) *KEL* genotyping showing the melting peaks at 59 and 66°C; (B) *FY* genotyping showing the melting peaks at 56 and 62°C; (C) *RHE/RHe* genotyping showing the melting peaks at 62 and 68°C.

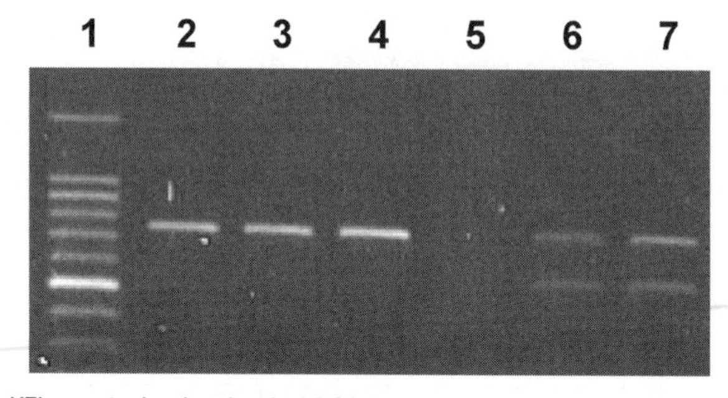

Fig. 3.2. *KEL* genotyping by classical PCR-RFLP (three fragments, with 740, 540, and 200 bp). Lane 1: 100 bp marker; lanes 2–4: PCR product from individuals homozygous *KEL1/KEL1* (one fragment with 740 bp); lane 5: negative control (water); lanes 6–7: PCR product after digestion with *Bsm* I, from heterozygous individuals *KEL1/KEL2* (two fragments, with 540 and 200 bp).

silent allele (*FY*) at 58°C (–33 T>C). In different runs, the positions and distances of the melting peaks are identical and differed by less than 1°C for the same allele (**Fig. 3.3**).

4. *RHD*: The use of SYBR Green I allows a simple identification of the alleles (**Fig. 3.4**).
5. *RHD* zygosity: The analysis of the results consistently shows three distinct populations: *RHD–/RHD–* (mean values of fluorescence ± sd: 0.31 ± 0.04), *RHD+/RHD–* (mean values of fluorescence ± sd: 0.16 ± 0.04) and *RHD+/RHD+* (values of fluorescence: <0.05) samples.
6. *RHCE*: The melting point of *RHE* is at 62°C and the *RHe* at 68°C. In different runs, the positions and distances of the melting peaks are identical and differed by less than 1°C for the same allele (**Fig. 3.1C**).
7. *DO1/DO2*: The melting point for *DO1* is 58 and 65°C for *DO2*. In different runs, the positions and distances of melting peaks for the same allele are identical, varying by less than 1°C.

3.3. Advantages and Precautions

The real-time PCR methodology has several practical advantages and a range of uses in clinical applications:

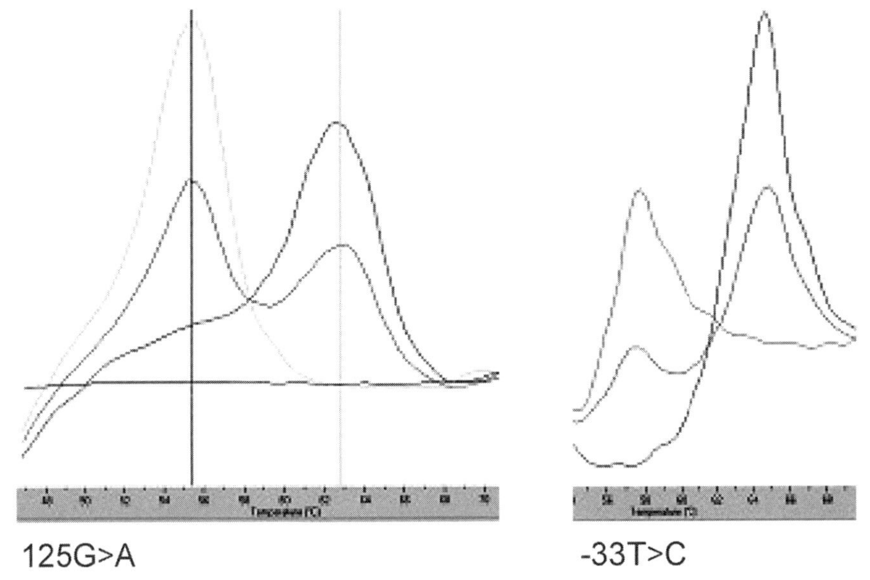

Fig. 3.3. *FY* genotyping using the hybridization probes protocols, showing the melting peaks of homozygous (one peak) and heterozygous (two peaks) individuals.

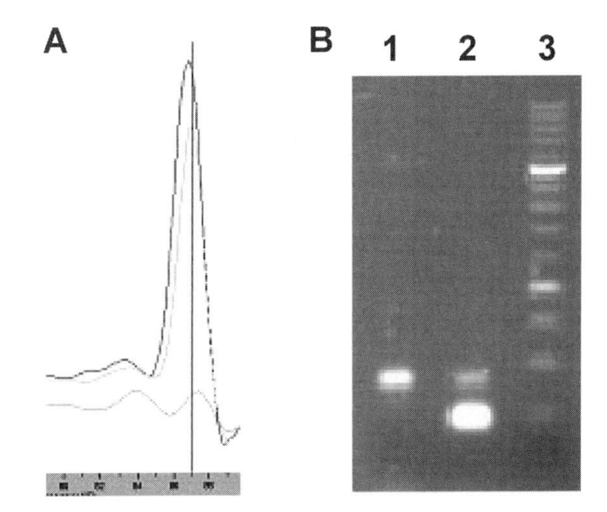

Fig. 3.4. Example of *RHD* genotyping for exon 7 using (**A**) the DNA binding dye SYBR Green I protocol (in this case for the exon 7 of the *RHD* gene) or (**B**) classical PCR. Lane 1: DNA from an individual *RHD*–negative; lane 2: DNA from an individual *RHD*–positive; lane 3: 50 bp marker.

1. Pre-natal determination of the *RHD* genotype through non-invasive methods: blood group genotyping techniques are particularly indicated in HDN, through the identification and characterization by molecular biology of fetal DNA circulating in the maternal plasma. The results, using the methodology described, indicated that fetal *RHD* genotyping is reliable from the 24th week of gestation. Nevertheless, more tests are

necessary in order to conclude for the adoption of this type of approach in the clinical routine.

2. HDN and the Kell blood group: in contrast with HDN due to the RhD antigen, the antibodies against Kell antigens cause fetal anemia specifically by the suppression of erythropoiesis, making the maternal antibody titres or bilirubin levels in the blood of the umbilical cord/amniotic fluid, not relevant indicators of the severity of the disease. Molecular biology methods helped prevent the use of invasive diagnostic techniques which could have placed the fetus at risk *(4)*.
3. Forensic medicine: the DNA-typing techniques may provide critical information in medico-legal situations where the phenotyping techniques have limitations.
4. Precautions when using commercial kits: blood group genotyping kits can make work easier in laboratories which have a high sample throughput or in routine labs where technical and human resources are limited. However, there are risks of false diagnosis, due to incorrect execution of the tests, faulty interpretation of the results obtained or even problems related to the development of the tests. This last situation is important, especially in cases where the molecular basis of the variants is not sufficiently known or when there is a reduced number of samples for validation of the kits.
5. Serological methods in transfusion medicine are reliable and adequate to the daily practice of the laboratories of the blood bank. However, there are situations in which they cannot provide an adequate response. The traditional PCR tests (PCR-SSP and PCR-RFLP) are slow, labor-intensive and sometimes difficult to interpret. Therefore, new molecular protocols were developed and validated using real-time PCR, specifically for the allele identification of the main blood groups.
6. The polymorphisms of a certain gene, its frequencies and distribution in the individuals studied, are of importance for the diagnostic application of the DNA typing.
7. However, when we intend to use the molecular genotyping results in clinical practice, we must be alert that in rare situations, the genotype may not correlate with the antigen expression on the red blood cells. A dominant inhibitor gene which is not linked to the locus analyzed, mutations in a location other than that being studied, mutations in the *GATA* box or the problem of hybrid genes, could generate false interpretations with clinical significance *(2)*. In addition, many DNA-based techniques still have to be evaluated: in a recent workshop report on the genotyping of blood cell allo-antigens, a 11% rate of mistyping in some systems was found, demonstrated that further efforts are needed to improve the precision of the genotyping techniques *(5)*.

4. Notes

1. The assay is performed on blood samples drawn by venipuncture of the antecubital vein, after informed consent.
2. DNA that is not used immediately should be frozen at –70°C until assayed.
3. In the case of *RHE/RHe*, the primers were designed in order to amplify only the *RHCE* gene and the probes to genotype the alleles.
4. For confirmation of the *KEL1/KEL2* genotypes and *RHD* zygosity, PCR-RFLP could be performed *(6–12)*, and for *JK1/JK2*, *FY1/FY2/FY* and *DO1/DO2*, PCR-SSP could be done, as described by others *(13–27)*. For confirmation of the results regarding *RHC/RHc*, *RHE/RHe*, and *RHD* genotypes, classical PCR reactions could be applied *(28–32)*.

Acknowledgments

To my Supervisor, J. Eduardo Guimarães, MD, PhD, Director of the Clinical Hematology Departments of the Oporto Faculty of Medicine and São João Hospital, for all the support, encouragement, and friendship during my training as a Doctor and in the carrying out of the thesis. To my Supervisor, Jill R. Storry, MSc, PhD, for her availability to orientate the thesis, for the criticism, suggestions, and knowledge shared, and for the certainty of being able to count on her support in future investigation assignments. To A. Rosa Araújo, MD, recently deceased and former Director of the Transfusion Medicine and Blood Bank Department of the São João Hospital, for having taught me the basic fundamentals of science and investigation. To the Technical Graduates in Clinical Analyses and Public Health, Fátima Monteiro, Isabel Henriques, Elsa Meireles, Aldina Cruz, Marina Ferreira, Gilberto Tavares, and Vitor Cunha, of the Molecular Biology Centre of the Transfusion Medicine and Blood Bank Department of the São João Hospital, for the indispensable help in carrying out the technical aspects of the assignment.

References

1. Cartron, J. P., Colin, Y. (2001) Structural and functional diversity of blood group antigens. *Transfus Clin Biol* 8, 163–199.
2. Reid, M. E., Rios M., Powell, V. I., Charles-Pierre, D., Malavade V. (2000) DNA from blood samples can be used to genotype patients who have recently received a transfusion. *Transfusion* 40, 48–53.
3. Rozman, P., Dovc, T., Gassner, C. (2000) Differentiation of autologous ABO, RHD, RHCE, KEL, JK and FY blood group genotypes by analysis of peripheral blood samples

of patients who have recently received multiple transfusions. *Transfusion* 40, 936–942.

4. Vaughan, J., Manning, M., Warwick, R., Letsky, E. A., Murray, N. A., Roberts, I. A. G. (1996) Inhibition of erythroid progenitor cells by anti-Kell antibodies in fetal alloimmune anemia. *N Engl J Med* 338, 798–803.
5. Daniels, G., van der Schoot, C., Olsson, M. (2005) Report of the First International Workshop on molecular blood group genotyping. *Vox Sang* 88, 136–142.
6. Lee, S., Wu, X., Reid, M., Zelinski, T., Redman, C. (1995) Molecular basis of the Kell (K1) phenotype. *Blood* 85, 912–916.
7. Hessner, M. J., McFarland, J. G., Endean, D. J. (1996) Genotyping of KEL1 and KEL2 of the human Kell blood group system by the polymerase chain reaction with sequence-specific primers. *Transfusion* 36, 495–499.
8. Lee, S., Wu, S., Son, S., Naime, D., Reid, M., Okubo, Y., Sistonen, P., Redman, C. (1996) Point mutations characterize KEL10, the KEL3, KEL4, and KEL21 alleles, and the KEL7 and KEL11 alleles. *Transfusion* 36, 490–494.
9. Lee, S. (1997) Molecular basis of Kell blod group phenotypes. *Vox Sang* 73, 1–11.
10. Wagner, F. F., Flegel, W. A. (2000) RHD gene deletion occurred in the Rhesus box. *Blood* 95, 3662–3668.
11. Matheson, K. A., Denomme, G. A. (2002) Novel 3′ Rhesus box sequences confound RHD zygosity assignment. *Transfusion* 42, 645–650.
12. Chiu, R. W. K., Murphy, M. F., Findler, C., Zee, B. C. Y., Wainscoat, J. S., Lo, Y. M. D. (2001) Determination of RhD zygosity: comparison of a double amplification refractory mutation system approach and a multiplex real-time quantitative PCR approach. *Clin Chem* 47, 667–672.
13. Heaton, D. C., McLoughlin, K. (1982) Jk(a-b-) red blood cells resist urea lysis. *Transfusion* 22, 70–71.
14. Frohlich, O., Macey, R. I., Edwards-Moulds, J., Gargus, J. J., Gunn, R. B. (1991) Urea transport deficiency in Jk(a-b) erythrocytes. *Am J Physiol* 260, C778–C783.
15. Olives, B., Merriman, M., Bailly, P., Bain, S., Barnett, A., Todd, J., et al. (1997) The molecular basis of the Kidd blood group polymorphism and its lack of association with type 1 diabetes susceptibility. *Hum Mol Genet* 6, 1017–1020.
16. Irshaid, N. M., Thuresson, B., Olsson, M. L. (1998) Genomic typing of the Kidd blood group locus by a single-tube allele specific primer PCR technique. *Br J Haematol* 102, 1010–1014.
17. Sidoux-Walter, F., Lucien, N., Nissinen, R., Moulds, J., Henry, S., Sistonen, P., et al. (2000) Molecular heterogeneity of the Jk-null phenotype: expression analysis of the Jk S291P mutation found in Finns. *Blood* 96, 1566–1573.
18. Irshaid, N. M., Henry, S. M., Olsson, M. L. (2000) Genomic characterization of the Kidd blood group gene: different molecular basis of the Jk(a-b-) phenotypes in Polynesians and Finns. *Transfusion* 40, 69–74.
19. Lucient, N., Sidoux-Walter, F., Olivés, B., Noulds, J., Le Pennec, P. Y., Cartron, J. P., Baily, P. (1998) Characterization of the gene encoding the human Kidd blood group/urea transporter protein. *J Biol Chem* 273, 12973–12980.
20. Iwamoto, S., Omi, T., Kajii, E., Ikemoto, S. (1995) Genomic organization of the glycoprotein D gene: duffy blood group system is associated with a polymorphism at the 44-amino acid residue. *Blood* 85, 662–666.
21. Mallinson, G., Soo, K. S., Schall, T. J., Pisacha, M., Anstee, D. J. (1995) Mutations in the erythrocyte chemokine receptor (Duffy) gene. The molecular basis of the Fya/Fyb antigens and identification of a deletion in the Duffy gene of an apparently healthy individual with Fy8a-b- phenotype. *Br J Haematol* 90, 823–829.
22. Tournamille, C., Le van Kim, C., Gane, P., Cartron, J. P., Colin, Y. (1995) Molecular basis and PCR-DNA typing of the Fya/Fyb polymorphism. *Hum Genet* 95, 407–410.
23. Gassner, C., Kraus, R. L., Dovc, T., et al. (2000) Fyx is associated with two missense point mutations in its gene and can be detected by PCR-SSP. *Immunohematology* 16, 61–67.
24. Tournamille, C., Colin, Y., Catron, J. P., Le Van Kim, C. (1995) Disruption of a GATA motif in the Duffy gene promoter abolishes erythroid gene expression in Duffy-negative individuals. *Nat Genet* 10, 224.
25. Parasol, N., Reid, M., Rios, M., Castilho, L., Harari, I., Kosower, N. (1998) A novel mutation in the coding sequence of the FY*B allele of the Duffy chemokine receptor gene is associated with an altered erythrocyte phenotype. *Blood* 92, 2237–2243.
26. Olson, M. L., Hansson, C., Avent, N. D., Akesson, I. E., Green, C. A., Daniels, G. (1998) A clinically applicable method for determining the three major alleles at the Duffy (FY) blood group locus using polymerase chain reaction with allele-specific primers. *Transfusion* 38, 168–173.

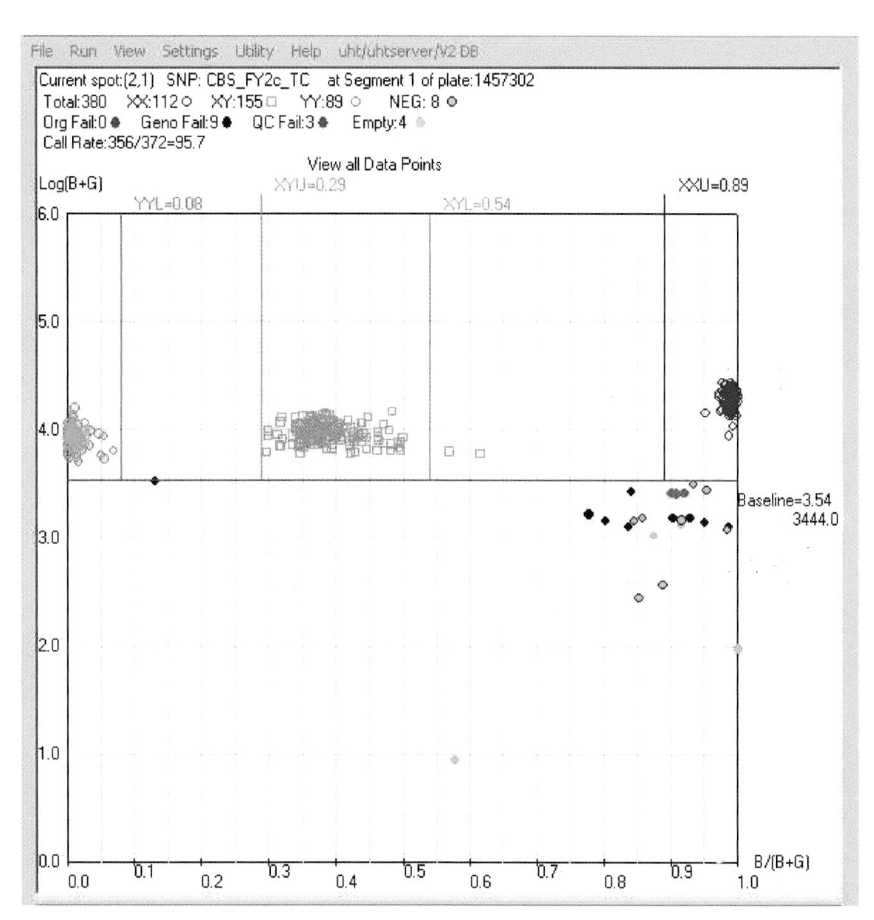

Fig. 2.2. Scatter plot diagram of the resultant BODIDY/TAMRA fluorescent output. The horizontal line represents baseline output (<3.54 units) and identifies genotype, QC, or origin failures and empty wells. The two vertical lines identify the genotypic groupings (XX, XY, and YY) and any outliers. Two outliers are identified in Fig. 2.2 and these samples can be selected for repeat analysis.

27. Reid, M. E. (2003) The Dombrock blood group system: a review. *Transfusion* 43, 107–114.
28. Cartron, J. P., Bailly, Le Van Kim C., Cherif-Zahar, B., Matassi, G., Bertrand, O., Colin, Y. (1998) Insights into the structure and function of membrane polypeptides carrying blood group antigens. *Vox Sang* 74 (2), 29–64.
29. Gassner, C., Schmarda, A., Kilga-Nogler, S., Jenny-Feldkircher, B., Rainer, E., Muller, T. H., Wagner, F. F., Flegel, W. A., Schonitzer, D. (1997) RHD/CE typing by polymerase chain reaction using sequence-specific primers. *Transfusion* 37, 1020–1026.
30. Le Van Kim, C., Mouro, I., Brossard, Y., Chavinié, J., Cartron, J. P. (1994) PCR-based determination of Rhc and RhE status of fetuses at risk of Rhc and RhE haemolytic disease. *Br J Haematol* 88, 193–195.
31. Singleton, B. K., Green, C. A., Avent, N. D., Martin, P. G., Smart, E., Daka, A., Narter-Olaga, E. G., Hawthorne, L. M., Daniels, G. (2000) The presence of an RHD pseudogene containing a 37 base pair duplication and a nonsense mutation in Africans with the Rh D-negative blood group phenotype. *Blood* 95, 8–12.
32. Simsek, S., Faas, B. H. W., Bleeker, P. M. M., Overbeeke, M. A. M., Cuijpers, H.T. M., van der Schoot C. E., von dem Borne A. E. G. (1995) Rapid Rh D genotyping by polymerase chain reaction-based amplification of DNA. *Blood* 85, 2975–2980.

Chapter 4

Real-Time PCR Assays for High-Throughput Human Platelet Antigen Typing

Simon E. McBride

Abstract

Most human platelet alloantigen (HPA) systems comprise biallelic single nucleotide polymorphisms in genes encoding major membrane glycoproteins. Genotyping for these systems is required in the investigation of patients with suspected HPA antibodies and for the provision of compatible blood products from HPA-typed donor panel populations.

Key words: SNP, TaqMan, HPA, high-throughput, platelet, glycoprotein, genotyping.

1. Introduction

A total of 24 serologically defined platelet alloantigens have been reported, 12 of which comprise the six well characterised and so called 'common' biallelic human platelet alloantigen (HPA) systems (HPA-1, -2, -3, -4, -5, and -15) *(1)*. Although HPA-4 is considered a minor allele in Caucasoid populations.

Twenty two of the 24 serologically defined HPAs have been identified at the molecular level, and with the exception of one system (HPA-14) all that differentiates between self and non-self in each case is a non-synonymous amino acid substitution resulting from a single nucleotide polymorphism (SNP) encoded in the relevant platelet glycoprotein (GP) gene *(2)*.

The HPAs are generally considered to be platelet specific due to their localisation on certain glycoproteins that are predominantly expressed on the surface of platelets (**Fig. 4.1**). Most of these membrane bound platelet GPs consist of molecules essential for cell to cell and cell to extracellular matrix interactions during

Peter Bugert (ed.), *DNA and RNA Profiling in Human Blood: Methods and Protocols, vol. 496*

DOI 10.1007/978-1-59745-553-4_4 Springerprotocols.com

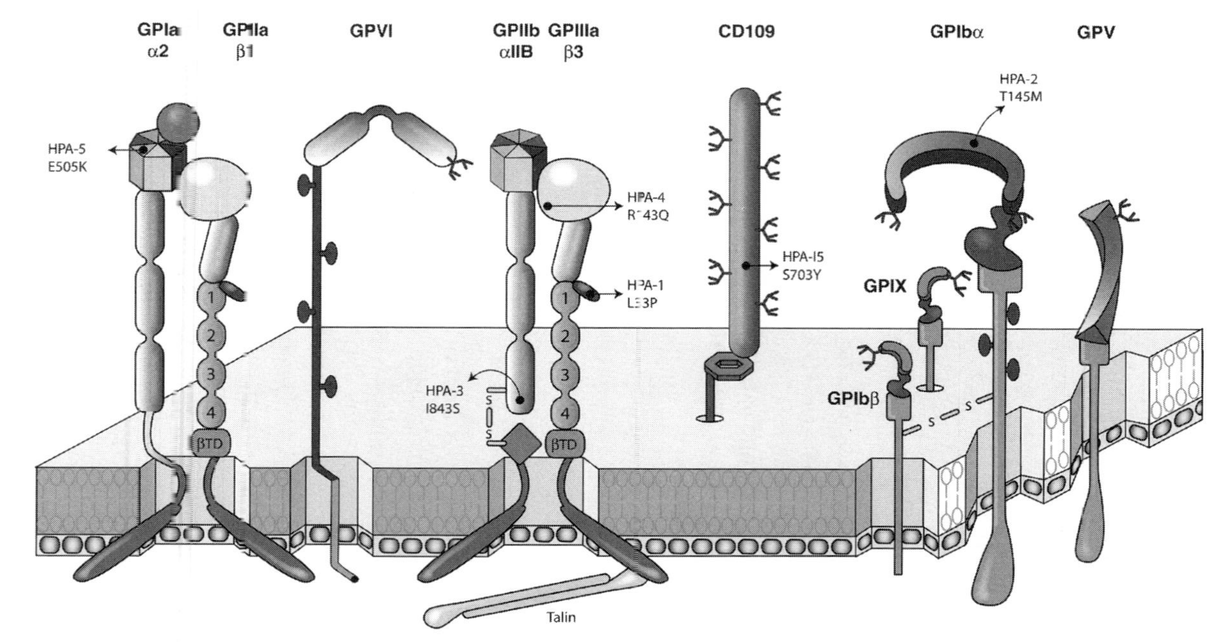

Fig. 4.1. Graphical representation of the major platelet membrane glycoproteins incorporating the positions and accepted nomenclature of the amino acid substitutions resulting in the HPA-1, -2, -3, -5, and -15 systems.

haemostasis, they comprise a family of adhesion receptors referred to as integrins *(3)*.

So far, the majority of known HPAs have been found on the GPIIb/IIIa complex, which as receptor for fibrinogen, von Willebrand factor, fibronectin, and vitronectin is a crucial molecule during platelet aggregation *(4)*. Other significant platelet GPs functioning as adhesion molecules include the GPIb/IX/V complex (a receptor for von Willebrand factor) as well as the GPIa/IIa complex, and CD109 (both receptors for collagen).

The TaqMan® (5′-nuclease) assay, is oligonucleotide hybridisation based and capable of detecting accumulation of PCR product in real time as a result of the release and subsequent detection (by laser) of a fluorophore during successful PCR. The underlying mechanism relies on the 5′-endonuclease activity of *Taq* polymerase for destruction of a hybridised probe oligonucleotide to which dye fluorophore and 'quencher' molecules are attached. Probe cleavage releases the dye from close proximity to the quencher, negating the Forster (or fluorescent) resonance energy transfer (FRET) effect and enabling excitation of the fluorophore by an argon laser *(5)*. It is therefore possible to design a flanking pair of primers with two 'internal' allele-specific probes complementary to a sequence incorporating a biallelic SNP. If each probe possesses a different fluorophore, primer extension leads to allele-specific 5′-endonuclease probe cleavage, fluorescence detection, and corresponding allelic discrimination *(6)*. This works because probe destruction is dependent on stringent probe hybridisation, the 'b' allele-specific probe will not hybridise to template DNA or PCR amplicons from an 'a/a' homozygous individual with enough efficiency to give increased fluorescence after 40 cycles of PCR. Although 'real-time' PCR allows quantifiable measurement of amplification at every cycle, the two-probe allelic discrimination TaqMan® 5′-nuclease assays only require end-point detection of increased fluorescence for genotype determination, a fact sometimes misrepresented by authors in the literature *(7)*. On completion of thermal cycling, TaqMan® allelic discrimination assay plates in 96- or 384-well format can be read in a real-time PCR instrument or other appropriate light amplification-based plate reader. Results are expressed as numerical values representing change in detectable fluorescence for the two 'reporter' dye fluorophores (typically FAM and VIC) and an internal positive control fluorophore (ROX). On the ABI PRISM 7900HT sequence detection system instrument platform providing at least two negative 'no template' control (NTC) wells are included on each plate, and a user-defined quality value is set, the analysis software will attempt to objectively assign genotypes using an algorithm based on the fluorescence ratio of the dye molecules in each sample. Results are displayed on a scatter plot and cluster according to genotype, each axis represents change in fluorescence for a different dye fluorophore; for example, HPA-1a and HPA-1b homozygous samples will exhibit only raised FAM or VIC fluorescence, respectively, whereas heterozygotes will have intermediate fluorescence values for both dyes. Samples may not autocall if the concentration or quality of DNA varies within a plate as this produces loose clusters on the scatter plot resulting in samples falling outside the defined confidence intervals. The facility exists to manually assign genotypes according to 'cluster' on the scatter graph. By also referring to the raw spectral data (a graph of fluorescence versus wavelength of light) operators can be confident of results defined in this way, with the possible exception of only very weak or outlying samples. The TaqMan® 5′-nuclease assay, a closed single tube 'homogeneous' genotyping technique offers the advantages of reduced carryover or cross contamination; no requirement for post-PCR sample manipulation; increased sensitivity *(8)*; fluorescence or fluorescence polarisation detection *(9, 10)*; 3′ minor groove binder (MGB) probes with non-fluorescent quenchers for increased specificity and detection *(11)*; standardised reagent design and thermal cycling protocol; an assays-by-design service *(12)*; the capability of objective allele calling for genotype determination and the export of results in various file formats for the electronic upload of data onto sample databases.

2. Materials

2.1. DNA Sample Dilution and Aliquoting

1. Low retention aerosol barrier tips, 1 ml (Molecular Bioproducts Inc., San Diego, CA, USA).
2. Low retention aerosol barrier tips, 200 μl (Molecular Bioproducts).
3. 'Deep' 96-well plate, 1.2 mL capacity (Alpha Laboratories, Eastleigh, UK).
4. Aluminium adhesive plate sealers (Beckman Coulter Inc., Fullerton, CA, USA).
5. Nuclease free water (Promega, Madison, WI).
6. Optical TaqMan 96-well plates (Applied Biosystems, Warrington, UK).

2.2. TaqMan Allelic Discrimination Assay

1. Sterile 'Universal', 25 ml (Sterilin).
2. PCR set-up dedicated pipettes 10, 100, 200, and 1000 μl.
3. Aerosol barrier pipette tips, 10, 100, 200 μl, and 1 ml (Alpha Laboratories).
4. Multidispensing pipette (Gilson).
5. Sterile individually wrapped multidispensing tips 10–1250 μl capacity (Gilson).
6. Flat optical strip caps (Applied Biosystems).

7. Cap sealing tool (Applied Biosystems).
8. TaqMan custom HPA assay primers, 100 μM (Applied Biosystems).
9. TaqMan custom HPA assay probes, 15 μM (Applied Biosystems).
10. TaqMan 2x Universal PCR Mastermix with UNG AmpErase (Applied Biosystems).
11. Nuclease free water (Promega). Store at 4°C once open.
12. MicroAmp base/s (Applied Biosystems).
13. Pre-aliquoted DNA in MicroAmp 96-well optical plates.
14. DNA samples of known genotype homozygous for the 'a' and 'b' alleles of the HPA systems to be assayed at a concentration of 5–10 ng/μl.

3. Methods

As with all PCR-based assays, particular care should be taken to avoid potential contamination of reagents and consumables at all times, ideally assay preparation should take place within a dedicated PCR set-up area or hood style workstation. Validation of the TaqMan 7900HT sequence detection platform for genotyping the five common human platelet alloantigen SNPs (HPA-1, -2, -3, -5, and -15) has been performed in direct comparison to a microarray-based approach with the use of an independently blinded, randomised and distributed set of test DNAs deliberately enriched for low-frequency 'b' allele homozygous samples *(13)*.

3.1. DNA Sample Dilution and Aliquoting

1. These directions assume the use of an entry-level automated liquid handling system such as the Beckman Biomek® 2000 robotic workstation, although a good quality multichannel pipette (and a steady arm) may suffice as a low cost alternative (*see* **Note 1**). DNA should be of a quality suitable for diagnostic PCR, any contaminants will likely impair optimal assay performance.
2. Prepare working dilution stocks of DNA samples (90 per plate) at a concentration of 5–10 ng/μl in the deep 96-well plate, leaving the last six wells of the bottom row empty. Final volume in each well of this plate should not exceed 600–700 μl to allow for sample displacement during aspiration. If not intended for immediate use seal the plate with an adhesive plate sealer (*see* **Note 2**). Repeat this step for every additional 90 samples to be genotyped.
3. For HPA-1, -2, -3, -4, -5, and -15 genotyping, create six replicates of each working dilution plate using the 96-well optical TaqMan plates. Aliquot 5μl of DNA into each well, checking to make sure there is DNA in all 90 sample positions of

every plate. Seal each plate with an adhesive plate sealer and store at 4°C for immediate use or –40°C long term storage. (*see* **Note 3**).

3.2. TaqMan Allelic Discrimination Assay

1. These instructions and the accompanying TaqMan primer and probe designs (including reaction conditions) assume the use of GeneAmp PCR system 9700 Thermal cycler. Use of other makes or types of thermal cycler may require some assay reoptimisation.
2. For each HPA system prepare sufficient reaction mix for the number of plates to be assayed, approximately 100 samples worth of mix per plate should be sufficient. Calculate the total volume of each reagent required using the values in **Table 4.1** (*see* **Note 4**). When making reaction mixes for multiple HPA assays it is best to prepare them one at a time to avoid the possibility of accidental cross-contamination of HPA-specific reagents. Store the mix at 4 °C until ready to use (*see* **Note 5**).
3. Select the required pre-aliquoted DNA plates (from step 3; **Section 3.1**) to genotype for each HPA assay. Centrifuge the plates at 3000 rpm for 1 min and briefly spin the appropriate control DNA samples in a microcentrifuge before use (*see* **Note 6**).
4. To each plate add 5μl of the 'a' and 'b' allele homozygous positive control samples to wells 91–92 (H7 and H8) and 93–94 (H9 and H10), respectively.
5. To each plate also add 5 μl of nuclease free water to wells 95–96 (H11 and H12) to provide 'no template' negative controls (NTCs).
6. Using a multidispensing pipette and a sterile 10–1250 μl capacity tip add 10 μl of reaction mix (from step 2) to each well of the corresponding plate/s.

Table 4.1
Stock solutions and volumes used per reaction

Reagent	Stock	Volume per reaction(μl)
Mastermix	2 ×	7.50
Forward primer	100 μM	0.14
Reverse primer	100 μM	0.14
'a' Allele probe	15 μM	0.20
'b' Allele probe	15 μM	0.20
Nuclease free water	NA	1.82

7. Seal the plates using optical strip caps and check to ensure all wells are completely sealed using the cap sealing tool (*see* **Note** 7).
8. Centrifuge the sealed plates at 3000 rpm for 1 min.
9. Place the prepared assay plate/s into thermal cycler/s.
10. Thermal cycle the plates using the following conditions: 50°C for 2 min, 95°C for 10 min, and then 40 cycles of 95°C for 15 s and 'X' °C for 1 min, followed by incubation at 4°C; where 'X' is 64°C for the HPA-2 assay, 60°C for the HPA-4 assay, and 56°C for the HPA-1, HPA-3, HPA-5, and HPA-15 assays when using the custom primer and probe sequences from **Table 4.2**.

Table 4.2
TaqMan custom primer and allele-specific probe sequences for HPA-1, -2, -3, -4, -5, and -15 genotyping

Assay	Primer sequences	MGB probe sequences[a] (3′ NFQ)
HPA-1	Forward primer:	HPA-1a probe:
	5′- CTG ATT GCT GGA CTT CTC TTT GG -3′	(5′ FAM labelled) 5′- CTG CCT CTG GGC TC -3′
	Reverse primer:	HPA-1b probe:
	5′- AGC AGA TTC TCC TTC AGG TCA CA -3′	(5′ VIC labelled) 5′- CTG CCT CCG GGC TC -3′.
HPA-2	Forward primer:	HPA-2a probe:
	5′-CTG AAA GGC AAT GAG CTG AAG AC- 3′	(5′ FAM labelled) 5′- CTC CTG ACG CCC ACA -3′
	Reverse primer:	HPA-2b probe:
	5′-CCA GAC TGA GCT TCT CCA GCT T- 3′	(5′ VIC labelled) 5′- CTC CTG ATG CCC ACA C -3′
HPA-3	Forward primer:	HPA-3a probe:
	5′- TGG GCC TGA CCA CTC CTT T -3′	(5′ FAM labelled) 5′- TGC CCA TCC CCA GCC -3′
	Reverse primer:	HPA-3b probe:
	5′- TGA TGG GCC GGG TGA A -3′	(5′ VIC labelled) 5′- CTG CCC AGC CCC AG -3′
HPA-4	Forward primer:	HPA-4a probe:
	5′- CAG AAC CTG GGT ACC AAG CT -3′	(5′ FAM labelled) 5′- CAG ATG CGA AAG CT -3′
	Reverse primer:	HPA-4b probe:
	5′- CAA TCC GCA GGT TAC TG -3′	(5′ VIC labelled) 5′- CAG ATG CAA AAG CT- 3′

(continued)

Table 4.2 (continued)

Assay	Primer sequences	MGB probe sequences[a] (3′ NFQ)
HPA-5	**Forward primer:**	**HPA-5a probe:**
	5′- GAC CTA AAG AAA GAG GAA GGA AGA GTC T -3′	(5′ FAM labelled) 5′- TTA CTA TCA AAG AGG TAA AAA -3′
	Reverse primer:	**HPA-5b probe:**
	5′- ATG CAA GTT AAA TTA CCA GTA CTA AAG CAA -3′	(5′ VIC labelled) 5′- TGT TTA CTA TCAAAA AGG TAA A -3′
HPA-15	**Forward primer:**	**HPA-15a probe:**
	5′- TGT ATC AGT TCT TGG TTT TGT GAT GTT -3′	(5′ FAM labelled) 5′- CTT CAG TTC CAG GAT TT -3′
	Reverse primer:	**HPA-15b probe:**
	5′- CCA AGA AGT GAT AGA ATC AGG TAC AGT TAC -3′	(5′ VIC labelled) 5′- CTT CAG TTA CAG GAT TT -3′

[a]Minor groove binding (MGB) modification is used as a non-fluorescent quencher (NFQ) at the 3′-end of each probe; underscores denote positioning of the complementary polymorphic nucleotide.

3.3. Analysis of TaqMan Allelic Discrimination Results for HPA Genotyping

1. These directions assume the use of an ABI Prism 7900HT sequence detection system using SDS software version 2.1.
2. Switch on the computer followed by the ABI Prism 7900HT instrument and allow to warm up for at least 30 min before use.
3. Start up the SDS software and create a new allelic discrimination plate document (96 well). Type or barcode in samples details corresponding to the equivalent wells of the plate grid, including control samples in wells 91–96 (H7–H12).
4. Create two 'detectors' (one for each probe, e.g., '1a' and '1b') and add these to a 'marker' corresponding to the assay plate to be analysed (e.g., HPA-1). Add this marker to the plate document.
5. Select all 96 wells of the plate grid and assign the marker to these samples, they will now be designated 'Unknown' and acquire a 'U' symbol on the plate grid graphic.
6. Select wells 95–96 of the plate grid and assign them as 'no template control' wells; they will now be designated 'NTC' and acquire an 'N' symbol on the plate grid.
7. Set the analysis parameters for allelic discrimination to a quality value of 95.0 and ensure the 'auto caller' is enabled.
8. Check the assay plate for bubbles, briefly spin the plate to remove all bubbles if there are any present.
9. Read the plate as a post-PCR run, then save the raw data file.

10. View the results as a scatter graph, check that the no template control wells and the positive control samples for both alleles have been amplified and correctly assigned by the software – an example is shown in **Fig. 4.2**.
11. Look at the clustering of the automatically assigned genotypes, check the raw spectra of any outlying or indeterminate

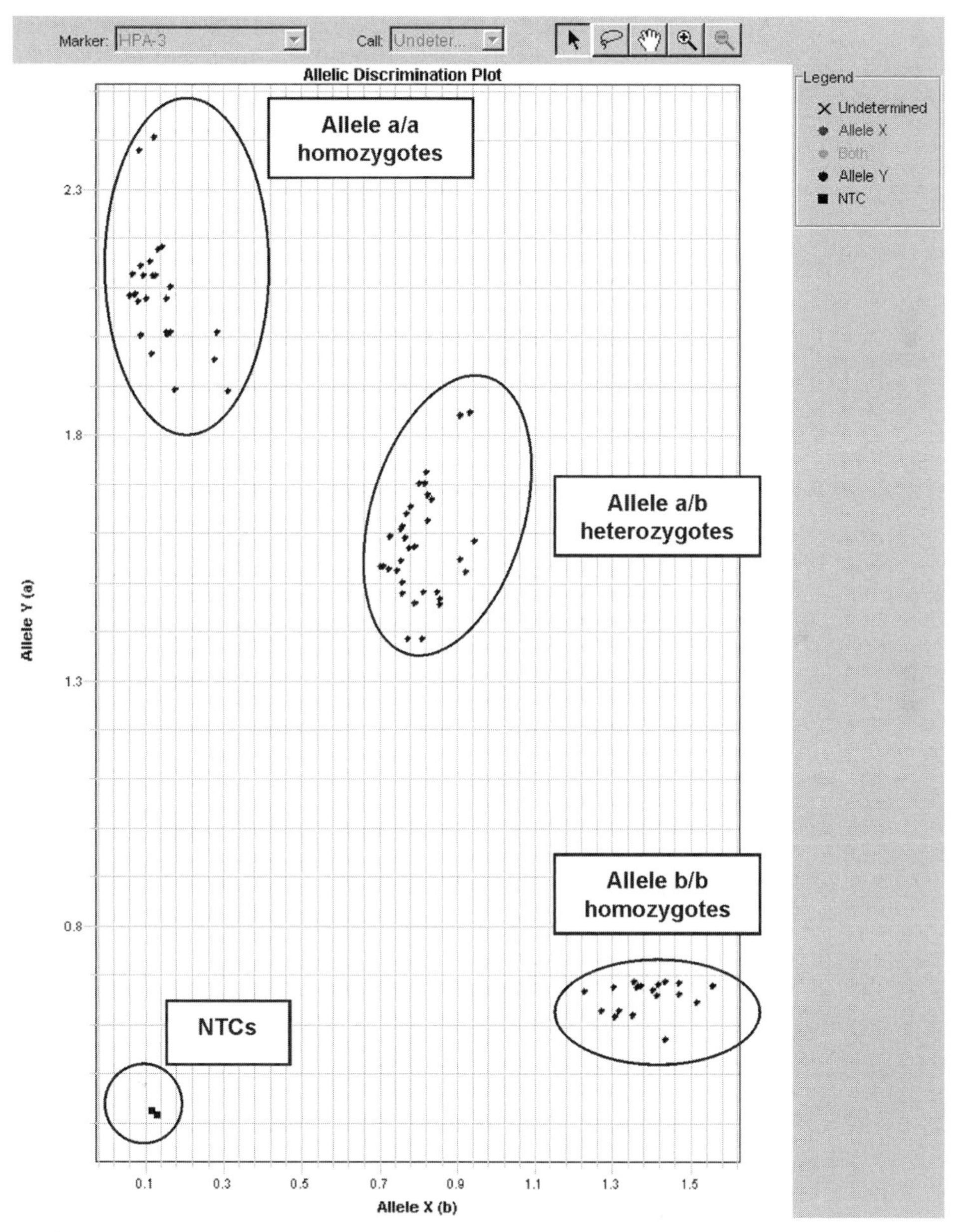

Fig. 4.2. Sample scatter plot showing HPA-3 genotype distribution for 90 samples of previously unknown type after post-PCR assignment of alleles by SDS software. Samples called as undetermined or having failed to amplify would be represented as 'x' (none present in this instance) (reproduced from Ref. *(13)* with permission from Blackwell Publishing).

results. If necessary, manually amend or reassign any erroneous autocalls (*see* **Note 8**).

4. Notes

1. The TaqMan 5′-nuclease assay for allelic discrimination is a fairly robust and sensitive assay and will work on samples of varying quality or concentration. However, avoiding human error and ensuring accurate pipetting is of paramount importance for high-throughput DNA amplification studies and the author highly recommends the use of a robot workstation if at all possible, especially when working in 384-well format.
2. Single-use disposable adhesive plate sealers are preferable to resealable flexible 'mat' style 96-well plate lids as they prevent potential cross contamination due to residual DNA retained on lids which are removed and replaced numerous times.
3. If the plates are stored long term at –40°C they will dry out over time until the wells of the plate are coated in desiccated DNA. This loss of volume will need to be taken into account during step 2 in Section 3.2 when calculating the volume of nuclease free water to add to the reaction mix. This can be used as a deliberate measure to maximise the reagent volume remaining when working in 384-well plate format.
4. Depending on the design of reagent reservoir it can be less wasteful to prepare larger batches of assay-specific reaction mix sufficient for as many plates as you can process at a time. We also found the use of a 'Distriman' multidispensing pipette which allowed us to minimise the volume of extra mix required.
5. We have stored TaqMan assay reaction mixes for up to 2 weeks at 4°C with no subsequent loss of reaction efficiency.
6. Centrifugation of the DNA plates and control samples is important to reduce the likelihood of cross-contamination due to aerosols.
7. It is essential to wear clean gloves and use a clean capping tool as any marks on the lids of reaction plates may impair plate reading or give rise to erroneous allele calls. To end this we recommend checking the heated lids of thermal cyclers for scorching or permanent marker stains; use of a clean rubber pressure mat easily solves this problem.
8. Comparing the raw spectra of unknown samples to those of known genotype is normally sufficient to resolve most discrepancies. However, care must be taken as outlying results may be due to the presence of mutation/s in the probe-binding region of the target sequence.

Acknowledgements

The author would like to thank Dr Willem Ouwehand for kind permission to reproduce the cartoon showing graphical representations of platelet glycoproteins in **Fig. 4.1**.

References

1. Metcalfe, P., Watkins, N. A., Ouwehand, W. H., Kaplan, C., Newman, P., Kekomaki, R., de Haas, M., Aster, R., Shibata, Y., Smith, J., Kiefel, V., Santoso, S. (2003) Nomenclature of Human Platelet Antigens (HPA). *Vox Sanguinis* 85, 240–245.
2. Santoso, S., Kiefel, V., Richter, I. G., Sachs, U. J., Rahman, A., Carl, B., Kroll, H. (2002) A functional platelet fibrinogen receptor with a deletion in the cysteine-rich repeat region of the beta (3) integrin: the Oe(a) alloantigen in neonatal alloimmune thrombocytopenia. *Blood* 99, 1205–1214.
3. Santoso, S. (2003) Human platelet alloantigens. *Transfusion Apheresis Sci* 28, 227–236.
4. Norton, A., Allen, D. L., Murphy, M. F. (2004) Review: platelet alloantigens and antibodies and their clinical significance. *Immunohematology* 20, 89–102.
5. Forster, V. Th. (1948) Zwischenmolekulare energiewanderung und fluoresenz. *Ann Phys (Leipzig)* 2, 55–75.
6. Livak, K. J. (1999) Allelic discrimination using fluorogenic probes and the 5′ nuclease assay. *Genetic Anal Biomol Eng* 14, 143–149.
7. Ficko, T., Galvani, V., Rupreht, R., Dovc, T., Rozman, P. (2004) Real-time PCR genotyping of human platelet alloantigens HPA-1, HPA-2, HPA-3 and HPA-5 is superior to the standard PCR-SSP method. *Trans Med* 14, 425–432.
8. Kalinina, O., Lebedeva, I., Brown, J., Silver, J. (1997) Nanoliter scale PCR with TaqMan detection. *Nucleic Acids Res* 25, 1999–2004.
9. Latif, S., Bauer-Sardina, I., Ranade, K., Livak, K. J., Kwok, P-Y. (2001) Fluorescence polarisation in homogeneous nucleic acid analysis II: 5′-nuclease assay. *Genome Res* 11, 436–440.
10. Kwok, P-Y. (2002) SNP Genotyping with fluorescence polarization detection. *Human Mutation* 19, 315–323.
11. Kutyavin, I. V., Afonina, I. A., Mills, A., Gorn, V. V., Lukhtanov, E. A., Belousov, E. S., Singer, M. J., Walburger, D. K., Lokhov, S. G., Gall, A. A., Dempcy, R., Reed, M. W., Meyer, R. B., Hedgpeth, J. (2000) 3′-Minor groove binder-DNA probes increase sequence specificity at PCR extension temperatures. *Nucleic Acids Res* 28, 655–661.
12. Higgins, M., Hughes, A., Buzzacott, N., Lown, J. (2004) High-throughput genotyping of human platelet antigens using the 5′-nuclease assay and minor groove binder probe technology. *Vox Sanguinis* 87, 114–117.
13. Bugert, P., McBride, S., Smith, G., Dugrillon, A., Kluter, H., Ouwehand, W., Metcalfe, P. (2005) Microarray-based genotyping for blood groups: comparison gene array and 5′-nuclease assay techniques with human platelet antigen as a model. *Transfusion* 45, 654–659.

Chapter 5

Multiplex ABO Genotyping by Minisequencing

Gianmarco Ferri and Susi Pelotti

Abstract

The minisequencing multiplex reaction to genotype ABO blood group system is described and discussed. This method was found to be a reproducible strategy to type 6 common alleles (A1, A2, B, 01, 01V, and 02) of the ABO blood group system, the high specificity and sensitivity make it suitable also in forensic science. It is more rapid than RFLP and SSCP analysis, resulting in unambiguous interpretation of ABO genotypes and newly discovered mutations are readily investigated by the addition of new extension primers in the minisequencing multiplex reaction.

Key words: ABO genotyping, polymorphism, minisequencing, polymerase chain reaction, single nucleotide polymorphism.

1. Introduction

Molecular analysis methods were introduced to transfusion medicine after cloning of the genes became routine. Single nucleotide polymorphisms (SNPs) inherited in a Mendelian manner are linked to many blood group antigens and the appropriate molecular methods to blood group testing are linked to the genetic basis for polymorphism and the development of PCR-based technology.

The ABO system is the most important blood group system in transfusion medicine, organ transplantation, and in forensic medicine as well, where the interpretation of forensic evidence materials requires an analysis method suitable for degraded DNA. In addition, ABO genotyping with the increasing number of alleles found continues to be well suited for paternity testing together with microsatellite typing. The cloning of the ABO gene in 1990

Peter Bugert (ed.), *DNA and RNA Profiling in Human Blood: Methods and Protocols, vol. 496*

DOI 10.1007/978-1-59745-553-4_5 Springerprotocols.com

(1) has led to the identification of up to 169 alleles *(2)* typed with different molecular typing methods.

The ABO gene codes for the glycosyltransferases that catalyze the biosynthesis of blood group A and B antigens. This gene maps to chromosome 9 and consists of seven exons and six introns ranging in size from 28 to 688 bp. Two mechanisms are considered to be responsible for genetic diversity at the ABO locus: point mutations and genetic recombination *(3)*, differing the common alleles in white individuals AB0*A101 (A1), AB0*A201 (A2), AB0*B101 (B1), AB0*O01 *(1)*, AB0*O02 (01V), and AB0*O03 *(2)* by few base positions in exons 6 and 7 *(4)*. Determination of ABO genotypes from DNA is complex, partly because of the large number of variants involved and partly because the ABO gene does not encode the blood group antigen directly, but rather encodes the transferases that catalyze their biosynthesis.

PCR–RFLP, alleles-specific PCR, sequence-specific PCR as single or multiplex assays, and real-time PCR have been applied to identify SNPs to complement routine serological typing *(5–7)*.

High-throughput techniques have been recently developed for the study of human SNPs; however, for medium/small clinical and research laboratory simple and inexpensive techniques are needed. The minisequencing method by SNaPshot kit is uncomplicated, does not require dedicated instrument(s), the samples are resolved by electrophoresis separation with DNA sequencer, and it can be easy multiplexed by the addition of different oligonucleotide tails to increase the size of minisequencing primers that are resolved and differentiated by length. In this assay, a DNA oligonucleotide or "primer" is extended by a single nucleotide unit. The identity of the extended base allows the sample to be accurately genotyped. In general, it is possible to separate single base extension (SBE) primers if the length differs by 3–5 nucleotides, even if there may be difficulties in separating some primers that are shorter than 30 nucleotides, because the electrophoretic mobility of short primers depends strongly on nucleotide composition and on fluorescent ddNTP addition. Therefore with a primer range from 20 to 100 nucleotides, about 30–40 SNPs can be detected, even if most of published SNaPshot assays identify approximately 10–25 SNPs, with a maximum of 52 in one reaction *(8)*.

To type the ABO blood group system, a method based on a duplex-PCR assay that amplifies 131/132 and 389 bp fragments on exons 6 and 7, respectively, of ABO gene followed by minisequencing using SNaPshot multiplex kit was developed. This method focuses on six relatively common ABO alleles (A1, A2, B, 01, 01V, and 02) by analyzing five SNPs at nucleotide positions 261 and 297 on exon 6 and 467, 646, and 703 on exon 7 *(9)*. New extension primers were designed and 70 blood donors with known serologically based predicted genotypes in family studies were tested, in addition, sensitivity assays were performed. Minisequencing has shown to be a reproducible strategy to type 6 common alleles (A1, A2, B, 01, 01V, and 02) of the ABO blood group system showing this method has several advantages: first, one reaction mixture is required for DNA template amplification and one for minisequencing analysis. Second, it is more rapid than RFLP and SSCP analysis, resulting unambiguous the interpretation of ABO genotypes with a table of reference. In addition, this easy and fast molecular approach may be very useful for personal identification as well as for paternity testing in forensic science, as it is characterized by both high specificity and sensitivity.

2. Materials

2.1. DNA Extraction

1. QIAamp DNA mini kit (Qiagen, Hilden, Germany).

2.2. PCR Amplification

1. PCR primers were designed to amplify the regions including nucleotide positions 261 and 297 in exon 6 and 467, 646, and 703 in exon 7 (*see* **Note 1**). All primers were designed with computer software (Primer3, H Rozen and J. Skaletsky, the code is available from: http://www.genome.wi.mit.edu/genome_software/other/primer3.html). Oligonucleotides were purchased from MWG-Biotech (Eberberg, Germany), desalted and lyophilized Primers sequences with final concentrations:

   AB0 6F 5'-GCCTCTCTCCATGTGCAGTA-3', 0.3 μM;
   AB0 6R 5'-AACCCAATGGTGGTGTTCTG-3', 0.3 μM;
   AB0 7F 5'-TGGCTTTCCTGAAGCTGTTC-3', 0.3 μM;
   AB0 7R 5'-GATGTAGGCCTGGGACTGG-3', 0.3 μM.

2. Reaction mixture with 1X gold buffer (12.5–25 μl), 2 mM $MgCl_2$, 200 μM each dNTP (Applera, Foster City, CA, USA), and 1.25 units AmpliTaqGold DNA polymerase (Applera).
3. PCR instrument: GeneAmp 9700 thermal cycler (Applera).

2.3. PCR Purification

1. *Escherichia coli* exonuclease shrimp alkaline phosphatase (ExoSAP) enzyme (USB Corporation, Cleveland, OH, USA).

2.4. Electrophoresis

1. Agarose (2%).
2. TBE: Tris (pH 8), boric acid, and EDTA.
3. Ethidium bromide.

2.5. Minisequencing

1. The extension primers were designed with Primer3 software, by the 3' end of primer that anneal one base before selected SNP (*see* **Note 2**): Extension primers were purchased from MWG-Biotech, desalted and lyophilized:
   261 F 5'-$(T)_8$GGAAGGATGTCCTCGTGGT-3', 0.15 μM;
   297 R 5'-GTTGAGGATGTCGATGTTGAA-3', 0.4 μM;
   467 F 5'-$(T)_{16}$TACTATGTCTTCACCGACCAGC-3', 0.6 μM;
   646 F 5'-$(T)_{24}$GCGTGGACGTGGACATGGAG-3', 0.4 μM;
   703 F 5'-$(T)_{17}$CGGCACCCTGCACCCC-3', 0.075 μM.
2. SNaPshot kit (Applera)
3. GeneAmp 9700 thermal cycler (Applera).

2.6. Post-Extension Treatment

1. Shrimp alkaline phosphatase (SAP) enzyme (USB).

2.7. Capillary Electrophoresis

1. ABI 310 or ABI 3130 DNA sequencers (Applera).
2. HiDi formamide (Applera).
3. Liz 120 standard (Applera).
4. Performance Optimizer Polymer 4 (Applera).
5. Array, 36 cm (Applera).

3. Methods

3.1. DNA Extraction

1. DNA was extracted with the DNA isolation kit (QIAamp DNA mini kit) from peripheral blood and buccal swabs from 70 unrelated healthy blood donors whose ABO phenotypes were determined by serologic testing and family studies.
2. Aliquots of 2, 1 ng, and 800 pg of genomic DNA from K562 cell line were prepared.

3.2. Amplification

1. PCR primers were designed to amplify the region including nucleotide positions 261 and 297 in exon 6 and 467, 646, and 703 in exon 7. The size of amplicons was 131/132 base pairs and 389 base pairs, respectively, for exon 6 and 7.
2. The simultaneous amplification of the two amplicons containing the polymorphic sites of interest is carried out in a total volume of 12.5 or 25 μL with 1–5 ng DNA templates, 1X Gold buffer, 2 mM $MgCl_2$, 200 μm each dNTP, and 1.25 units AmpliTaqGold DNA polymerase.
3. Place in a PCR cycler and cycle under the following conditions:

Initial denaturation	95°C; 10 min
34 cycles of:	94°C, 1 min; 59°C, 1 min; 72°C, 1 min
Final extension:	72°C, 7 min.

4. Run PCR products in 2% agarose gels containing ethidium bromide.

3.3. PCR Purification

Primers and unincorporated dNTPs are removed by incubation of 1.5 μl of PCR product with 1.5 μl of ExoSAP enzymes for 15 min at 37°C followed by inactivation by heating at 80°C for 15 min.

3.4. Minisequencing

1. Minisequencing multiplex reaction are performed by using SNaPshot kit with 1 μl of purified PCR product in a 8 μl of total reaction volume with 2.5 μl of SNaPshot buffer, and SBE primers at the final concentration (Section 2.5) (*see* **Note 3**).
2. Place in a PCR cycler and cycle under the following conditions: 25 cycles of: 96°C, 10 s; 50°C, 5s;60°C, 30 s.

3.5. Post-extension Treatment

1. Add 1 U of SAP directly in SBE reaction tubes and incubate at 37°C for 45–60 min (*see* **Note 4**).
2. Inactivate enzymes by incubation at 80°C for 15 min.

3.6. Capillary Electrophoresis and Analysis of Minisequencing Products

SBE purified products are resolved by electrophoresis separation in an ABI 310 or in an ABI 3130 DNA sequencer.

1. For preparation of samples to electrophoresis separation, 1 μl of SNaPshot purified extension product is mixed with 12–15 μl of formamide and 0.5 μl of size standard Liz 120.
2. Samples are run on an ABI 310 or 3130 genetic analyzer using POP 4.
3. Data are analyzed either using GenScan v 3.1 or GeneMapper v 3.7 (*see* **Note 5**).

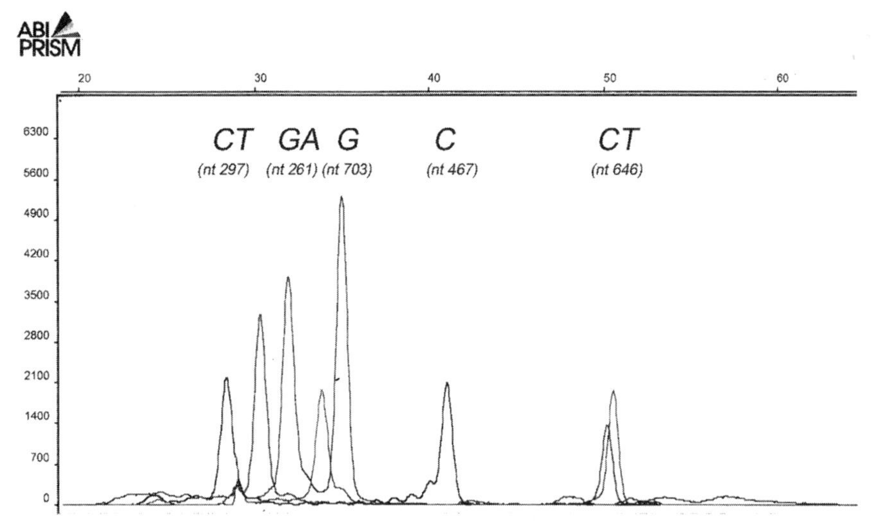

Fig. 5.1. Representative electropherogram showing SNaPshot minisequencing of the ABO A1/01V genotype.

Table 5.1
Reference table for ABO alleles assignment based on minisequencing analysis of five SNPs

Allele	R 297(22 bp)	F 261(28 bp)	F 703(34 bp)	F 467(39 bp)	F 646(45 bp)
A1	T	G	G	C	T
A2	T	G	G	T	T
B	C	G	A	C	T
01	T	A	G	C	T
01V	C	A	G	C	A
02	C	G	G	C	T

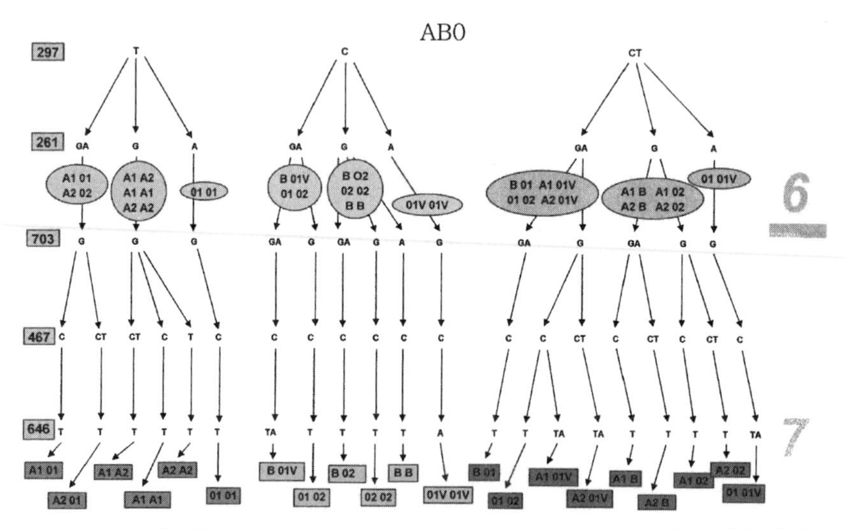

Fig. 5.2. Reference diagram for ABO genotype assignment based on minisequencing analysis of five SNPs.

4. The interpretation of the electrophoretic pattern from minisequencing (**Fig. 5.1**) with a table of reference (**Table 5.1**, **Fig. 5.2**) allows to assign the ABO genotype without difficulties (*see* **Note 6**).

4. Notes

1. To achieve unambiguous genotyping some optimizations were necessary due to unspecific annealing of extension primers overcoming by several attempts of primers set design and each extension primer was first tested by uniplex reaction. The use

of software like Primer 3 or similar allows to test the designed primers for the best annealing temperature and hairpin and secondary structures avoiding.

2. The tails of the detection primers must not contain nucleotides sequences that influence the hybridization process and this can be gained by the addition of random nucleotide sequences unspecific for human DNA and, or by polynucleotide sequences (polyA, polyT, polyC, and polyG).
3. The addition of 20 mM ammonium sulfate to the SBE reaction, as suggested elsewhere, had no significant effects for suppression of nonspecific peaks.
4. Some artifact peaks can appear in the electropherogram if the SAP digestion does not remove all the ddNTP. In this case, the purification can be repeated on the same minisequencing product.
5. Alternative nucleotide incorporation for each SNP affects the migration size of extension primer. Before starting, the multiplex assay is strongly recommended to test extension primers in a singleplex minisequencing reaction to know the real-migration size of the two alternative nucleotide incorporation. Consequently, building a reference table (see **Table 5.1**) with specific migration size of alternative nucleotide for each SNP site is also suggested to avoid mutations swap and/or misinterpretation, supporting the electropherograms reading.
6. The occurrence of hybrid or subgroup ABO alleles has emphasized the need of better genotyping protocols to safe phenotype prediction and a profound knowledge of the genetic diversity at this locus is required to design genotyping kits. The minisequencing technique described was developed to screen common alleles, but since new alleles are apparently the result of gene conversion events with a hotspot in and around exon 6 and are characterized by new arrangements of known mutations, a minisequencing, genotyping assay that investigates each known polymorphic site of exons 6 and 7, or in ambiguous cases of the entire intron/exon regions, makes possible resolving of novel ABO alleles containing known sequence motifs in new combinations. It is impossible to investigate every single mutation that leads to the 169 different types of ABO alleles, but a situation of potential discrepancies in some case of hybrid alleles originated by unequal crossing over can be resolved. The potential mistyping due to hybrid alleles may be resolved in most cases by the minisequencing approach, e.g., heterozygous sample A1/01-B hybrid, phenotype A1. Also a serological discovery of the weak aggregation with monoclonal reagents requests a DNA investigation of the ABO alleles. Besides, newly discovered mutations are readily investigated by the addition of new extension primers in the minisequencing multiplex reaction.

Acknowledgments

Authors would like to thank the Blackwell Publishing and the Transfusion Journal for the permission to use parts of the article: Ferri et al. (2004) ABO genotyping by minisequencing analysis. *Transfusion* 44, 943–944.

References

1. Yamamoto, F., Clausen, H., White, T., Marken, J., Hakomori, S. (1990) Molecular genetic basis of the histo-blood group ABO system. *Nature* 345, 229–233.
2. http://www.ncbi.nlm.nih.gov/projects/mhc/xslcgi.fcgi?cmd=bgmut/systems_info&system=abo
3. Yip, S. P. (2000) Single-tube multiplex PCR-SSCP analysis distinguishes 7 common ABO alleles and readily identifies new alleles. *Blood* 95, 1487–1492.
4. Seltsam, A., Hallensleben, M., Kollmann, A., Blasczyk, R. (2003) The nature of diversity and diversification at the ABO locus. *Blood* 102, 3035–3042.
5. Olsson, M. L., Irshaid, N. M., Hosseini-Maaf, B., Hellberg, A., Moulds, M. K., Sareneva, H., Chester, M. A. (2001a) Genomic analysis of clinical samples with serologic ABO blood grouping discrepancies: identification of 15 novel A and B subgroup alleles. *Blood* 98, 1585–1593.
6. Olsson, M. L., Chester M. A. (2001b) Polymorphism and recombination events at the ABO locus: a major challenge for genomic ABO blood grouping strategies. *Transfus Med* 11, 295–313.
7. Yazer, M. H. (2005) What a difference 2 nucleotides make: a short review of ABO genetics. *Transfus Med Rev* 19, 200–209.
8. Sanchez, J. J., Phillips, C., Børsting, C., Balogh, K., Bogus, M., Fondevila, M., Harrison, C. D., Musgrave-Brown, E., Salas, A., Syndercombe-Court, D., Schneider, P. M., Carracedo, A., Morling, N. (2006) A multiplex assay with 52 single nucleotide polymorphisms for human identification. *Electrophoresis* 27, 1713–1724.
9. Ferri, G., Bini, C., Ceccardi, S., Pelotti, S. (2004) ABO genotyping by minisequencing analysis. *Transfusion* 44, 943–944

Chapter 6

Multiplex Genotyping for Thrombophilia-Associated SNPs by Universal Bead Arrays

Susan Bortolin

Abstract

This chapter describes a method for the multiplex analysis of six biallelic single nucleotide polymorphisms (SNPs) associated with thrombophilia. The method may, however, be adapted for the simultaneous analysis of up to 100 markers (50 biallelic SNPs) in a single reaction. In the method described, the targets of interest are amplified by single-tube multiplex PCR using six primer sets followed by single-tube multiplex allele-specific primer extension using 12 universally tagged genotyping primers. Labeled extension products are sorted using the xTAG™ universal bead-based array and detected on the Luminex xMAP® system. The 12 universal tag sequences used in the assay derive from a set of 100 universal tags which have been designed to be isothermal and have been empirically validated to show that mismatch hybridization events are minimal. The method is suitable for cost-effective high-throughput clinical genotyping applications.

Key words: SNP, multiplex PCR, multiplex allele-specific primer extension, universal bead array, thrombophilia, genotyping.

1. Introduction

Venous thromboembolism is a classic example of a complex, multifactorial disorder involving multiple genetic abnormalities *(1, 2)*. The factor V Leiden polymorphism (G1691A) is the most common cause of inherited thrombophilia followed by the factor II (prothrombin) G20210A polymorphism *(3, 4)*. A third independent risk factor, although controversial, is the methylenetetrahydrofolate reductase (MTHFR) C677T polymorphism *(5)*. Three additional SNPs, MTHFR A1298C, factor XIII val34leu, and tissue factor pathway inhibitor (TFPI) C536T, have been included in the thrombophilia assay described in this chapter

Peter Bugert (ed.), *DNA and RNA Profiling in Human Blood: Methods and Protocols, vol. 496*

DOI 10.1007/978-1-59745-553-4_6 Springerprotocols.com

(6–8). Individually, these additional polymorphisms have little or no independent effect on venous thrombosis *(9–14)*, but may act synergistically with other genetic or acquired risk factors resulting in a more than additive effect or, in the case of factor XIII, a protective effect. It may be that polymorphisms believed to have no clinical utility were characterized as such simply because each SNP was studied in isolation and not as part of a comprehensive panel. Since the risk of thromboembolism increases when multiple variant genes are present, comprehensive panel testing on large sample populations may prove useful in better defining the genetic component of this complex disorder. To date, the tools required for high-throughput, cost-effective SNP panel testing were unavailable limiting the number of SNPs being analyzed in combination. The multiplexed genotyping method described here addresses this issue and may be easily modified to allow newly identified SNPs with potential clinical relevance to be incorporated.

A generalized overview of the method is given in **Fig. 6.1**. The four basic steps include target amplification by multiplex PCR, genotyping by multiplex allele-specific primer extension (ASPE) chemistry, universal array sorting by hybridization, and detection on the Luminex xMAP® flow cytometer. The first two steps are specific to the 6-SNP thrombophilia panel being analyzed while the universal array sorting step remains constant and can be adapted to any SNP panel.

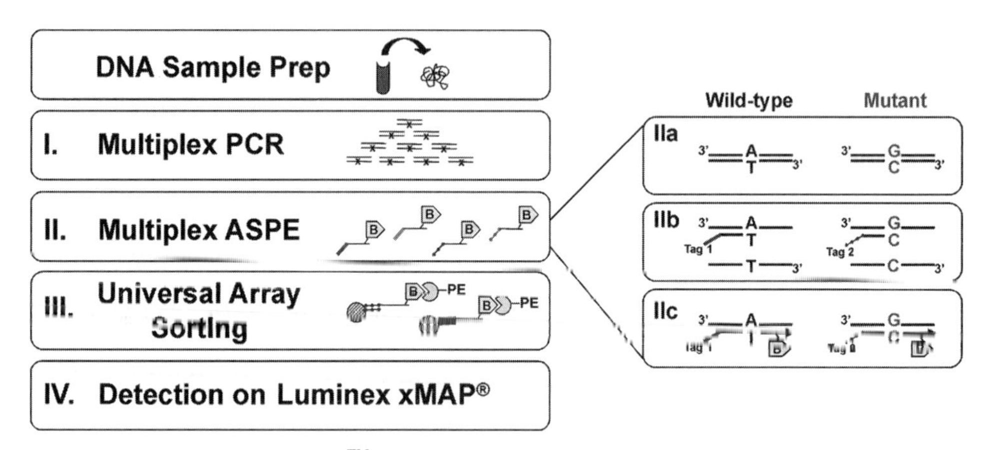

Fig. 6.1. Generalized overview of the xTAG™ microsphere-based universal array genotyping platform. Following extraction, sample DNA is PCR-amplified. Each SNP site of interest is simultaneously queried using two universally tagged allele-specific primers whose 3′ ends define the alleles. A thermophilic DNA polymerase is used for label incorporation into extended products. Since ASPE primers overlap the SNP site in the PCR-amplified target DNA, only the correctly hybridized primer(s) will be extended. Allelic discrimination occurs as a result of the high sequence specificity of the polymerase and its inability to extend 3′ mismatches. Labeled extension products are then sorted on the microsphere-based universal array and detected on the Luminex xMAP® instrument. The output data is then interpreted and presented as a sample-specific genotype (reproduced from Ref. *(20)* with permission from The American Association for Clinical Chemistry).

Multiplex PCR is carried out to enrich target populations of interest *(15)*. The multiplex reaction overcomes the need to set up multiple individual reactions, greatly improving throughput and cost. Genomic DNA extracted from a variety of sample types (usually whole blood) is used as the starting material. For multiplex PCR to be successful, optimization of multiple parameters is required *(16)*.

Multiplex genotyping is performed via ASPE chemistry *(17, 18)*. In ASPE, each SNP in the panel is simultaneously queried using two universally tagged allele-specific primers whose 3′ ends define the alleles. A thermophilic DNA polymerase is used for label incorporation (biotin–dCTP) into extended products. Since ASPE primers overlap the SNP site in the PCR-amplified target DNA, only the correctly hybridized primer(s) will be extended. Allelic discrimination occurs as a result of the high sequence specificity of the polymerase and its inability to extend 3′ mismatches. In other words, ASPE incorporates two levels of allelic discrimination, one occurring at the hybridization step and the other at the enzymatic extension step. As ASPE is a linear amplification method, extension products are single-stranded. This facilitates their capture onto the universal array as the universal tag sequences are single-stranded and competition by strand complements is avoided.

Following ASPE, single-stranded biotin-labeled extension products are captured via their 'tag' sequences on the xTAG™ universal bead array which may consist of up to 100 spectrally distinct polystyrene microsphere populations each bearing a unique 'anti-tag' sequence. The tag/anti-tag sequences have been designed to be isothermal and minimally cross-hybridizing as a set. This 'universal' approach overcomes the differences in the sequence-dependent thermodynamic properties of gene-specific sequences which make it difficult to select a single set of hybridization conditions suitable for all probes. Universal arrays are advantageous in that, once the optimal tag/anti-tag hybridization conditions have been established, no further assay-specific hybridization optimization is required. Bead-based arrays have the added benefits of faster reaction kinetics and greater flexibility with respect to array composition.

Finally, hybridization products are detected using the Luminex xMAP® flow cytometer *(19)*. Microspheres are individually interrogated by two lasers as they pass single file through a rapidly flowing fluid stream. The instrument's red laser identifies the bead's unique spectral address (and hence its corresponding anti-tag) while the green laser determines whether a hybridization event has occurred on the bead's surface by quantifying a fluorescent reporter. In this case, streptavidin–phycoerythrin is the fluorescent reporter molecule used to indirectly detect the biotin–dCTP incorporated through ASPE.

The multiplexed 6-SNP genotyping assay for thrombophilia described here, together with its validation study have previously been published *(20)*. Commercial versions of the assay are available from Luminex Molecular Diagnostics. As such, the methods described here relate to the 6-SNP thrombophilia panel but are also described in general terms as they would apply to the multiplex design and analysis of an alternate SNP panel.

2. Materials

2.1. Multiplex PCR

1. Oligonucleotide primers (Integrated DNA Technologies, Coralville, IA, USA).
2. Platinum *Taq* with supplied 10× buffer and 50 mM $MgCl_2$ (Invitrogen, Burlington, ON, Canada).
3. Individual dNTP stocks, 100 mM (Roche Diagnostics, Laval, QC, Canada).

2.2. PCR Clean-Up

1. Shrimp alkaline phosphatase (USB Corporation, Cleveland, OH, USA).
2. Exonuclease I (EXO) (USB Corporation).

2.3. Multiplex ASPE

1. Universally tagged allele-specific oligonucleotide primers (Integrated DNA Technologies).
2. Platinum *Tsp* with supplied 10× buffer and 50 mM $MgCl_2$ (Invitrogen).
3. Individual dNTP stocks, 100 mM (Roche).
4. Biotin–dCTP, 0.4 mM (Invitrogen).

2.4. Universal Array Sorting and Detection

1. Luminex xMAP® flow cytometer system (Luminex Corporation, Austin, TX, USA).
2. xTAG™ universally tagged fluorescent microspheres (Luminex Corporation)
3. Hybridization buffer: 0.22 M NaCl; 0.11 M Tris–HCl, pH 8.0; 0.088% (v/v) Triton X-100.
4. Wash buffer: 0.2 M NaCl; 0.1 M Tris–HCl, pH 8.0; 0.08% (v/v) Triton X-100 (see **Note 1**).
5. Streptavidin-conjugated phycoerythrin, 1 mg/mL (Invitrogen). Store conjugate at 4°C in the dark. Do not freeze.

3. Methods

The methods described here will reflect the use of xTAG™ microspheres which represent the commercially available, open-access universal array. The universal sequences coupled to xTAG™ microspheres are a different set than those previously described *(20)* but are equivalent given that validation of both arrays generated almost identical results.

3.1. Multiplex PCR (6-plex)

After identifying the target genes of interest, PCR primers were designed using Primer3 software (http://fokker.wi.mit.edu/primer3/input.htm). For multiplex PCR to be successful, the critical parameter is primer melting temperature. For the 6-SNP thrombophilia panel, the 12 PCR primers were designed to have melting temperatures between 57 and 63°C. Primers were designed to be 18–22 bases in length with an average GC content of 50%. Where possible, PCR primers were designed to generate amplicons that were relatively small (<200 bp) with the SNP of interest being located towards the center (*see* **Note 2**). Additionally, primers were designed so that each amplicon within the multiplex reaction was distinguishable by size to allow for gel resolution. PCR primer sequences and amplicon sizes for the thrombophilia panel are given in **Table 6.1** (*see* **Note 3**).

PCR primers were synthesized on a 250 nmol scale by Integrated DNA Technologies. PCR primers were unmodified and were purified using standard desalting procedures. Individual, lyophilized PCR primers were reconstituted to about 200 μM in sterile ddH_2O. Exact oligo concentrations were determined spectrophotometrically using extinction coefficients provided by the supplier. Reconstituted oligos were scanned between 200 and 800

Table 6.1
PCR primer sequences (5′ to 3′) and amplicon sizes *(20)*

Gene		Sequence	Amplicon size (bp)
factor V	Forward	CGCCTCTGGGCTAATAGGAC	154
	Reverse	GCCCCATTATTTAGCCAGGA	
factor II	Forward	GAACCAATCCCGTGAAAGAA	112
	Reverse	CCAGAGAGCTGCCCATGA	
MTHFR 677	Forward	CTTTGAGGCTGACCTGAAGC	101
	Reverse	CAAAGCGGAAGAATGTGTCA	
MTHFR 1298	Forward	AGGAGCTGCTGAAGATGTGG	105
	Reverse	CTTTGTGACCATTCCGGTTT	
TFPI	Forward	TCTATTTTAATTGGCTGTATTTTTTTC	97
	Reverse	TGCGGAGTCAGGGAGTTATT	
factor XIII	Forward	TCTAATGCAGCGGAAGATGA	117
	Reverse	TGTGCCTGGACCCAGAGT	

nm on a Beckman DU640 spectrophotometer and absorbance was measured at 260 nm to calculate concentration. PCR primers were aliquoted and stocks were stored at –20°C.

1. Multiplex PCR is carried out using 25 ng genomic DNA in a 25 μL final volume.
2. A 'no target' PCR negative control is included with each assay run (*see* **Note 4**).
3. Each reaction consits of 1.5× buffer (30 mM Tris–HCl, pH 8.4; 75 mM KCl), 2 mM $MgCl_2$, 200 μM each dNTP, 1.25 units platinum *Taq*, with factor II primers each at 0.5 μM, factor V and TFPI primers each at 0.3 μM, and all others present at 0.2 μM (*see* **Notes 5, 6**).
4. Samples are cycled in an MJ Research PTC-200 thermocycler with cycling parameters as follows:

Initial denaturation:	95°C, 5 min
30 cycles:	95°C, 30 s; 58°C, 30 s; 72°C, 30 s
Final extension:	72°C, 5 min.
Hold temperature	4°C until further use (*see* **Note 7**).

5. While optimizing the multiplex PCR for buffer composition, cycling parameters, annealing temperature, and primer concentrations, PCR reactions were separated electrophoretically to verify that all six desired bands were present and were of the correct size and to ensure that nonspecific amplification was absent (both in samples and negative control). The gel image shown in **Fig. 6.2** clearly demonstrates the high specificity of the PCR primers used in the multiplex reaction under optimized conditions. Since PCR represents the first step in the assay, it is essential that the reaction be as specific as possible to minimize potential complications downstream (*see* **Note 8**).

3.2. PCR Clean-Up

Prior to the ASPE reaction, each PCR reaction is treated with shrimp alkaline phosphatase (SAP) to inactivate remaining nucleotides (particularly dCTP) so that biotin–dCTP may be efficiently incorporated during the primer extension reaction. Each PCR reaction is also treated with EXO to degrade remaining PCR primers in order to avoid any interference with the tagged ASPE primers and the extension reaction itself.

1. To each 25 μL PCR reaction directly add 2 μL SAP (= 2 units) and 0.5 μL EXO (= 5 units) (*see* **Note 9**).
2. Samples are incubated at 37°C for 30 min followed by a 15 min incubation at 99°C to inactivate the enzymes.
3. Treated PCR samples are then added directly to the ASPE reaction (*see* **Note 10**).

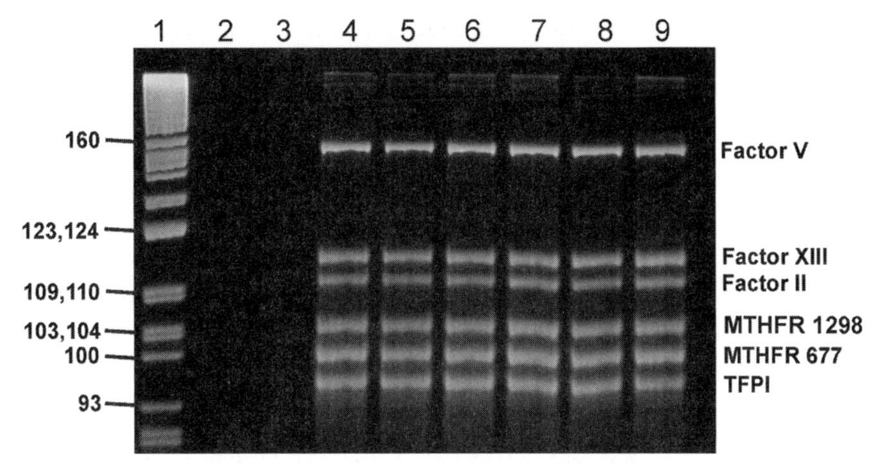

Fig. 6.2. Gel image of multiplex PCR products representing patient samples separated electrophoretically on a Helixx SuperGel 150 in 1× Tris–acetate–EDTA buffer. Lane 1, markers; Lane 2, empty; Lane 3, PCR negative control; Lanes 4–9, patient samples. The multiplex PCR reaction was highly specific for the six desired amplimers with no spurious amplification products detected. The amplimers were of the expected size (97–154 bp) and the PCR negative control showed no evidence of primer dimer formation (reproduced from Ref. *(20)* with permission from The American Association for Clinical Chemistry).

3.3. Multiplex ASPE

Given that ASPE is a linear amplification reaction and that the SNP site is pre-defined, design of genotyping primers for multiplex ASPE is much simpler than multiplex PCR primer design. For each SNP in the multiplex panel, a wild-type and a mutant primer, usually differing only in their 3′ base, are required (*see* **Note 11**). These two primers may be selected to prime off of either the 5′ or 3′ strand of the PCR product but both must be in the same direction. Some factors, which should be considered when selecting which strand to prime off, include the presence of underlying polymorphisms in the ASPE primer binding region and secondary structure within the 3′ end of the primer. Melting temperature (T_m) and primer length should also be taken into consideration when designing ASPE primers; however, a wider T_m range is permitted since linear ASPE is less impacted by T_m as compared to multiplex PCR. For the thrombophilia panel, the gene-specific sequence of each ASPE primer ranged from 18 to 22 bases in length with melting temperatures ranging from about 52 to 65°C.

Once the allele-specific primers have been designed, they must be tagged with universal sequences complementary to those coupled to the xTAG^{TM} microspheres making up the bead array. Universal xTAG^{TM} sequences as well as information on how to assign tags and select microsphere sets are available on the Luminex website (http://www.luminexcorp.com/uploads/data/xTAG%20Protocols%20FAQs/xtagmicrosphere71008.pdf).

For the thrombophilia panel, chimeric ASPE primers consisting of the 24mer universal tag sequence 5′ to the allele-specific (18–22mer) sequence were synthesized on a 250 nmol scale by Integrated DNA Technologies. Primers were unmodified and were purified by polyacrylamide gel electrophoresis (PAGE) (*see* **Note 12**). Individual, lyophilized ASPE primers were reconstituted to approximately 100 μM in sterile ddH_2O. Reconstituted primers were scanned between 200 and 800 nm on a Beckman DU640 spectrophotometer and absorbance was measured at 260 nm to calculate oligo concentration using extinction coefficients provided by the supplier. ASPE primer stocks were stored at –20°C. ASPE primer sequences are given in **Table 6.2**.

1. Multiplex ASPE is carried out using 5 μL of treated PCR product in a final volume of 20 μL.
2. Each reaction consists of 1× buffer (20 mM Tris–HCl, pH 8.4, 50 mM KCl), 1.25 mM $MgCl_2$, 5 μM biotin–dCTP, 5 μM each of dATP, dGTP and dTTP, 1.5 units Platinum *Tsp* and 25 nM ASPE primer pool (i.e., each ASPE primer present at 500 fmol/reaction) (*see* **Note 13**).
3. The ASPE reactions are incubated in a PCR cycler under the following conditions:

Initial denaturation:	96°C, 2 min
40 cycles at:	94°C, 30 s; 54°C, 30 s; 74°C, 60 s.
Hold temperature:	4°C until use (*see* **Note 14**).

Table 6.2
ASPE primer sequences (5′ to 3′) ***(20)***

Gene	Allele	Sequence
factor V	Wild-type	Tag-GGACAAAATACCTGTATTCCTC
	Mutant	Tag-GGACAAAATACCTGTATTCCTT
factor II	Wild-type	Tag-CAATAAAAGTGACTCTCAGCG
	Mutant	Tag-CAATAAAAGTGACTCTCAGCA
MTHFR 677	Wild-type	Tag-GAGAAGGTGTCTGCGGGAGC
	Mutant	Tag-GAGAAGGTGTCTGCGGGAGT
MTHFR 1298	Wild-type	Tag-ACAAAGACTTCAAAGACACTTT
	Mutant	Tag-ACAAAGACTTCAAAGACACTTG
TFPI	Wild-type	Tag-GGCTGTATTTTTTTCCAGC
	Mutant	Tag-GGCTGTATTTTTTTCCAGT
factor XIII	Wild-type	Tag-GACGCCCCGGGGCACCAC
	Mutant	Tag-GACGCCCCGGGGCACCAA

4. The ASPE component of the assay was optimized for several parameters affecting specificity and signal output. These parameters included cycling conditions, annealing temperature, buffer conditions, ASPE primer concentrations, and PCR reaction volume added to the ASPE reaction. However, the main optimization required at the ASPE step is the balancing of median fluorescence intensity (MFI) signals generated for each allele of a given SNP following array hybridization. This is achieved by redesigning ASPE primers either by lengthening them at the 5′ end of the gene-specific portion to increase MFI or shortening them (at their 5′ end) to reduce nonspecific MFI generated for an absent allele. Typical MFIs are usually >1000 for present alleles and, as a rough measure, absent alleles should generate MFIs that are <10% the specific MFI generated for the present allele of any given SNP.

3.4. Universal Array Sorting and Detection

1. Prepare the universal array by combining the 12 xTAG™ microsphere populations selected to correspond to the 12 ASPE primers. Resuspend the beads in hybridization buffer (0.22 M NaCl, 0.11 M Tris (pH 8.0), and 0.088% (v/v) Triton X-100) so that each bead population is present at 56 beads/μL (*see* **Note 15**).
2. Briefly vortex and sonicate the bead mix. For each hybridization reaction, combine 45 μL of the bead mix (approximately 2500 beads of each of the 12 anti-tag bearing bead populations) with 5 μL of the ASPE reaction in a microwell plate and seal.
3. Heat samples to 96°C for 2 min in an MJ Research PTC-200 followed by a 1 h incubation at 37°C.
4. To remove excess biotin–dCTP and excess primers following hybridization, the reactions are filtered using a 1.2 μm Durapore® Membrane (Millipore Corp). A vacuum manifold is required for this step. Prepare the membrane by first washing twice with 200 μL wash buffer (0.2 M NaCl, 0.1 M Tris (pH 8.0), and 0.08% (v/v) Triton X-100).
5. To the hybridized samples, add 100 μL wash buffer and then transfer the entire 150 μL volume to the filter plate. Apply the vacuum to filter the reactions. Again, wash with 200 μL of wash buffer.
6. Resuspend the beads in 150 μL reporter solution (*see* **Note 16**) and incubate for 15 min at room temperature.
7. Transfer the filter plate to the Luminex xMAP® for detection. Before reading the samples, ensure that the instrument has been calibrated. Setup a run file by selecting and naming the 12 required bead populations. Ensure that a gate setting has been established and set acquisition parameters to measure 100 events per bead population and a 100 μL sample volume (*see* **Note 17**).

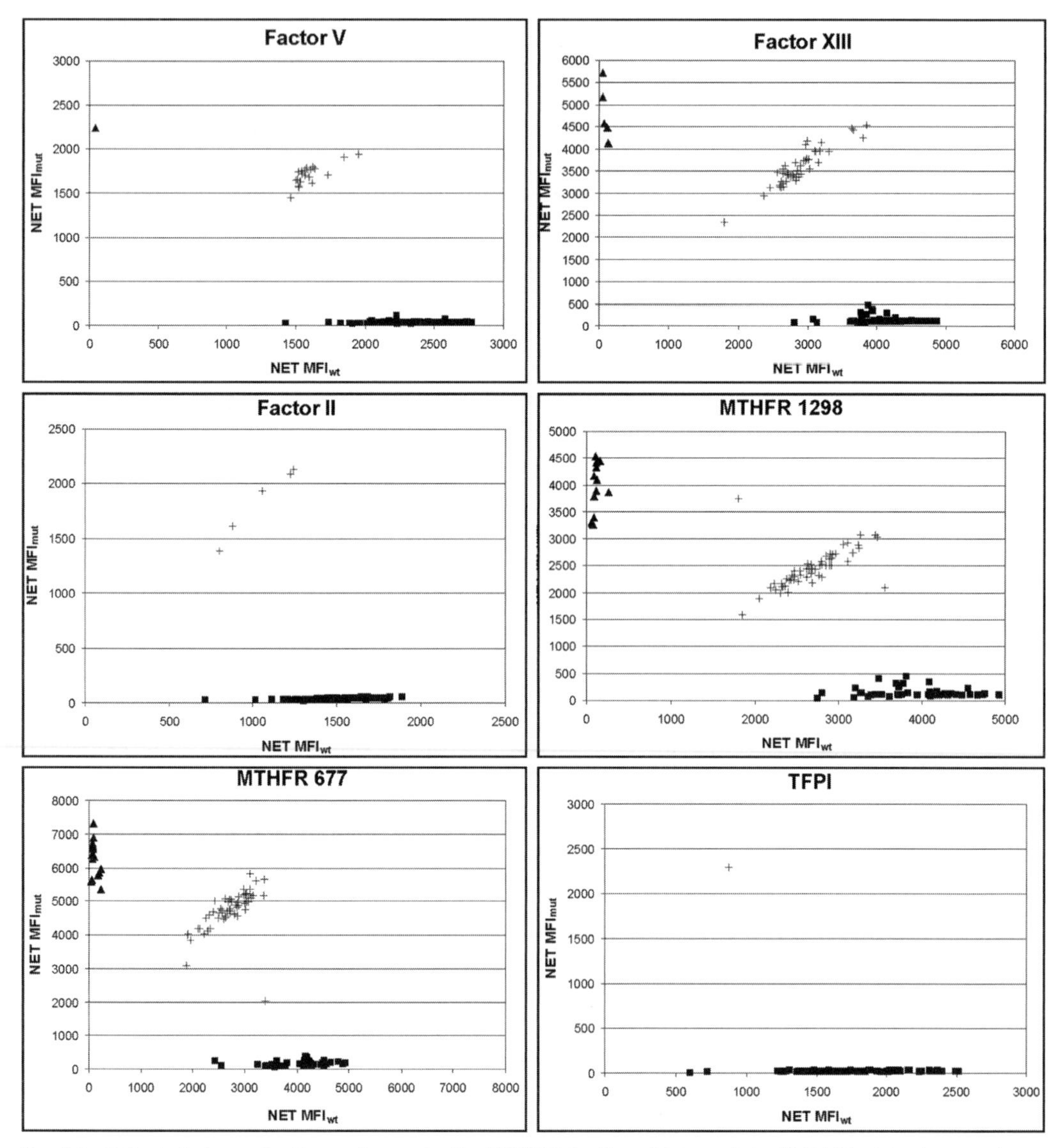

Fig. 6.3. Patient study results. For each sample, the NET MFI for the wild-type allele (NET MFI$_{wt}$) is given on the x-axis and the NET MFI for the mutant allele (NET MFI$_{mut}$) is given on the y-axis. Based on allelic ratio, the assay easily discriminates wild-type (■), mutant (▲), and heterozygous (+) samples with the large separation between signals greatly reducing the risk of miscalls. Although two MTHFR 1298 calls and one MTHFR 677 call appear as outliers, the calculated mutant allelic ratios for these samples fell within defined ranges for heterozygous calls. The variation in signal intensities between the different SNPs is likely due to the different priming efficiencies of the ASPE primers and/or the number of incorporated biotin–dCTP nucleotides. The variation observed between samples with the same call for any given SNP is attributed to differences in DNA sample quality or the presence of underlying polymorphisms (reproduced from Ref. *(20)* with permission from The American Association for Clinical Chemistry).

3.5. Data Analysis

For each DNA sample tested, MFI units are collected by the Luminex xMAP® instrument for each of the 12 bead populations corresponding to each allele within the assay. The data collected requires very minimal processing and genotype calls may easily be made for each SNP directly from the raw data. However, for each allele, NET MFI may be calculated by subtracting the 'no target' (PCR negative control) MFI values from the MFI values for each allele of a given sample. The genotype may then be determined based on the mutant allelic ratio:

$$\text{Mutant allelic ratio} = \frac{(\text{NET MFI})_{\text{mutant allele}}}{(\text{NET MFI})_{\text{mutant allele}} + (\text{NET MFI})_{\text{wild-type allele}}}$$

In other words, the mutant allelic ratio represents the fraction of the total MFI signal for a given SNP attributed to the presence of the mutant allele. By setting cut-off values, the allelic ratio is used to discriminate homozygous wild-type, homozygous mutant and heterozygous SNP calls. Cut-off values are empirically determined for each individual SNP. Typically, the mutant allelic ratio ranges from 0.00 to 0.10 for homozygous wild-type calls, 0.30–0.70 for heterozygous calls and 0.90–1.00 for homozygous mutant calls (*see* **Note 18**). The results of a patient study generated using the 6-SNP thrombophilia genotyping assay appear in **Fig. 6.3**.

4. Notes

1. A 10× buffer stock solution (2 M NaCl, 1 M Tris (pH 8.0), and 0.8% (v/v) Triton X-100) may be prepared and then diluted to 1.1× for the hybridization buffer and 1.0× for the wash buffer. The 10× buffer stock and the 1.1× hybridization buffer may be stored long-term at 4°C. The 1.0× wash buffer should be maintained at room temperature.
2. Short amplicons are amplified with higher efficiencies so that shorter cycling times may be used. By designing the amplicon so that the SNP is in the center, ASPE primers may be designed in either direction with enough length to incorporate multiple biotin labels.
3. In addition to checking individual primer pairs for 3′ sequence complementarity, all PCR primers within a multiplex reaction should be examined (either visually or using appropriate software) for potential alignments.
4. For the 'no target' control, water is used. The control monitors carryover contamination and gives an indication as to whether PCR primers are being consumed in nonspecific interactions such as primer dimer formation.
5. Before combining all PCR primers into a single multiplex reaction, test each primer pair individually first to ensure

that a single product of the correct size is being generated. For this purpose, basic PCR reaction conditions (1× buffer, 2 mM $MgCl_2$, 200 μM dNTPs, 0.2 μM PCR primers, 1 unit Pt *Taq*) and standard thermal cycling will suffice since singleplex reactions are relatively robust. Following amplification, run the products out on an agarose gel to verify size and confirm that only a single-specific product is being generated for each primer pair. BLAST searches and other sequence alignment programs are extremely useful tools in determining whether nonspecific sequence complementarity exists. However, empirical determination of PCR primer specificity is even more powerful since it is assessed in the context of the reaction conditions being used.

6. For multiplex PCR, all primer pairs are initially combined at equimolar concentrations. For example, begin with all primer pairs in the multiplex at 0.2 μM. Run the multiplex reactions out on a gel and begin adjusting primer concentrations either up or down (by about 0.1 μM increments) to even out band intensities. However, it should be noted that band intensity on a gel is not always directly proportional to the median fluorescence signal generated following hybridization and detection by the Luminex xMAP®. At this point, primer concentrations should only be adjusted roughly to either intensify an extremely weak band or weaken an extremely strong band which is dominating the multiplex reaction.
7. At this point, PCR products may proceed to the next step or may be held at 4°C until use. It is recommended that PCR products not be stored longer than 48 h prior to use.
8. Electrophoresis may be carried out using either commercially available high-resolution gel systems or in-house prepared high-resolution agarose gels. Gel electrophoresis is only used during work-up of the assay and is therefore amenable to any variety of procedures. In the case of the thrombophilia panel, the only requirement is that the gel system used be capable of resolving four basepair differences. The Helixx SuperGel 150 system used here is capable of resolving single basepair differences within products.
9. Due to the small volumes of highly viscous enzymes being pipetted, a master mix containing both SAP and EXO may be prepared. Combine enough of each enzyme to cover the total number of reactions being run. Mix the two enzymes by very brief vortexing and spin down briefly. Add 2.5 μL of the enzyme mixture to each reaction. Ensure that the added enzymes are thoroughly mixed with PCR reactions.
10. Treated PCR products should be used for ASPE as soon as possible. Long-term storage of SAP/EXO- treated products is not recommended.

11. For the initial round of ASPE primer design, the wild-type, and mutant primers for any given SNP are usually of equal length.
12. As PAGE purification is not performed by all oligo houses, alternate purification methods suitable for 40–50mer sequences may be used provided they guarantee utmost purity and an intact 3′ base. The success of the ASPE reaction is highly dependent on primer quality.
13. The enzyme used in the ASPE reaction must be suitable for genotyping. In other words, it should lack 3′–5′ exonuclease activity and have minimal terminal transferase activity. The safest method of selecting an alternate enzyme for ASPE is empirical testing using known wild-type, mutant, and heterozygous samples.
14. At this point, ASPE products may proceed to the next step or may be held at 4°C until use. It is recommended that ASPE products not be stored longer than 48 h prior to use.
15. xTAG™ microspheres are supplied at two concentrations, either at 2.5×10^5 or 6.25×10^6 beads/mL. While preparing the array, protect the microspheres from light as much as possible. A large volume of the liquid array may be prepared and stored at 4°C. As indicated in the method, the bead array is added in a 45 μL volume and contains 2500 beads of each of the 12 populations (56 beads/μL). To prepare 4.5 mL of the bead array (enough for 100 reactions), 2.5×10^5 beads are required. For each bead population, pipette either 1 mL of the less concentrated stock or 40 μL of the more concentrated stock into individual microcentrifuge tubes. Centrifuge the beads at 10 000 × *g* for 1 min to pellet the beads. Remove the supernatant, being careful not to disturb the pellet, and resuspend each bead population in a small volume of hybridization buffer (about 100 μL). Combine the 12 bead populations into a single vesicle and bring volume up to 4.5 mL with hybridization buffer. Store the bead mix at 4°C in the dark until use.
16. The reporter solution is prepared just prior to the end of the hybridization reaction by diluting the 1 mg/mL stock solution 1000× in wash buffer to a final concentration of 1 μg/mL. Given that streptavidin has a very high affinity for plastic, the phycoerythrin conjugate should be diluted in glass and kept in the dark until use.
17. Details regarding instrument calibration, file set-up and gate setting are provided in the User's manual supplied with the Luminex xMAP® instrument.
18. As with any hybridization-based assay, underlying polymorphisms occurring in the allele-specific primer binding regions (other than those being queried) may affect the fluorescence signals generated and consequently, the calls made.

References

1. Rosendaal, F. R. (1999) Venous thrombosis: a multicausal disease. *Lancet* 353, 1167–1173.
2. Franco, R. F., Reitsma, P. H. (2001) Genetic risk factors of venous thrombosis. *Hum Genet* 109, 369–384.
3. Bertina, R. M., Koeleman, B. P., Koster, T., Rosendaal, F. R., Dirven, R. J., de Ronde, H., et al. (1994) Mutation in blood coagulation factor V associated with resistance to activated protein C. *Nature* 369, 64–67.
4. Poort, S. R., Rosendaal, F. R., Reitsma, P. H., Bertina, R. M. (1996) A common genetic variation in the 3′-untranslated region of the prothrombin gene is associated with elevated plasma prothrombin levels and an increase in venous thrombosis. *Blood* 88, 3698–3703.
5. Frosst, P., Blom, H. J., Milos, R., Goyette, P., Sheppard, C. A., Matthews, R. G., et al. (1995) A candidate genetic risk factor for vascular disease: a common mutation in methylenetetrahydrofolate reductase. *Nat Genet* 10, 111–113.
6. van der Put, N. M. J., Gabreels, F., Stevens, E. M. B., Smeitink, J. A. M., Trijbels, F. J. M., Eskes, T. K. A. B., et al. (1998) A second common mutation in the methylenetetrahydrofolate reductase gene: an additional risk factor for neural-tube defects? *Am J Hum Genet* 62, 1044–1051.
7. Mikkola, H., Syrjala, M., Rasi, V., Vahtera, E., Hamalainen, E., Peltonen, L., Palotie, A. (1994) Deficiency in the A-subunit of coagulation factor XIII: two novel point mutations demonstrate different effects on transcript levels. *Blood* 84, 517–525.
8. Girard, T. J., Eddy, R., Wesselschmidt, R. L., MacPhail, L. A., Likert, K. M., Byers, M. G., et al. (1991) Structure of the human lipoprotein-associated coagulation inhibitor gene. *J Biol Chem* 266, 5036–5041.
9. Franco, R. F., Morelli, V., Lourenco, D., Maffei, F. H., Tavella, M. H., Piccinato, C. E., et al. (1999) A second mutation in the methylenetetrahydrofolate reductase gene and the risk of venous thrombotic disease. *Br J Haematol* 105, 556–559.
10. Franco, R. F., Reitsma, P. H., Lourenco, D., Maffei, F. H., Morelli, V., Tavella, M. H., et al. (1999) Factor XIII Val34Leu is a genetic factor involved in the etiology of venous thrombosis. *Thromb Haemost* 81, 676–679.
11. van Hylckama Vlieg, A., Komanasin, N., Ariens, R. A. S., Poort, S. R., Grant, P. J., Bertina, R. M., Rosendaal, F. R. (2002) Factor XIII Val34Leu polymorphism, factor XIII antigen levels and activity and the risk of deep vein thrombosis. *Br J Haematol* 119, 169–175.
12. Kleesiek, K., Schmidt, M., Gotting, C., Brinkmann, T., Prohaska, W. (1998) A first mutation in the human tissue factor pathway inhibitor gene encoding [P151L]TFPI. *Blood* 92, 3976–3977.
13. Kleesiek, K., Schmidt, M., Gotting, C., Schwenz, B., Lange, S., Muller-Berghaus, G., et al. (1999) The 536C→T transition in the human tissue factor pathway inhibitor (TFPI) gene is statistically associated with a higher risk for venous thrombosis. *Thromb Haemost* 82, 1–5.
14. Evans, G. D., Langdown, J., Brown, K., Baglin, T. P. (2000) The C536T transition in the tissue factor pathway inhibitor gene is not a common cause of venous thromboembolic disease in the UK population. *Thromb Haemost* 83, 511.
15. Mullis, K. B., Faloona, F. A. (1987) Specific synthesis of DNA in vitro via a polymerase-catalyzed chain reaction. *Methods Enzymol* 155, 335–350.
16. Markoulatos, P., Siafakas, N., Moncany, M. (2002) Multiplex polymerase chain reaction: a practical approach. *J Clin Lab Anal* 16, 47–51.
17. Ugozzoli, L., Wahlqvist, J. M., Ehsani, A., Kaplan, B. E., Wallace, R. B. (1992) Detection of specific alleles by using allele-specific primer extension followed by capture on solid support. *Genet. Anal Tech Appl* 9, 107–112.
18. Ye, F., Li, M., Taylor, J. D., Nguyen, Q., Colton, H. M., Casey, W. M., et al. (2001) Fluorescent microsphere-based readout technology for multiplexed human single nucleotide polymorphism analysis and bacterial identification. *Hum Mutat* 17, 305–316.
19. Dunbar, S. A. (2006) Applications of Luminex® xMAP™ technology for rapid, high-throughput multiplexed nucleic acid detection. *Clin Chim Acta* 363, 71–82.
20. Bortolin, S., Black, M., Modi, H., Boszko, I., Kobler, D., Fieldhouse, D., et al. (2004) Analytical validation of the Tag-It high-throughput microsphere-based universal array genotyping platform: application to the multiplex detection of a panel of thrombophilia-associated single-nucleotide polymorphisms. *Clin Chem* 50, 2028–2036.

Chapter 7

Pyrosequencing of Toll-Like Receptor Polymorphisms of Functional Relevance

Parviz Ahmad-Nejad

Abstract

Inflammation is becoming increasingly recognized and discussed as an important pathobiochemical origin in many disease entities such as atherosclerosis, cancer, or infections and genetically determined susceptibility to danger signals may influence the development of inflammatory diseases. Members of the 'toll-like receptor' (TLR) family are pivotal molecules in the activation of the innate immune system and specifically recognize structurally conserved pathogen-associated molecular patterns. Since their discovery a growing number of single nucleotide polymorphisms (SNPs) have been identified, functionally characterized and in part linked to multiple diseases. Here we report several protocols for Pyrosequencing® approaches to genotype functionally relevant SNPs in TLR-genes and further molecules of the innate immune system.

Key words: toll-like receptor; lipopolysaccharide; pathogen-associated molecular pattern; Pyrosequencing®; single nucleotide polymorphism; genotyping.

1. Introduction

The delineation of members of the toll-like receptor (TLR) family within the recent years has lead to a new level in understanding the innate immune systems *(1)*. TLRs sense pathogen-associated molecular patterns (PAMPs). To date ten members of the TLR family are described in man *(1)*. While TLR4 complexed with MD-2 senses lipopolysaccharide (LPS) of gram-negative microbes as a bacterial danger signal *(2)*, TLR2 recognizes bacterial lipoproteins (BLP) *(3)*, heat shock proteins (HSP) *(4–6)*, and fungal-derived structures such as zymosan/mannane-activated *(7)*. For TLR9, it was demonstrated that it recognizes bacterial unmethylated CpG-DNA *(8)*, and bacterial flagellin is considered

Peter Bugert (ed.), *DNA and RNA Profiling in Human Blood: Methods and Protocols, vol. 496*

DOI 10.1007/978-1-59745-553-4_7 Springerprotocols.com

as ligand for TLR5 *(9)*. More recent publications show that specific stimuli for TLR7 and 8 exist. Specifically the low-molecular antiviral substance R848 triggers immune cells signals via human TLR7 and 8 *(10–12)*. TLR engagement leads to the production and secretion of proinflammatory cytokines like IL-6, IL-12, and TNFα in cells of the innate immune system. Furthermore mitogen-activated protein kinases (MAPK) and transcription factors like NF-κB and AP-1 are activated. Thus TLRs are directly involved in the very first recognition of microbial danger signals. As with many other genes single nucleotide polymorphisms (SNPs) have been reported in public databases *(13)* for members of this ancient receptor family and many of this polymorphisms have already been functionally characterized *(14–17)*. A few of these SNPs have been linked to disease entities, but for the majority no genotyping methods are currently available.

In the recent years Pyrosequencing® has emerged as a valuable technique in molecular diagnostics *(18–20)*. Pyrosequencing® is a bioluminometric DNA sequencing method based on the sequencing-by-synthesis principle. This real-time DNA sequencing approach employs a cascade of four enzymatic reactions producing sequence peak signals. In this report we describe genotyping methods based on the Pyrosequencing® technology for all currently known, functionally relevant TLR sequence variations and further important SNPs in innate immune system genes (Fig 7.1).

2. Material

2.1. PCR

1. Water, 500 ml (Sigma-Aldrich, Deisenhofen, Germany), store at 4°C (*see* **Note 1**).
2. dNTPs (10 mM) (Fermentas MBI, St Leon-Rot, Germany), store at –20°C.
3. GoTaq Flexi DNA polymerase Kit (Promega, Mannheim, Germany), store at –20°C), contains: GoTaq polymerase 500 U, $MgCl_2$ (25 mM), 5x Flexi-buffer green (*see* **Note 2**).
4. PCR-SOFTSTRIPS (0.2 ml) (Biozym, Oldendorf, Germany).
5. Oligonucleotides (**Table 7.1**).
6. PCR-cycler (Mastercycler Ep gradient S; Eppendorf, Hamburg, Germany).

2.2. Pyrosequencing

1. Pyro-Gold-Star Kit 5x96 (Biotage, Uppsala, Sweden) includes enzymes and substrates
2. Streptavidin Sepharose™ HP (Amersham Biosciences, Uppsala, Sweden).
3. PSQ-Reagent-Cartridge (Biotage).
4. LiChrosolv (Merck, Darmstadt, Germany).
5. PCR-plates (Eppendorf).

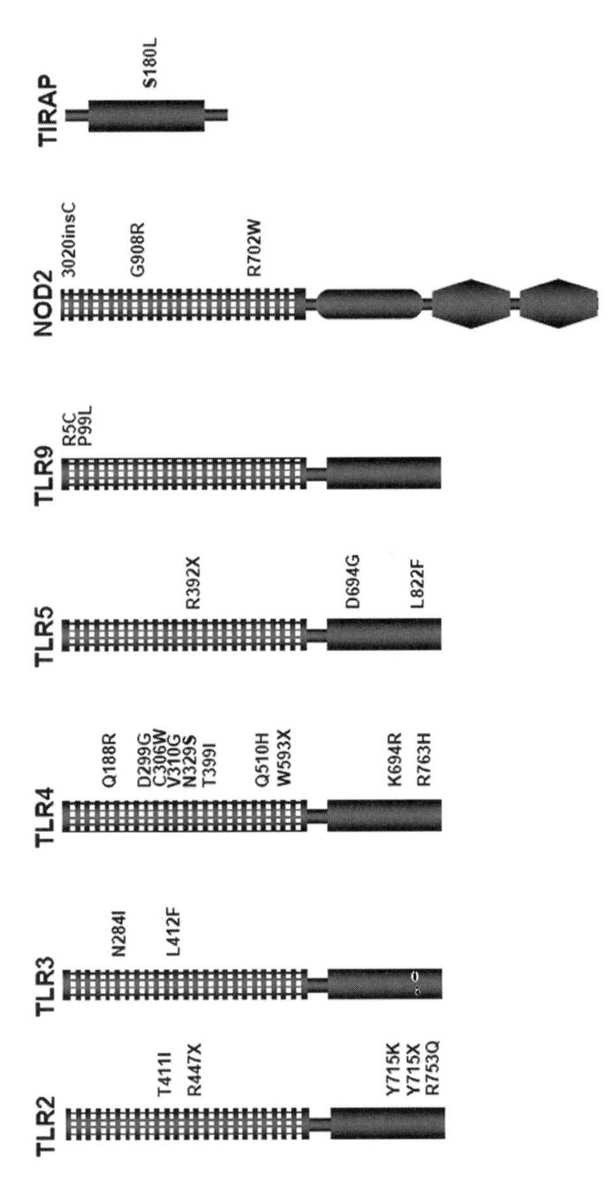

Fig. 7.1. Molecules of the innate immune system and SNPs of functional relevance.

Table 7.1
Primer sequences, $MgCl_2$ -concentration and annealing temperature recommended for PCR to genotype functional relevant SNPs

Name	Sequence[a]	$MgCl_2$ (mM)/annealing temperature (°C)	Sequence to analyze
TLR2			
T2-R753Q-for	B-CCATTCCCCACCGCTTCT	1.5/60	C/TGCAGCTTGCAGAAGCGCTGGGGAAT
T2-R753Q-rev	TGGGCCACTCCAGGTAGGT		
T2-R753Q-seq	TCTTGGTGTTCATTATCTTC		
T2-R447x-for	B-TGCCTGAAACTTGTGAGTGG	3/58	TCA/GTGTGCTGGATAAGTTCAAATATTTC
T2-R447x-rev	GGGAATGCAGCCTGTTACACT		
T2-R447x-seq	CAGCCTGTTACACTGTGTAT		
TLR3			
T3-N284I-for	GGGACTAAAGTGGACAAATCTCA	3/58	CAA/TCAACTTAAATGTGGTTGGTAACGAT
T3-N284I-rev	B-AAGCCAAGCAAAGGAATCGT		
T3-N284I-seq	ATGCTCGATCTTTCCTA		
T3-Y307D-for	TTTGCTTGGCTTCCACAAC	3/58	T/GATAATAATATACAGCATTTGTTTTC
T3-Y307D-rev	B-CCGTGCAAAGAGTGAGAAAAC		
T3-Y307D-seq	TAGAATATTTCTTCCTAGAG		
T3-L412F-for	GCGAACTTTGACAAATGAAACA	3/58	ATAC/TTCAACCTAACCAAGAATAAAATCT
T3-L412F-rev	B-TTTCATTAAGGCCCAGGTCA		
T3-L412F-seq	GCTCATTCTCCCTTACAC		

(continued)

Table 7.1 (continued)

Name	Sequence[a]	$MgCl_2$ (mM)/annealing temperature (°C)	Sequence to analyze
TLR4			
T4-Q188R-for	B-TTGGACCTTTCCAGCAACAA	3/58	C/TGAATCTTGTTGCTGGAAAGGTCCAA
T4-Q188R-rev	AGGGGCATTTGATGTAGAACC		
T4-Q188R-seq	AGTCTGTGCAATAAATACTT		
T4-D299G-for	TGACCATTGAAGAATTCCGATTA	3/58	A/GTATTATTGACTTATTTAATTGTTTG
T4-D299G-rev	B-CACCAGGGAAAATGAAGAAACA		
T4-D299G-seq	CTTAGACTACTACCTCGATG		
T4-C306W-V310G-for	GACCATTGAAGAATTCCGATTAGC	3/58	T/GTTGACAAATG
T4-C306W-V310G-rev	B-CCCTTTCAATAGTCACACTCACCA		
T4-C306W-V310G-seq	ATTATTGACTTATTTAATTG		
T4-N329S-for	B-TTTTCCCTGGTGAGTGTGACTAT	3/58	AAAC/TTATAAGAAAAGTCTTTTACCCTTT
T4-N329S-rev	ATGTGGGAAACTGTCCAAATTTA		
T4-N329S-seq	AAATGTTGCCATCCG		
T4-T399I-for	TTCTCAAAGTGATTTTGGGACAA	3/58	AC/TCAGCCTAAAGTATTTAGATCTGAGC
T4-T399I-rev	B-CCCAAGAAGTTTGAACTCATGGTA		
T4-T399I-seq	AAAGTGATTTTGGGACA		
T4-Q510H-for	CTTGACCTTCCTGGACCTCTCTC	3/58	CAG/TTTGTCTCCAACAGCATTTAACTCAC
T4-Q510H-rev	B-CCTGGAGGGAGTTCAGACACTTAT		
T4-Q510H-seq	CTCAGTGTCAACTGGAG		

(continued)

Table 7.1 (continued)

Name	Sequence[a]	$MgCl_2$ (mM)/annealing temperature (°C)	Sequence to analyze
T4-W593x-for	B-CTTCCAAAAAACAGGAACTACAGC	3/58	C/TCATTGCAGGAAACTCTGGTGTTCAC
T4-W593x-rev	CGTTCAACTTCCACCAAGAGC		
T4-W593x-seq	GCCTCTGGTCCTTGAT		
T4-K694R-for	AGGACTGCGGTAAGGAATGAGC	1.5/58	A/GAGAATTTAGAAGAAGGGGTGCCTCC
T4-K694R-rev	B-GGAGGCACCCCCTTCTTCTAA		
T4-K694R-seq	GGGTAAGGAATGAGCTAGTA		
T4-R763H-for	TTGCTCAGACCTGGCAGTTTC	3/58	G/ATGCTGGTATCATCTTCATTGTCCTG
T4-R763H-rev	B-CACTGTCCCTCCCACTCCAGGTAA		
T4-R763H-seq	GCAGTTTCTGAGCAGTC		
TLR5			
T5-R392x-for	B-TGTCATATAACCTTCTGGGGGAAC	1.5/60	TCA/GGAGATCCAAGGTCTGTAATTTTTCC
T5-R392x-rev	AGAAGATATCGGGTATGCTTGG		
T5-R392x-seq	GGTTGTAAGAGCATTGTC		
TIRAP			
TIRAP-S180L-for	GCACCATCCCCCTGCTGT	1.5/58	C/TGGGCCTCAGCAGAGCTG CCTACCCA
TIRAP-S180L-rev	B-CCCCTGCCATCGACGTAGTA		
TIRAP-S180L-seq	CCATCCCCCTGCTGT		

(continued)

Table 7.1 (continued)

Name	Sequence[a]	$MgCl_2$ (mM)/annealing temperature (°C)	Sequence to analyze
NOD2/CARD15			
R702W-F	TTCCTGGCAGGGCTGTT	3/64	AGGCCCTGCTCC/TGGCGCCAGGCCTGTGCCCGC TGGTG
R702W-R	B-CCAGACACCAGCGGGC		
CARD15 702seq	GTGCCAGACATCTGAGA		
G908R-F	TAGAGGGAGGAGGACTGTTAGTT	3/64	TCTGGG/CGCAACAGAGTGGGTGACGAGGGGGC
G908R-R	B-CCCCCTCGTCACCCACT		
Card15 908seq	TTTTGGCCTTTTCAGAT		
CARD15forw1007	TAGGGGCAGAAGCCCTCC	3/64	C[C]TTGAAAGGAATGACACCATCCTGGA
CARD15rev1007	B-CAGACTTCCAGGATGGTGTCA		
CARD15seq 1007	AGCCCTCCTGCAGGCC		

[a]B, biotin-labeled at 5′-end.

6. PSQ 96 Plate Low (Biotage).
7. Annealing buffer (Biotage).
8. Binding buffer (Biotage).
9. Ethanol (70%).
10. Denaturation buffer, 500 ml (Biotage).
11. Washing buffer, 200 ml (Biotage).
12. Oligonucleotides for sequencing (**Table 7.1**).
13. Laboratory shaker (e.g., Variomag Monoshake, H+P Labortechnik AG, Oberschleißheim, Germany)
14. Pyrosequencing system (Biotage).

3. Methods

3.1. Isolation of Human DNA from Peripheral Blood

The PCR-assays which will be described in the following report will work on human DNA extracted either manually using DNA extraction kits by various manufacturers or automated techniques. Generally, it is sufficient to isolate DNA from 200 μl of peripheral, K-EDTA anticoagulated blood to achieve a final concentration between 10 and 40 ng/μl. Purity of the isolated DNA should be verified once by photometry. The aim is to achieve a ratio OD260/280 between 1.8 and 2.0.

3.2. PCR

Except for the DNAse-free water, which is stored at 4°C, all other components should be stored at –20°C. Generally, setting up of the PCR is to be separated locally from the subsequent sequencing reaction to strictly avoid contaminations in the following runs. All pipetting is to be performed with DNAse- and RNAse-free filter tips and it is strongly recommended to wear laboratory coats and using vinyl or latex hand gloves for pipetting the PCR-master mixes. All pipetting must be prepared on ice.

No matter if single-plex or duo-plex assays will be prepared, labeled, and unlabeled lyophilized oligonucleotides will be resuspended with DNAse-free water to a final concentration of 10 pmol/μl. In each PCR run at least one no-template control must be included to monitor PCR-contamination. Furthermore a heterozygous sample should also be included as a positive control (*see* **Note 3**).

Note that the $MgCl_2$ concentration varies from PCR to PCR depending on the primers used. All primers were designed using the Pyrosequencing® Assay Design Software (version 1.0.6). The concentration determined for each TLR-PCR is listed in **Table 7.1** (*see* **Note 4**).

1. Prepare 48 μl of a master mix containing: 10 μl 5x Flexi-buffer; X μl $MgCl_2$ (25 mM) (calculate the volume to obtain final concentration given in Table 7.1); 1.0 μl Primer 1 (forward); 1.0 μl Primer 2 (reverse); 1.0 μl dNTPs (10 mM each);

0.2 μl GoTaq polymerase; X μl H_2O (depending on the volume of $MgCl_2$).
2. Vortex and spin down quickly.
3. Transfer master mix to a labeled PCR-tube and add either 2 μl of DNA or 2 μl H_2O for the no template control.
4. Samples are cycled under the following conditions (*see* **Note 5**):

Initial denaturation:	95°C, 3 min
45 cycles:	95°C, 30 s; X°C, 45 s; 72°C, 60 s
Final extension:	72°C, 5 min
Hold temperature	4°C until further use.

Please note that the annealing temperature varies depending on the primers used. The best annealing temperature for each PCR is listed in **Table 7.1**.
5. The PCR-product can now be used for a Pyrosequencing® run directly or might be stored at –20°C.

3.3. Pyrosequencing

3.3.1. Preparation for the Sequencing Reaction

It is strongly recommended performing the sequencing reaction in different and separate room from the pre-PCR-area (*see* **Note 6**). All buffers and reagents necessary should be stored at 4°C until usage. For the sequencing reaction all reagent should be available at room temperature.

1. Resuspend enzyme and substrate components of the Pyro-Gold-Star Kit in 620 μl water at room temperature.
2. Thaw and spin down the sequencing primer at a concentration of 10 pmol/μl.
3. Fill the vacuum-prep-station with 70% ethanol, NaOH (0.2 M), washing buffer, and water.
4. Turn on the vacuum-prep-pump and the heating plate.

3.3.2. Preparation of the PSQ-plate low

1. For a single-plex assay mix 1.6 μl sequencing primer and 38.4 μl annealing buffer per analyzed sample/patient and transfer into a well of the PSQ-plate low (*see* **Note 7**).
2. For a duo-plex assay mix 1.6 μl sequencing primer 1, 1.6 μl sequencing primer 2 and 36.8 μl annealing buffer per analyzed sample/patient and transfer into a well of the PSQ-plate low.
3. Transfer 40 μl annealing buffer without any sequencing primer in one well of the PSQ-plate low to control the sequencing reaction for false positive signals.
4. Place the plate to the right-hand side of the vacuum-prep-station.

3.3.3. Preparation of the PCR-plate

1. Mix the following reagents in each cavity of a separate 96-well PCR-plate: 20 μl H_2O; 40 μl sepharose beads mixture

consisting of 3 μl beads (well-mixed) and 37 μl binding buffer; 20 μl PCR-product

2. Prepare one well with 40 μl H_2O and 40 μl sepharose beads mixture to monitor unspecific signals generated by the sequencing primer.
3. Shake the PCR-plate on a laboratory shaker for at least 5–10 min at room temperature.
4. Place the PCR-plate to the left-hand side of the vacuum-prep-station.

3.3.4. Denaturation and Annealing for the Sequencing Reaction

1. Take the vacuum-prep-tool, place it into the PCR-plate and turn it on. EacH$_2$Of the following steps should be done for 5 s to guarantee an optimal preparation for the subsequent sequencing reaction.
2. Wash in 70% ethanol for 5 s.
3. Let dry for 5 s.
4. Denature for 5 s.
5. Let dry for 5 s.
6. Incubate in washing buffer.
7. Let dry for 30 s.
8. Place the vacuum-prep-tool carefully above the PSQ-plate low. Keeping a distance of approximately 0.5 mm between the vacuum-prep-tool and the bottom of the well of the PSQ-plate turn the vacuum-off. The beads will now be falling into the designated PSQ-plate low wells.
9. Carefully shake the vacuum-prep-tool to remove all aspirated beads.
10. The PSQ-plate low which is now containing the beads and the single-stranded PCR-product is now heated to 80°C for 2 min.
11. Take the PSQ-plate low from the heater and let it stand for 5 min at room temperature.
12. During this time prepare the reagent cartridge as recommended by the manufacturer.
13. Sequencing can now be performed with the PSQ™ 96 MA (Biotage, Uppsala, Sweden). The determination of peaks and the generation of pyrogram® will be done by the Pyrosequencing Software (PSQ96MA, Version 2.1.1, Biotage, Uppsala, Sweden) automatically but the results should be validated for each probe and the controls by the investigator (**Figs. 7.2** and **7.3**) (*see* **Note 8**).

3.4. Troubleshooting

Concerning troubleshooting with new tests based on Pyrosequencing® technology the following controls can be performed to check for unspecific signals:

1. Primer-dimers: Sequencing primer + annealing buffer without PCR-product biotinylated PCR-primer + annealing buffer

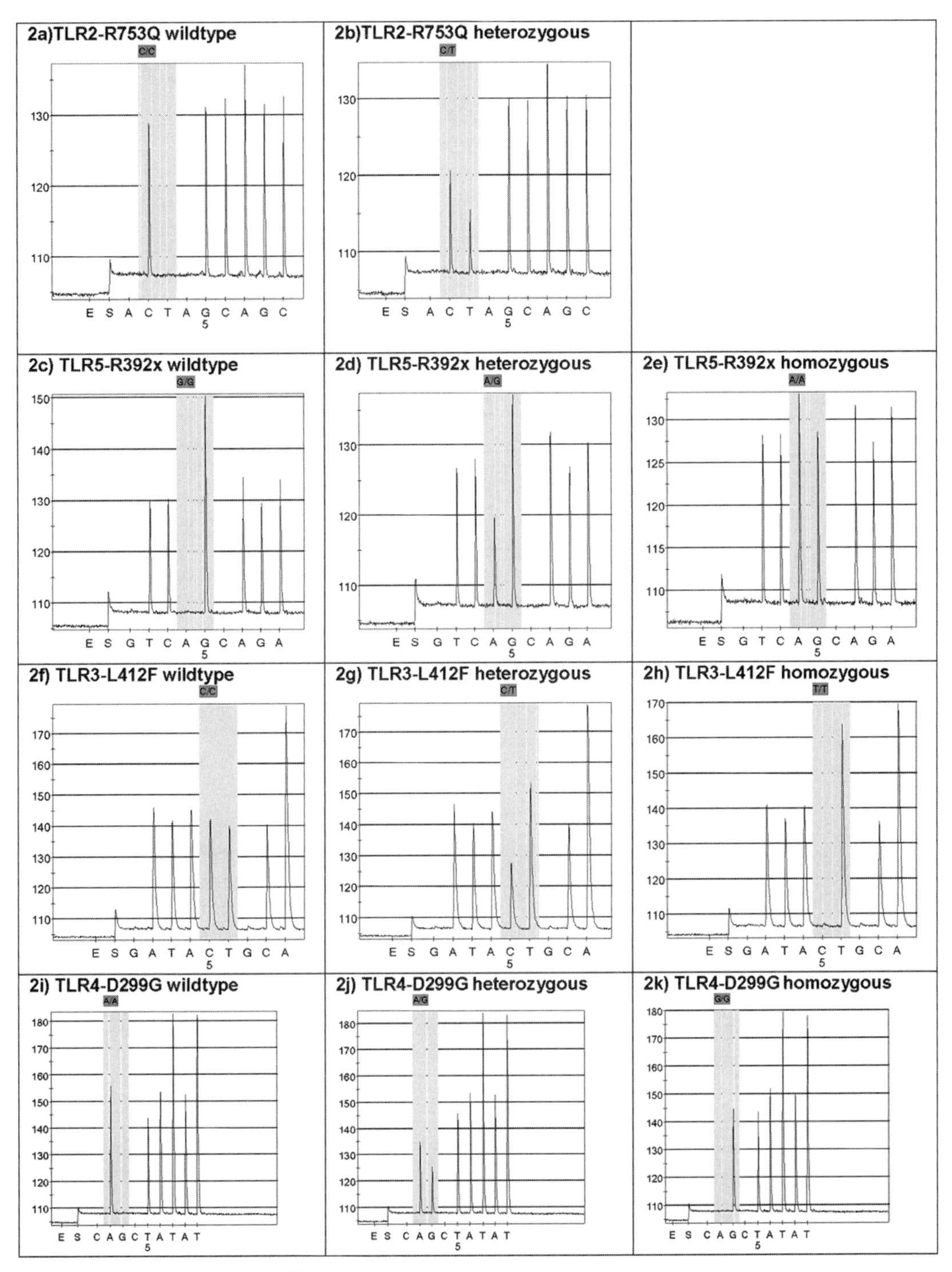

Fig. 7.2. Representative pyrogram® results for different functionally relevant TLR-SNPs (2a–m).

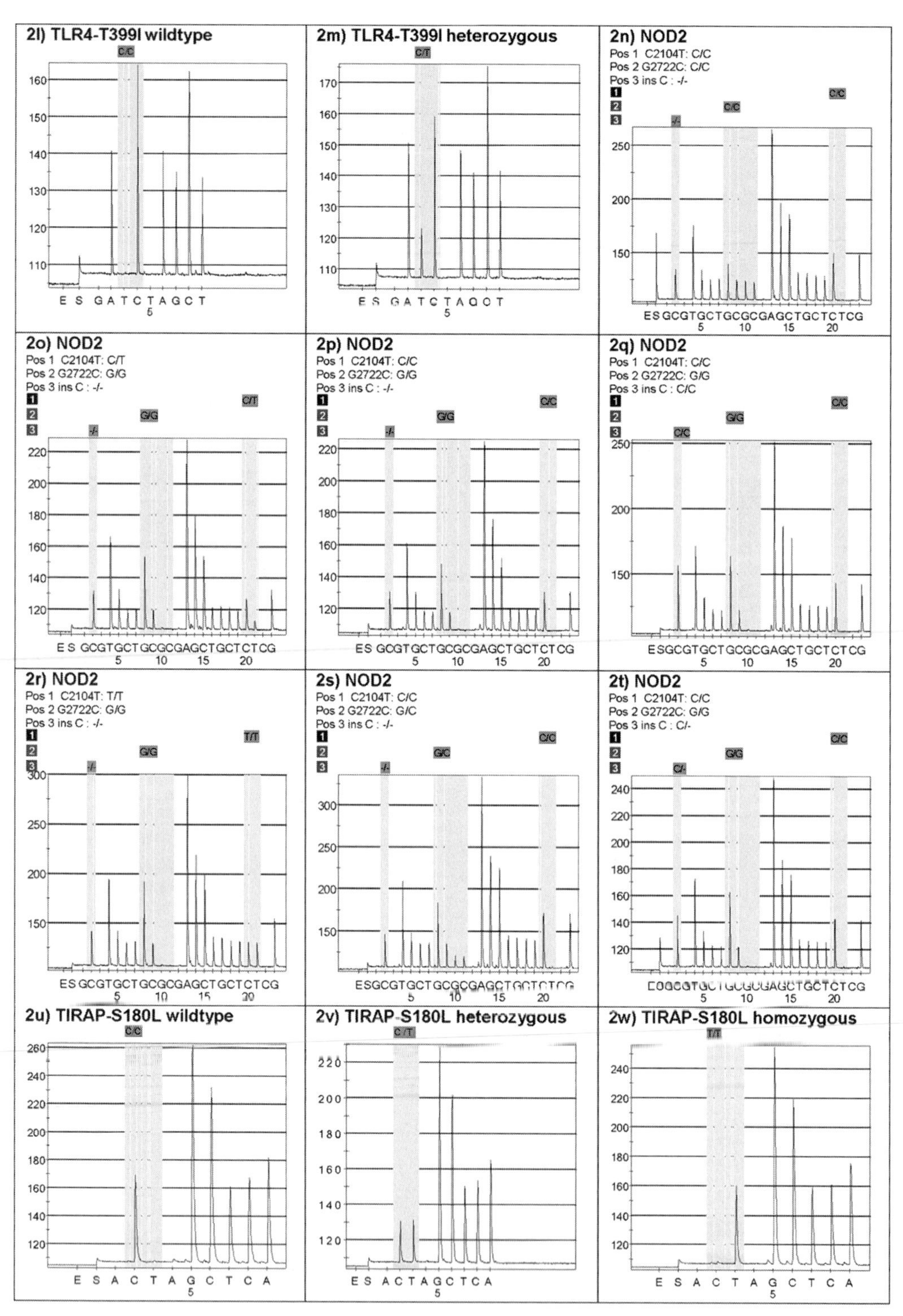

Fig. 7.3. Representative pyrogram® results for different functionally relevant sequence variations in NOD2 (CARD15) (2n–t) and TIRAP (2u–w).

without PCR-product sequencing primer + biotinylated PCR-primer + annealing buffer without PCR-product

2. Potential annealing of the biotinylated PCR-primer: Biotinylated PCR-primer + annealing buffer with PCR-product
3. 3′-end loops of the specific PCR-product:
   PCR-product + annealing buffer
4. PCR-background:
   PCR-negative control + sequencing primer + annealing buffer
   PCR-negative control + biotinylated PCR-primer + annealing buffer
   PCR-negative control + annealing buffer

4. Notes

1. Unless stated otherwise, all solutions should be prepared in water that has a resistivity of 18.2 MΩ cm and total organic content of less than five parts per billion. This standard is referred to as "water" in the text.
2. Hot-start DNA polymerases such as the GoTaq DNA polymerase (Promega, Mannheim, Germany) reduce the generation of unspecific PCR-products. The green indicator/loading dye does not interfere with downstream techniques and the bioluminometric Pyrosequencing® procedure.
3. For rare mutations the use of positive controls makes sense to monitor false negative determinations. To generate such positive controls plasmids might be used but should be handled with great care to avoid contaminations.
4. All oligonucleotides should be designed carefully using the manufacturer's software accompanying the instrument. Primer design should strictly avoid mispriming or hairpin structures which might be generated in the PCR or the following sequencing reaction.
5. In general PCRs for genotyping assays on a Pyrosequencing® instrument have three important characteristics: (1) generation of short PCR-amplicons (usually <200 bp); (2) including one biotinylated primer; (3) running a high number of cycles within the PCR (>40 cycles).
6. Since the PCR-tube will be opened for Pyrosequencing® after the PCR-amplification the technique generally bears a great risk for PCR-contamination as compared to other genotyping methods. A strict separation of pre- and post-amplification laboratory area as described in reference literature is therefore an indispensable prerequisite to avoid false determinations *(21)*.
7. Pyrosequencing enables the investigator to perform duoplex or tri-plex genotyping assays out of one multiplex-PCR.

Table 7.2
Different master mixes to optimize the PCR-conditions for subsequent pyrosequencing

Component	Master mix (μl) I	II	III
H_2O	31.8	29.8	30.8
5x Buffer	10	10	10
$MgCl_2$ (25 mM)	3	5	3
Primer forward (10 μM)	1	1	1.5
Primer reverse (10 μM)	1	1	1.5
dNTPs (each 10 mM)	1	1	1
goTaq (500 U)	0.2	0.2	0.2
Genomic DNA (10–40 ng/μl)	2	2	2

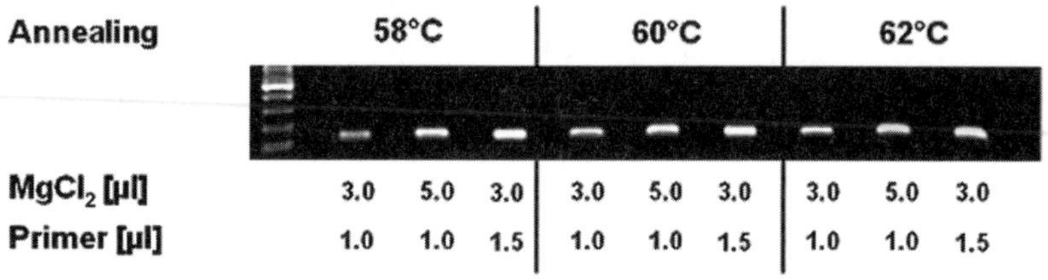

Annealing 58°C 60°C 62°C
$MgCl_2$ [μl] 3.0 5.0 3.0 3.0 5.0 3.0 3.0 5.0 3.0
Primer [μl] 1.0 1.0 1.5 1.0 1.0 1.5 1.0 1.0 1.5

Fig. 7.4. Nine different PCR-protocols have been tested simultaneously in a MasterCycler EP gradient S.

Besides the fact that multiplex PCR must be validated we recommend to start with single-plex runs first and subsequently try to establish multiplex-assays by merging both assays.

8. Like in capillary sequencing Pyrosequencing® assays are PCR-based. Therefore it is important to optimize the PCR as a basic technique first in setting up genotyping methods. It is recommended to determine the concentration of $MgCl_2$, primers, and the annealing temperature necessary to yield the best results and document the run by agarose gel-electrophoresis (**Table 7.2** and **Fig. 7.4**).

Acknowledgments

The author would like to thank Prof. Dr Michael Neumaier for his advice and encouragement and Heidi Prunes for excellent and outstanding technical assistance. This work was supported by the Fritz Thyssen Stiftung Az. 10.03.2.115.

References

1. Takeda, K., Kaisho, T., Akira, S. (2003) Toll-like receptors. *Annu Rev Immunol* 21, 335–376.
2. Poltorak, A., He, X., Smirnova, I., Liu, M. Y., Huffel, C. V., Du, X., Birdwell, D., Alejos, E., Silva, M., Galanos, C. et al. (1998) Defective LPS signaling in C3H/HeJ and C57BL/10ScCr mice: mutations in Tlr4 gene. *Science* 282, 2085–2088.
3. Aliprantis, A. O., Weiss, D. S., Zychlinsky, A. (2001) Toll-like receptor-2 transduces signals for NF-kappa B activation, apoptosis and reactive oxygen species production. *J Endotoxin Res* 7, 287–291.
4. Vabulas, R. M., Ahmad-Nejad, P., da Costa, C., Miethke, T., Kirschning, C. J., Hacker, H., Wagner, H. (2001) Endocytosed HSP60s use toll-like receptor 2 (TLR2) and TLR4 to activate the toll/interleukin-1 receptor signaling pathway in innate immune cells. *J Biol Chem* 276, 31332–31339.
5. Vabulas, R. M., Ahmad-Nejad, P., Ghose, S., Kirschning, C. J., Issels, R. D., Wagner, H. (2002) HSP70 as endogenous stimulus of the Toll/interleukin-1 receptor signal pathway. *J Biol Chem* 277, 15107–15112.
6. Vabulas, R. M., Braedel, S., Hilf, N., Singh-Jasuja, H., Herter, S., Ahmad-Nejad, P., Kirschning, C. J., Da Costa, C., Rammensee, H. G., Wagner, H. et al. (2002) The endoplasmic reticulum-resident heat shock protein Gp96 activates dendritic cells via the Toll-like receptor 2/4 pathway. *J Biol Chem* 277, 20847–20853.
7. Underhill, D. M., Ozinsky, A., Hajjar, A. M., Stevens, A., Wilson, C. B., Bassetti, M., Aderem, A. (1999) The Toll-like receptor 2 is recruited to macrophage phagosomes and discriminates between pathogens. *Nature* 401, 811–815.
8. Hemmi, H., Takeuchi, O., Kawai, T., Kaisho, T., Sato, S., Sanjo, H., Matsumoto, M., Hoshino, K., Wagner, H., Takeda, K. et al. (2000) A Toll-like receptor recognizes bacterial DNA. *Nature* 408, 740–745.
9. Hayashi, F., Smith, K. D., Ozinsky, A., Hawn, T. R., Yi, E. C., Goodlett, D. R., Eng, J. K., Akira, S., Underhill, D. M., Aderem, A. (2001) The innate immune response to bacterial flagellin is mediated by Toll-like receptor 5. *Nature* 410, 1099–1103.
10. Jurk, M., Heil, F., Vollmer, J., Schetter, C., Krieg, A. M., Wagner, H., Lipford, G., Bauer, S. (2002) Human TLR7 or TLR8 independently confer responsiveness to the antiviral compound R-848. *Nat Immunol* 3, 499.
11. Heil, F., Ahmad-Nejad, P., Hemmi, H., Hochrein, H., Ampenberger, F., Gellert, T., Dietrich, H., Lipford, G., Takeda, K., Akira, S. et al. (2003) The Toll-like receptor 7 (TLR7)-specific stimulus loxoribine uncovers a strong relationship within the TLR7, 8 and 9 subfamily. *Eur J Immunol* 33, 2987–2997.
12. Hemmi, H., Kaisho, T., Takeuchi, O., Sato, S., Sanjo, H., Hoshino, K., Horiuchi, T., Tomizawa, H., Takeda, K., Akira, S. (2002) Small anti-viral compounds activate immune cells via the TLR7 MyD88-dependent signaling pathway. *Nat Immunol* 3, 196–200.
13. Riva, A., Kohane, I. S. (2002) SNPper: retrieval and analysis of human SNPs. *Bioinformatics* 18, 1681–1685.
14. Hawn, T. R., Verbon, A., Lettinga, K. D., Zhao, L. P., Li, S. S., Laws, R. J., Skerrett, S. J., Beutler, B., Schroeder, L., Nachman, A. et al. (2003) A common dominant TLR5 stop codon polymorphism abolishes flagellin signaling and is associated with susceptibility to legionnaires' disease. *J Exp Med* 198, 1563–1572.
15. Lorenz, E., Mira, J. P., Cornish, K. L., Arbour, N. C., Schwartz, D. A. (2000) A novel polymorphism in the toll-like receptor 2 gene and its potential association with staphylococcal infection. *Infect Immun* 68, 6398–6401.
16. Merx, S., Neumaier, M., Wagner, H., Kirschning, C. J., Ahmad-Nejad, P. (2007) Characterization and investigation of single nucleotide polymorphisms and a novel TLR2 mutation in the human TLR2 gene. *Hum Mol Genet* 16, 1225–1232.
17. Merx, S., Zimmer, W., Neumaier, M., Ahmad-Nejad, P. (2006) Characterization and functional investigation of single nucleotide polymorphisms (SNPs) in the human TLR5 gene. *Hum Mutat* 27, 293.
18. Fakhrai-Rad, H., Pourmand, N., Ronaghi, M. (2002) Pyrosequencing: an accurate detection platform for single nucleotide polymorphisms. *Hum Mutat* 19, 479–485.
19. Ahmadian, A., Ehn, M., Hober, S. (2006) Pyrosequencing: history, biochemistry and future. *Clin Chim Acta* 363, 83–94.
20. Langaee, T., Ronaghi, M. (2005) Genetic variation analyses by Pyrosequencing. *Mutat Res* 573, 96–102.
21. Neumaier, M., Braun, A., Wagener, C. (1998) Fundamentals of quality assessment of molecular amplification methods in clinical diagnostics. International Federation of Clinical Chemistry Scientific Division Committee on Molecular Biology Techniques. *Clin Chem* 44, 12–26.

Chapter 8

Multiplex *HLA*-Typing by Pyrosequencing

Ying Lu, Julian Boehm, Lynn Nichol, Massimo Trucco, and Steven Ringquist

Abstract

Class I and II loci of the human leukocyte antigens (*HLA*) represent the most polymorphic region of the genome. Evolutionary pressure has resulted in a large number of allelic variants of these loci ensuring the high frequency of heterozygous genotypes observed in human populations. Molecular techniques, including sequencing, are capable of precisely defining *HLA* alleles. Sequencing by synthesis methodology employed by pyrosequencing represents a complementary approach to other molecular methods of *HLA* genotyping. Out-of-phase sequencing of *HLA* alleles by pyrosequencing can resolve certain *cis/trans* ambiguities that would otherwise require the sequencing of cloned DNA. Genotyping of *HLA* loci for the presence of specific amino acid variants is beneficial for proper matching of organ donor to recipient, the monitoring of *HLA* associated genetic risk to autoimmune diseases, population genetic studies, as well as evaluation of the genetics of human host–human pathogen interaction.

Key words: DNA sequencing, histocompatibility antigens, immunogenetics, sequence-based typing, transplantation.

Abbreviations: Expressed Sequence Tag, EST; Human Leukocyte Antigen, *HLA*; Polymerase Chain Reaction, PCR; Single Nucleotide Polymorphism, SNP; Single Stranded Binding Protein, SSBP

1. Introduction

Human leukocyte antigens (*HLA*) class I and class II loci are located in a roughly four megabase segment on chromosome 6 at cytogenetic location 6p21.3 *(1, 2)*. The *HLA* encoding region can be subdivided into three physical regions encoding the class I, class III, and class II molecules *(3)* (**Fig. 8.1**). The physiologic function of the HLA proteins is to present antigenic peptides from the cytoplasm onto the cell surface. The genes encoding the

Peter Bugert (ed.), *DNA and RNA Profiling in Human Blood: Methods and Protocols, vol. 496*

DOI 10.1007/978-1-59745-553-4_8 Springerprotocols.com

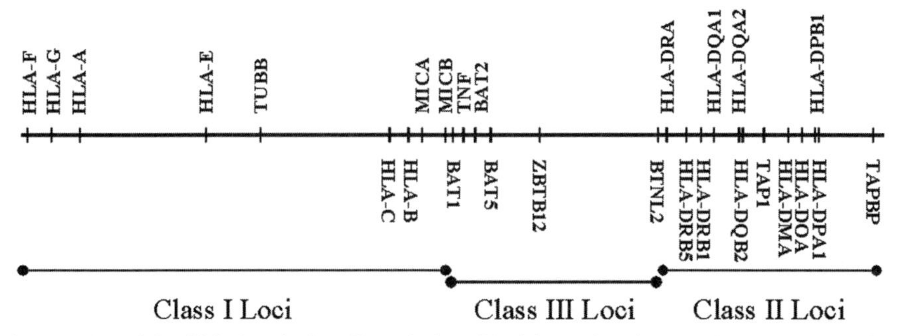

Fig. 8.1. Organization of the *HLA* class I, class III, and class II loci located on human chromosome 6p. Selected loci are shown. The class I and class II regions encode the major immunogenetic loci *HLA-A*, *-B*, *-C* and *HLA-DR*, *-DQ*, *-DP*, respectively. Loci located in the class III region are not related to class I or class II loci structurally or functionally.

HLA class I and class II proteins constitute the most highly polymorphic genetic system in humans (**Table 8.1**). Allelic variations have been identified for most of the *HLA* loci with *HLA-B* being the most polymorphic having over 900 reported alleles *(4)*. The numbers of confirmed *HLA* alleles is consistently rising *(4)*, there being as many as 1,910 and 889 alleles reported for the class I and class II *HLA* loci, respectively.

Determination of genetic polymorphisms within the *HLA* loci is essential for successful transplantation of tissue during solid organ and bone marrow transplantation. In transplant genetics, matching donor and recipient for bone marrow transplantation

Table 8.1
***HLA* allele and protein variants**

HLA locus	Allelic variants	Protein variants
Class I loci		
HLA-A	580	462
HLA-B	921	789
HLA-C	312	249
Class II loci		
HLA-DRA	3	2
HLA-DRB1	501	423
HLA-DQA1	34	25
HLA-DQB1	86	64
HLA-DPA1	23	14
HLA-DPB1	127	114

Numbers of allelic and protein variants were obtained from the IMGT/HLA database *(4)*.

generally means determination of an appropriate immunological antigen match based on each of two alleles at *HLA* class I loci *HLA-A* and *-B* as well as class II loci *HLA-DRB1*, *-DQB1*, and *-DPB1* *(5)* (**Table 8.2**). In human population genetics, *HLA* allele frequency differences have been observed when comparing geographically distinct human populations and have identified probable patterns associated with ancient human migration patterns as well as recent admixture events *(6)*. Changes in the frequency of *HLA* alleles are currently being observed due to selective pressure from viral pathogens, such as, HIV and susceptibility to AIDS *(7, 8)*. Genome-wide association scans as well as population studies associate *HLA-B*5701* with AIDS resistance observed in some HIV positive patients *(9)*. The frequency of this allele has been reported to be increasing in populations where HIV prevalence is endemic *(8)*. Evidence for a role of *HLA* alleles has also been observed in preference for mate selection observed

Table 8.2
Frequently occurring *HLA* haplotypes observed in Caucasians

Selected *HLA* class I haplotypes				Selected *HLA* class II haplotypes			
HLA-A	*HLA-B*	*HLA-C*	Frequency	*HLA-DRB1*	*HLA-DQA1*	*HLA-DQB1*	Frequency
*0101	*0801	*0701	0.070	*1501	*0102	*0602	0.142
*0201	*4402	*0501	0.054	*0301	*0501	*0201	0.131
*0301	*0702	*0702	0.041	*0701	*0201	*0202	0.111
*0201	*0702	*0702	0.033	*0101	*0101	*0501	0.091
*0201	*4001	*0304	0.025	*1101	*0501	*0301	0.056
*0101	*5701	*0602	0.022	*1301	*0103	*0603	0.056
*0201	*0801	*0701	0.020	*0401	*0302	*0301	0.053
*2601	*3801	*1203	0.017	*0401	*0301	*0302	0.042
*0201	*1501	*0303	0.015	*0404	*0301	*0302	0.039
*0301	*3501	*0401	0.015	*0701	*0201	*0303	0.037
*3101	*4001	*0304	0.015	*1302	*0102	*0604	0.034
*1101	*3501	*0401	0.013	*1104	*0501	*0301	0.027
*2902	*4403	*1601	0.013	*0801	*0401	*0402	0.022
*2402	*3501	*0401	0.011	*1401	*0104	*0503	0.020
*0201	*3501	*0401	0.011	*0102	*0101	*0501	0.014
				*1201	*0501	*0301	0.011

HLA class I and II haplotypes with observed frequencies greater than 1% are shown. Haplotype frequencies are summarized from Cao et al. *(15)* and Klitz et al. *(16)*.

in humans *(10)*. People as well as other animals have the ability to recognize HLA peptides as olfactory cues that enable social recognition *(11)*. Selective pressure to maintain population-wide high frequencies of *HLA* heterozygous individuals has been connected with human population survival in the face of otherwise lethal infectious diseases *(12, 13)*.

The basis for *HLA* polymorphism resides in stable inherited variations of the genomic DNA sequence present in the coding regions of the *HLA* genes (**Fig. 8.2**). Polymorphisms are clustered into discrete hypervariable regions directing changes in the amino acid sequence of the protein *(14)*. While most *HLA* alleles occur in all ethnic groups, they vary in frequency among these populations *(15,16)*. Numerous ambiguities occur when genotyping *HLA* alleles by sequence-based typing *(17)*. This is primarily due to the presence of *cis/trans* combinations of alleles that occur in certain allelic pairs in heterozygous individuals *(4, 18)*. As a result of the method's intrinsic sequencing by synthesis approach, pryosequencing yields out-of-phase sequence information from individual alleles *(19)*. Out-of-phase pyrosequencing allows many of the *cis/trans* pairs to be resolved into their individual haplotypes. Thus, providing high-quality data, focused genotyping results, as well as improved *HLA* resolution *(18, 20–22)*.

2. Materials

2.1. Collection and Purification of DNA from Whole Blood

1. Vacutainer tubes with EDTA anticoagulant may be purchased from Becton-Dickinson Inc., Franklin Lakes, NJ. A trained phlebotomist should be available to perform blood draws from volunteers. This step may require approval by the local Institutional Review Board.
2. QIAamp DNA blood mini kit (Qiagen, Inc., Valencia, CA) is used to extract DNA from whole blood. This kit may also be used for preparing DNA from purified cells.
3. Dry blood spots are prepared by spotting whole blood onto S&S 903 Paper (Schleicher and Schuell Inc., Keene, NH).
4. Biotech grade ethanol and methanol are purchased from Thermo Fisher Scientific (Waltham, MA).
5. Punches of 1.5 mm in diameter can be obtained using a hand-operated paper puncher or a Wallac Delfia DBS Puncher (Perkin-Elmer Life Sciences Inc., Turku, Finland).

2.2. Estimation of DNA Concentration

1. PicoGreen dsDNA quantification kit is obtained from Invitrogen Corp. (Carlsbad, CA). The fluorescence-based assay will provide precise measurement of DNA concentration from whole blood-isolated genomic DNA.

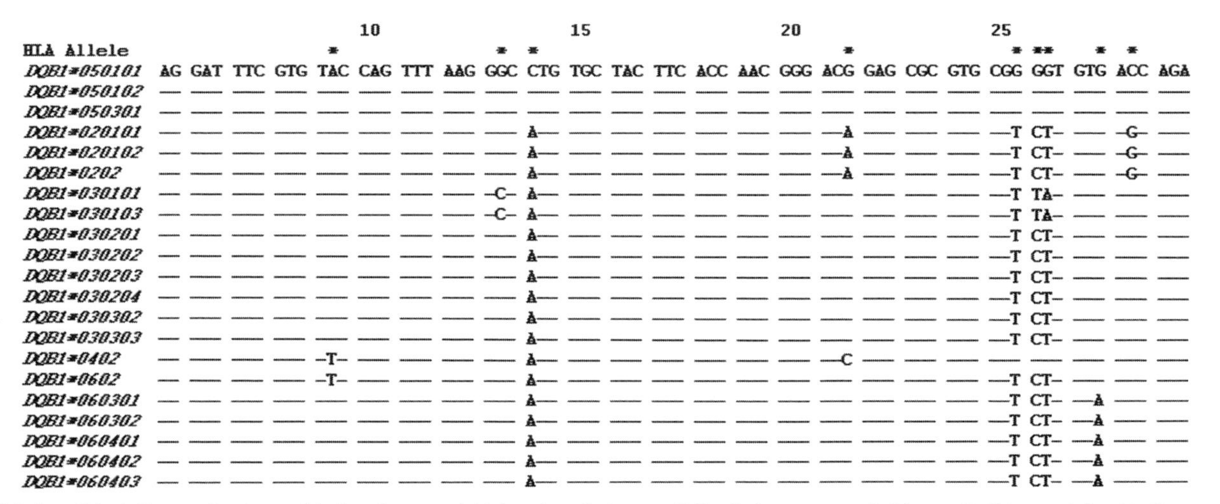

Fig. 8.2. In the case of *HLA* class II loci, the most polymorphic (i.e., hypervariable) regions between allelic chains are encoded by nucleotides contained in the second exon of the gene. The nucleotides in the sequences are shown grouped in triplets to reflect the amino acid coding structure. In the example, the allelic nucleotides of *HLA-DQB1* are shown beginning at the second nucleotide residue of codon 5 (corresponding to the beginning of exon 2) through codon 30. Hypervariable residues are indicated by asterisk symbols.

2. 10x TE buffer (pH 8) is prepared by dissolving in 1 l distilled water, 12.1 g Tris base and 3.72 g disodium EDTA. Adjust pH with concentrated HCl to pH 8 (*see* **Notes 1** and **2**).

2.3. Polymerase Chain Reaction (PCR)

1. 10× Taq buffer consists of 500 mM KCl, 100 mM Tris–HCl (pH 8.3), 0.01% (w/v) gelatin and can be obtained from Applied Biosystems (Foster City, CA). Store frozen at –20°C.
2. Magnesium chloride solution is prepared as a 25 mM solution by dissolving 0.5 g $MgCl_2$–hexahydrate in 100 ml distilled water. The solution should be transferred as 1 ml aliquots and stored frozen at –20°C.
3. Deoxynucleotide triphosphate solution consists of 10 mM dNTP mixture of dATP, dCTP, dGTP, and TTP can be purchased from Applied Biosystems (Foster City, CA). Store frozen at –20°C.
4. *Taq* DNA polymerase (5 units/μl) can be purchased from Applied Biosystems (Foster City, CA). Store frozen at –20°C.
5. DNA oligonucleotide primers for PCR can be purchased from Integrated DNA Technologies, Inc. (Ames, IA), Invitrogen (Carlsbad, CA), Operon (Alameda, CA), or another provider of high-quality DNA oligonucleotides. Desalted primers should be dissolved in 1× TE buffer (pH 8) at 10 μM concentration. Store in 1× TE buffer (pH 8) at 4°C.
6. Biotinylated DNA oligonucleotide primers containing a 5′-end biotin-TEG modification can be purchased from Integrated DNA Technologies, Inc. (Ames, IA), Invitrogen (Carlsbad, CA), Operon (Alameda, CA), or another provider of high-quality DNA oligonucleotides. HPLC-purified biotinylated primers should be dissolved in 1× TE buffer (pH 8) at a concentration of 10 μM. Store in 1× TE buffer (pH 8) at 4°C.

2.4. Agarose Gel Electrophoresis

1. Agarose (low electroendosmosis) can be obtained from Roche, Inc. (Indianapolis, IN).
2. One liter of a solution of 5× TBE buffer should be prepared using 54 g Tris base, 27.5 g boric acid, and 3.72 g disodium EDTA. 5x TBE may be stored at room temperature for several months.
3. Ethidium bromide solution (10 mg/ml) can be purchased from Sigma Chemical Company (Saint Louis, MO) (*see* **Note 3**).

2.5. Pyrosequence-Based Typing

1. Streptavidin-coated beads (GE Healthcare Bio-Sciences, Corp., Piscataway, NJ).
2. 96-well MultiScreen-HV 0.45 μm hydrophilic, low protein binding, Durapore membrane (Millipore Corp., Bedford, MA).

3. Binding buffer is 10 mM Tris–HCl (pH 7.6), 2 M NaCl, 1 mM EDTA, 0.1% Tween 20. This solution is prepared by mixing 1.2 g Tris base, 116.9 g NaCl, and 0.34 g disodium EDTA in 900 ml distilled water. Adjust the solution to pH 7.6 with concentrated HCl. Prepare the final solution by addition of 1 ml Tween 20 and adjust the volume to 1 l with distilled water. Store the solution at room temperature.
4. Denaturation solution is 0.2 M NaOH. Prepared by dissolving 8 g NaOH in 1 l distilled water. Store at room temperature.
5. Washing buffer consists of 10 mM Tris–acetate (pH 7.6). The solution is prepared by dissolving 1.2 g Tris base in 800 ml distilled water and is titrated to pH 7.6 with 4 M acetic acid. The final volume is then adjusted to 1 l with distilled water. Store the washing buffer at room temperature.
6. Annealing buffer is composed of 20 mM Tris–acetate (pH 7.6), 2 mM magnesium acetate. Prepare by dissolving 2.42 g Tris base and 0.43 g magnesium acetate-tetrahydrate in 900 ml distilled water. Adjust pH to 7.6 using 4 M acetic acid. Add distilled water to adjust the final volume to 1 l. Store at room temperature.
7. *E. coli* single-stranded DNA binding protein can be obtained from Promega Corp., Madison, WI. Store frozen at –20°C.
8. 96-well tray formatted MultiScreen-HV 0.45 μm hydrophilic, low-protein binding, Durapore membranes are purchased from Millipore Corp., Bedford, MA.
9. Pyrosequencing reagents (PSQ 96 MA sample preparation kit and PSQ 96 SQA reagent kit) and instrumentation may be purchased from Biotage, AB (Charlottesville, NC).

3. Methods

Pyrosequencing was developed for the analysis of expressed sequence tags (ESTs) in order to enable accurate sequencing of short stretches of DNA. The methodology has found increasing acceptance as a method for analysis of single nucleotide polymorphisms (SNPs), insertion/deletion polymorphisms, and for investigating sites of DNA methylation *(23–25)*. Pyrosequencing allows increased resolution of allelic pairs due to its ability to accurately resolve heterozygous nucleotides by enabling out-of-phase sequencing *(26)*. Performed by stepwise addition of dNTPs, pyrosequencing enables synthesis of the nascent nucleotide chain that is extended one nucleotide residue per dispensation event. Sequencing by DNA synthesis is performed through a chain of enzymatic reactions involving DNA polymerase, apyrase, ATP sulfurylase, and luciferase *(27, 28)*. Each nucleotide is dispensed independently to the reaction mix. The graphic display depicting the incorporation of a particular nucleotide is in the form of nucleotide dispensation event (x-axis) versus the intensity of emitted light (y-axis) and is referred to as a pyrogram (**Fig. 8.3**). The pyrosequencing reaction is quantitative, in that increased light intensity is produced upon incorporation of multiple nucleotides. Thus, DNA sequence is determined by examination of light intensity emitted immediately after nucleotide dispensation. Polymorphic residues are identified from nucleotide incorporation patterns consistent with known genotype possibilities (e.g., homozygous AA, heterozygous AB, and homozygous BB) (*see* **Note 4**).

3.1. Collection and Storage of Whole Blood Obtained by Venous Puncture from Human Subjects

1. Collection of whole blood by vein puncture. Sterile vein puncture can be used as a method of obtaining whole blood suitable for DNA extraction and pyrosequencing. This step should be performed by a trained phlebotomist on volunteers who have been recruited in a manner specified by the local Institutional Review Board (*see* **Note 5**).
2. Whole blood (at least 5 ml) can be drawn into Vacutainer tubes containing EDTA as an anticoagulant.
3. Samples can be processed immediately or aliquots can be stored frozen at –80°C. When transferring blood samples biohazard safety precautions should be observed, such as working within a biohazard laminar flow hood, wearing latex gloves, and eye protection. Disposal of used laboratory materials should follow the local regulations of the institution in which the work is being performed.

3.2. Collection and Storage of Dry Blood Spot from Human Subjects

1. An alternative approach to collection of human sample is to transfer whole blood directly to filter paper to generate dry blood spots for storage and DNA purification. This process uses a sterile needle to deliver a finger prick in order to induce a minimum amount of bleeding, 20–50 μl is sufficient. This step should be performed by a trained phlebotomist on volunteers who have been recruited in a manner specified by the local Institutional Review Board (*see* **Note 5**).
2. Whole blood from the site of bleeding can be transferred directly by pressing the finger onto S&S 903 paper.
3. The blood spotted S&S 903 paper is allowed to dry overnight in a biosafety hood.
4. Dry blood spots are sealed from moisture and stored at room temperature. A zip lock plastic bag is sufficient for this step. The sealed samples can be stored in #10 sized mailing envelopes. This is a convenient method to separate and avoid contamination of samples that have been collected from different individuals.

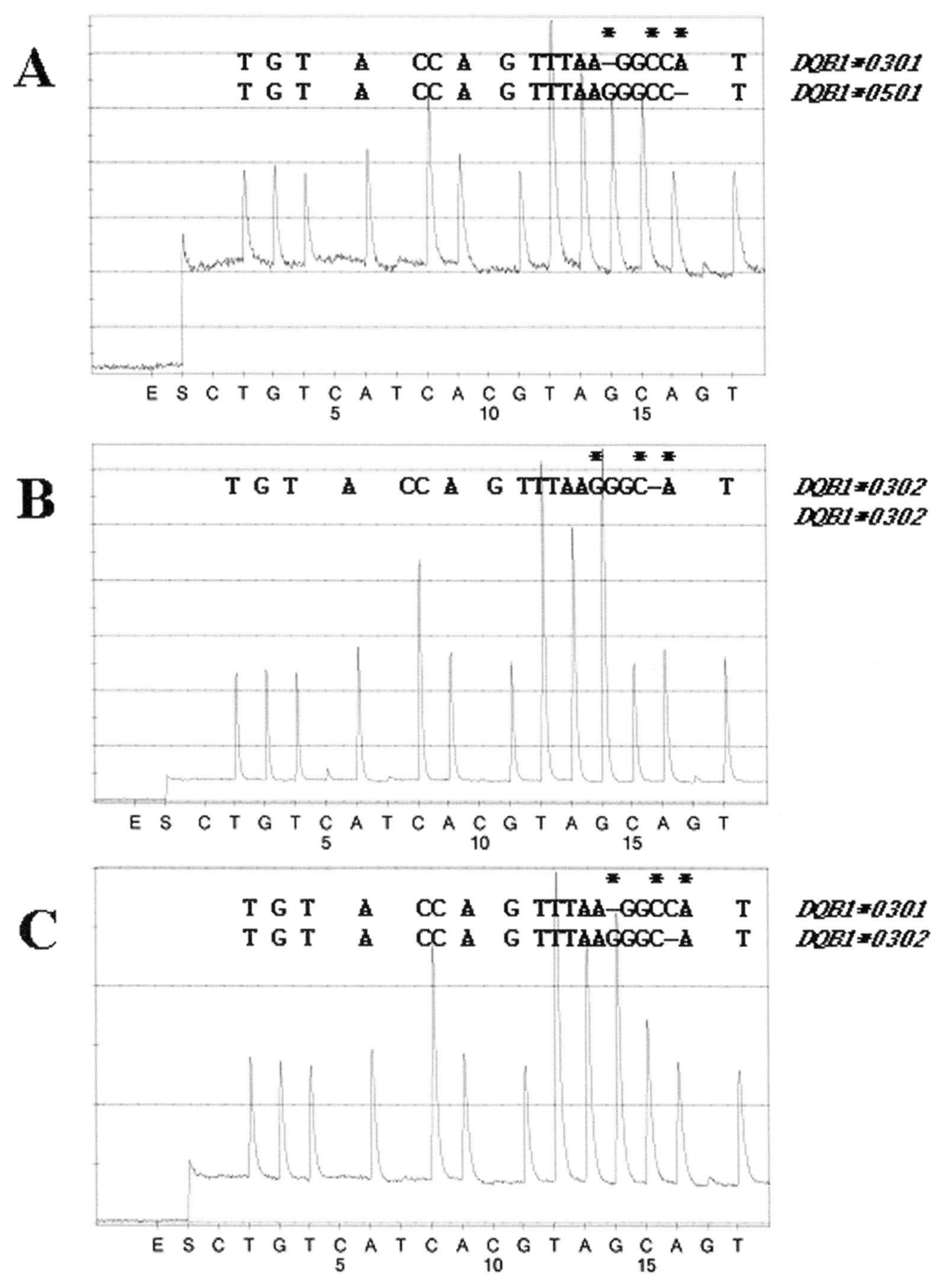

Fig. 8.3. Pyrograms of *HLA-DQB1* alleles obtained from DNA isolated from a family trio pedigree. A. *HLA-DQB1*0301+*0501*, paternal DNA. B. *HLA-DQB1*0302+*0302*, maternal DNA. C. *HLA-DQB1*0301+*0302*, offspring DNA. DNA was prepared from whole blood and PCR amplified using *HLA-DQB1*-specific primer SR25 and SR24. The sequencing primer was SR26 and the pyrosequence step used the nucleotide dispensation order indicated in **Table 8.9**. The residues marked by an asterisk symbol indicate the position of hypervariable residues unique to the three genotypes.

3.3. Purification of DNA from Whole Blood Collected by Venous Puncture

1. DNA can be extracted from fresh as well as frozen whole blood. QIAamp DNA mini kit has been used for DNA isolation. The method typically yields greater than 0.5 μg of purified genomic DNA per 75 μl of whole blood. The DNA isolation protocol is modified from the QIAamp DNA Blood Mini Kit Handbook provided by Qiagen, Inc., for use with whole blood collected with EDTA as an anticoagulant. The following items are recommended precautions before starting the DNA isolation protocol:
   a. Equilibrate whole blood samples to room temperature before starting.
   b. Prepare a 56°C heating block.
   c. Equilibrate buffer AE to room temperature.
   d. Prepare buffers AW1, AW2, and QIAGEN protease according to the instructions in the QIAamp DNA Mini Kit Handbook.
2. In a 1.5 ml centrifuge tube, thoroughly mix 20 μl QIAGEN protease with 75–200 μl whole blood. If less than 200 μl whole blood is used, add an appropriate volume of phosphate-buffered saline solution.
3. Add 200 μl buffer AL and vortex for 15 s.
4. Incubate the mixture for 10 min at 56°C followed by brief centrifugation to remove droplets from the inside of the lid.
5. Add 200 μl of absolute ethanol to the protease-treated blood sample and mix by vortexing. Briefly centrifuge in order to remove droplets from the inside of the lid.
6. Transfer the solution to a QIAamp mini column placed in 2 ml collection tubes. Seal the tubes and centrifuge at 6,000 x *g* for 1 min.
7. Transfer the QIAamp mini column to a new 2 ml collection tube. Seal the tube and centrifuge at 6,000 x *g* for 1 min.
8. Add 500 μl buffer AW1 to the column. Seal the column and centrifuge at 6,000 x *g* for 1 min. Place the column into a clean 2 ml collection tube.
9. Add 500 μl buffer AW2 to the QIAamp mini column. Seal the column and centrifuge at 20,000 x *g* for 3 min. Place in new collection tube and centrifuge at 20,000 x *g* for 1 min.
10. Repeat step 10 and transfer the QiAamp mini column to a new 1.5 ml centrifuge tube.
11. Add 50 μl of distilled water by transferring directly onto the center of the membrane of the column. Seal the column and incubate at room temperature for 5 min. Centrifuge at 6,000 x *g* for 1 min.
12. Store purified DNA samples either refrigerated or frozen at –20°C. Prior to use in PCR amplification the concentration of the DNA samples should be determined. This will be helpful when optimizing the amount of DNA template required

during the PCR step, allowing careful management of limited samples.

3.4. Isolation of DNA from Dried Blood Spots

1. Purification of genomic DNA from dried blood spot samples follows the protocol outlined by the Qiagen QIAamp DNA Mini Kit and QIAamp DNA Blood Mini Kit Handbook. Multiple 3.2 mm diameter punches can be prepared from each dried blood spot. This can be accomplished by using a hand-operated paper puncher or a motor-operated Wallac Delfia DBS Puncher. The hand-operated paper puncher and Wallac Delfia DBS Puncher are equipped with a 3.2 mm punch head. Specimens can be punched directly into sample collection trays. Six 3.2 mm diameter punches per tube are sufficient for each extraction.
2. The following items are recommended before starting the DNA isolation protocol:
   a. Prepare 85, 70 and 56°C heating blocks.
   b. Equilibrate buffers AW1 and AW2 to room temperature.
3. Using a 3.2 mm diameter punch head transfer six punches of the dried blood spot to a 1.5 ml centrifuge tube. Add 180 μl buffer ATL.
4. Place the tube in an 85°C heating block for 10 min. Following the incubation briefly centrifuge to remove droplets from the lid.
5. Add 20 μl proteinase K and mix thoroughly. Incubate the sample at 70°C for 10 min. Briefly centrifuge to remove droplets from the inside of the lid.
6. Add 200 μl absolute ethanol. Mix thoroughly and centrifuge to remove droplets from the lid.
7. Transfer the sample to the QIAamp spin column and place the column into a 2 ml collection tube. Seal the column and centrifuge at 6,000 x *g* for 1 min. Transfer the column to a clean 2 ml collection tube and discard the filtrate.
8. Open the column and add 500 μl buffer AW1 to the center of the membrane. Seal the column and centrifuge at 6,000 x *g* for 1 min. Transfer the column to a clean 2 ml collection tube and discard the filtrate.
9. Open the column and add 500 μl buffer AW2 to the center of the membrane. Seal the column and centrifuge at 20,000 x *g* for 3 min.
10. Transfer the column to a clean 1.5 ml centrifuge tube. Open the column and transfer 100 μl buffer AE to the center of the membrane. Centrifuge at 6,000 x *g* for 1 min in order to elute the DNA sample from the column.
11. Store purified DNA samples either refrigerated or frozen at –20°C. Prior to use in PCR amplification the concentration of the DNA samples should be determined. This will allow the laboratory to optimize the amount of DNA template

required during the PCR step and enable management of limited samples.

3.5. Determining the Concentration of Purified DNA

1. DNA concentration can be determined using the PicoGreen dsDNA quantitation kit (Invitrogen, Carlsbad, CA). The method allows estimation of DNA concentration by comparison of fluorescent signal obtained from each sample against the dilution series of a DNA standard. Signal can be measured with a fluorescent microplate reader using excitation wavelength 484 nm, emission wavelength 538 nm, and a 530 nm bandpass filter. The DNA quantification protocol is modified from the method provided by Invitrogen Corp., Carlsbad, CA for use in a 96-well tray.
2. The following items compose the recommended precautions to be observed before starting the DNA concentration protocol:
3. Allow the PicoGreen reagents, DMSO stock solution, and DNA samples to warm to room temperature.
4. Prepare TE buffer (10 mM Tris–HCl, 1 mM EDTA, pH 7.5) for use in diluting DNA samples and assay reagents.
5. Use plastic rather than glass containers to prepare samples and reaction mixtures to avoid possible absorption of the fluorescent dye to glass surfaces.
6. Protect the solutions from exposure to light as the PicoGreen reagent is susceptible to photobleaching.
7. Use all solutions within a few hours of their preparation.
8. Freshly prepare an aqueous working solution of the PicoGreen reagent by preparing a 200-fold dilution of the concentrated DMSO solution in 1X TE (pH 8). For 20 assays, mix 0.1 ml PicoGreen dsDNA quantitation reagent with 19.9 ml 1X TE (pH 8).
9. Prepare two stock solutions of lambda DNA. Using the 0.1 mg/ml lambda DNA standard provided with the PicoGreen dsDNA quantitation kit the high-range standard stock solution is prepared by mixing 30 μl lambda DNA with 1.47 ml 1x TE (pH 7.5). The low-range standard stock solution is prepared by mixing 3 μl of the high-range standard with 1.5 ml 1x TE (pH 7.5).
10. The high-range standard DNA curve for a range of 1 ng/ml to 1 μg/ml can be prepared from the 2 μg/ml high-range DNA standard stock solution as given in **Table 8.3**.
11. The low-range standard DNA curve for a range of 250 pg/ml to 25 ng/ml can be prepared from the 50 ng/ml low-range DNA standard stock solution as given in **Table 8.4**.
12. Preparation of the purified DNA samples for quantification is performed using each of the purified DNA samples as given in **Table 8.5**.

Table 8.3
Preparation of the high-range standard DNA curve for a range of 1 ng/μl to 1 μg/ml

Concentration	1× TE buffer	High-range DNA standard (2 μg/ml)	Diluted picogreen reagent (ml)
1 μg/ml	–	1 ml	1
100 ng/ml	900 μl	100 μl	1
10 ng/ml	990 μl	10 μl	1
1 ng/ml	999 μl	1 μl	1
No DNA control	1 ml	–	1

Table 8.4
Preparation of the low-range standard DNA curve for a range of 250 pg/μl to 25 ng/ml

Concentration	1× TE buffer	High-range DNA standard (50 ng/ml)	Diluted picogreen reagent (ml)
25 ng/ml	–	1 ml	1
2.5 ng/ml	900 μl	100 μl	1
250 pg/ml	990 μl	10 μl	1
No DNA control	1 ml	–	1

13. DNA prepared for the standard curve and the dilutions of sample DNA are mixed thoroughly and 0.2 ml of each is transferred to a 96-well microplate suitable for use in a fluorescence plate reader. Incubate these reactions for 2–5 min at room temperature.
14. Fluorescent signals are measured in a fluorescent microplate reader. The excitation wavelength is 484 nm, emission wavelength is 538 nm, and the instrument should use a 530 nm bandpass filter.

Table 8.5
Preparation of the purified DNA samples for quantification

Dilution factor	1× TE buffer (μl)	DNA sample (μl)	Diluted picogreen reagent (μl)
1:2	50	50	100
1:20	90	10	100
1:200	99	1	100

15. Linear regression can be used to analyze the background corrected signals. Concentration of the sample DNAs are estimated from the relationship of concentration and signal obtained from the standard curves (see Section 3.6.). If none of the sample signals correspond to the signal range measured for the standard then appropriate fresh dilutions of the sample should be prepared. In the event that more than one of the sample dilutions is within range of the standard curves their estimated concentrations may be reported as the mean concentration.
16. Adjust the concentration of the samples to 2 ng/μl using 1× TE buffer. Store diluted DNA samples in cryotubes frozen at –20°C.

3.6. Estimation of Sample DNA Concentration

The concentration of an unknown, along with the associated confidence limits, can be estimated from linear regression of the relationship between a set of known concentration standards and the fluorescent signal obtained from the PicoGreen assay *(29)*. The calculations are performed in two parts: (A) linear regression analysis of the concentration (*x*-axis) and observed signal (*y*-axis) followed by estimation of the concentration of the unknown; and (B) determination of the 95% confidence limits of this estimate. The mathematical steps used are described in the following:

Part A: Estimating the concentration of an unknown using a standard curve:

Examples of experimental data obtained from the PicoGreen assay.

Background corrected sample data for the DNA standard:

DNA concentration (X_i)	1,000	100	10	1	25	2.5	0.25
Fluorescent signal (Y_i)	4,707.0	456.2	60.5	20.6	138.8	28.0	17.7

Background corrected sample data for the unknown DNA sample:

Fluorescent signal of the unknown (Y_{UK}) = 100.0

1. Step 1A: Determination of sample size, sums, means, sum of squares, and sum of products:

Sample size (N) = 7	
Sum of X (X_S) = 1,301	Sum of Y (Y_S) = 6,204
Mean X (X_M) = 162.68	Mean Y (Y_M) = 775.5
Sum of squares X (X_{SS}) = 825,482	Sum of squares Y (Y_{SS}) = 18,180,979
Sum of products (XY_{SP}) = 3,873,640	

2. Step 2A: Calculate the regression coefficient (b_{YX}):
   $b_{YX} = XY_{SP}/X_{SS}$
   $b_{YX} = 3{,}873{,}640 / 825{,}482 = 4.69$
3. Step 3A: Calculate the y-intercept (a):
   $a = Y_M - (b_{YX} * X_M)$
   $a = 775.5 - (4.69)*(162.68) = 12.53$
4. Step 4A: Determine the explained sum of squares (E_{SS}).
   $E_{SS} = (XY_{SP})^2/(X_{SS})$
   $E_{SS} = (3,873,640)^2/8,25,482 = 18,177,364$
5. Step 5A: Determine the unexplained sum of squares (U_{SS}):
   $U_{SS} = \Sigma\, [Y_i - a + (b_{YX} * X_i)]^2 = 780.01$
6. Step 6A: Estimate the concentration of the unknown sample (E_{UK}):
   $E_{UK} = (Y_{UK} - a) / b_{YX}$
   $E_{UK} = (100.0 - 12.53) / 4.69 = 18.65$

Part B: Determine the 95% confidence limits:

7. Step 1B: Determine the value of $t_{0.05[n-2]}$:
   The value of $t_{0.05[n-2]}$ is determined using a look-up table of critical t-values *(30)* (partially reproduced in **Table 8.6**). For the sample data specified in Part 1 the degrees of freedom is $(N - 2) = 5$ and therefore $t_{0.05[n-2]} = 2.571$.
8. Step 2B: Calculate the unexplained error around the regression line s^2_{YX}: $s^2_{YX} = U_{SS} / (N - 2)$
   $s^2_{YX} = (780.01 / 5) = 156$
9. Step 3B: Calculate the standard error of the regression coefficient s_b: $s_b = (s^2_{YX} / X_{SS})^{1/2}$
   $s_b = (156 / 825{,}482)^{1/2} = 0.01375$

Table 8.6
Critical *t*-values for the determination of the $t_{0.05[n-2]}$ value (partially reproduced from *30*)

Degrees of freedom	Alpha (0.05) 2-tailed *t*-test
1	12.706
2	4.303
3	3.182
4	2.776
5	**2.571**
6	2.447
7	2.365
8	2.306
9	2.262
10	2.228

10. Step 4B: Calculate the value of D:
    $D = b^2_{YX} t^2_{0.05[n-2]} * s^2_b$
    $D = (4.69)^2 - (2.571)^2 * (0.01375)^2 = 21.99$
11. Step 5B: Calculate the value of H:
    $H = (t_{0.05[n-2]}/D) * (s^2_{YX} * (D * ((N+1)/N) + (Y_{UK} - Y_M)^2/X_{SS}))^{1/2}$
    $H = (2.571/21.99) * (156 * (21.99 * ((7+1)/7) + (100.0 - 775.5)^2/825,482))^{1/2}$
    $H = 0.1169 * (156 * (25.1314 + 0.5528))^{1/2}$
    $H = 0.1169 * (4,006.7352)^{1/2} = 7.3996$
12. Step 6B: Calculate the 95% confidence limits $L1$ and $L2$:
    $L1 = X_M + (b_{YX} * (Y_{UK} - Y_M)/D) - H$
    $L1 = 162.68 + (4.69 * (100.0 - 775.5)/21.99) - 7.3996$
    $L1 = 162.68 + (-144.07) - 7.3996 = 11.21$
    $L2 = X_M + (b_{YX} * (Y_{UK} - Y_M)/D) + H$
    $L2 = 162.68 + (4.69 * (100.0 - 775.5)/21.99) + 7.3996$
    $L2 = 162.68 + (-144.07) + 7.3996 = 26.01$
13. The estimated concentration of the unknown is reported as $E_{UK} = 18.65$ (see Step 6A) with 95% confidence intervals $L1 = 11.21$ and $L2 = 26.01$ (as reported in Step 6B).

3.7. Polymerase Chain Reaction

1. Amplification of *HLA* loci follows the protocol described by Ringquist et al. *(18, 20)*. The reactions are performed in 50 μl volumes using the PCR forward and biotinylated reverse primers indicated in **Table 8.7**.
2. The following solutions should be prepared for use during PCR amplification of *HLA* loci and can be stored at –20°C until needed:
   a. 10× *Taq* buffer: 500 mM KCl, 100 mM Tris–HCl (pH 8.3), 0.01% (w/v) gelatin.
   b. Magnesium chloride solution: 25 mM $MgCl_2$.
   c. Deoxynucleotide triphosphate solution. 10 mM dNTP mixture.
   d. Forward primer: 10 μM forward PCR primer in TE (pH 8) buffer.
   e. Biotinylated primer: 10 μM biotinylated reverse PCR primer in TE (pH 8) buffer.
3. Thaw reagents in an ice-water bath or by gentle warming to room temperature. Excessive incubation at warmer temperatures may reduce the shelf life of the reagents. Mixtures are prepared using DNA purified from whole blood or extracted from dried blood spots (**Table 8.8**).
4. Thermal cycling conditions are 96°C incubation for 3 min followed by 32 cycles at 96, 55, and 72°C incubated for 30 s at

Table 8.7
DNA oligonucleotide primers for PCR amplification of *HLA-DQB1* alleles

Name	Sequence (written 5′ to 3′)[a]	Length (bp)	Comment[b]
Forward strand PCR amplification			
SR25	TTTGACCCCGCAGAGGA TTTCGTG	283	-*DQB1* forward primer use with SR22 or SR24
SR22	Biotin-TEG-CTCTCC TCTGACRGATCCC		-*DQB1* reverse primer, alleles **05/06* use with SR25
SR24	Biotin-TEG-CTCGCC GCTGCAAGGTCGT		-*DQB1* reverse primer, alleles **02/03/04* use with SR25
Reverse strand PCR amplification			
YL22	CTCTCCTCTGACRGATCCC	283	-*DQB1* forward primer for **05/06*
YL24	CTCGCCGCTGCA AGGTCGT		-*DQB1* forward primer for **02/03/04*
YL25	Biotin-TEG-TTTGACCCCGC AGAGGATTTCGTG		-*DQB1* reverse primer used with yl22 and yl24

[a]In PCR primers SR22 and YL22 nucleotide residue 13 is designated R corresponding to the presence of equal molar amounts of A and G at this position. This can be accomplished by ordering the oligonucleotides as separate syntheses or by incorporation of mixed residues at this position during a single synthesis;
[b]Oligonucleotide primers SR22 / YL22 and SR24 / YL24 are used for amplification of *HLA-DQB1* group-specific alleles from **05/06* and **02/03/04*, respectively.

Table 8.8
Mixture for PCR amplification of *HLA* loci

Solution	DNA purified from whole blood (μl)	DNA purified from dried blood spot (μl)
10× Taq buffer	5	5
$MgCl_2$ (25 mM)	4	4
dNTP (10 mM)	4	4
Forward primer (10 μM)	1	1
Biotinylated primer (10 μM)	1	1
Distilled water	29.8	14.8
Taq DNA polymerase (5 U/μl)	0.2	0.2
DNA sample (2 ng/μl)	5	–
Dry blood spot extract	–	20

each step. It is recommended that thermal cycling be followed by a final incubation for 10 min at 72°C. Samples can be stored at 4°C or frozen prior to pyrosequencing (*see* **Note 6**).

3.8. Viewing the Results of PCR Amplification by Agarose Gel Electrophoresis

1. Quality control of PCR amplification of *HLA* loci can be assessed by gel electrophoresis using 1.5% agarose gel electrophoresis. The following protocol for agarose gel electrophoresis corresponds to that described by Sambrook et al. *(31)*. Begin by sealing the edges of a clean electrophoresis tray with laboratory tape. Place the sealed gel tray onto a level surface.
2. Prepare a mixture of 1× TBE and powered agarose to make a sufficient volume of gel solution to fill the tray to a level of 50–100 mm. For example, for a preparation of 1.5% agarose in 100 ml 1× TBE use 1.5 g powered agarose. The 1× TBE buffer is 0.9 M Tris–HCl (pH 8), 0.9 M boric acid, and 0.002 mM EDTA. A 5× solution of TBE can be prepared ahead of time, stored at room temperature, and diluted with distilled water prior to use.
3. Add a magnetic stir bar and heat the slurry in an Erlenmeyer flask, loosely sealed with aluminum foil, using an electric stirrer-hot plate. Make certain that the agarose has completely dissolved. It is common for agarose gel solutions containing greater than 1.2% agarose that this process may take a few minutes of constant boiling.
4. A microwave oven may also be used to dissolve the agarose. Caution should be observed when handling the heated agarose solution in that it can become super-heated and create a safety hazard by unexpectedly boiling over when the container is handled. When using the microwave oven the aluminum foil and magnetic stir bar should be omitted.
5. The appearance of undissolved agarose in the mixture can be determined by observing the presence of small clumps of semitransparent material when the mixture is stirred. This material must be completely dissolved by heating prior to using it to prepare a gel.
6. The dissolved agarose mixture is cooled to 60°C and ethidium bromide is added to a final concentration 0.5 μg/ml. Thorough mixing of the ethidium bromide into the agarose solution is required.
7. The agarose–ethidium bromide solution is poured into the sealed gel tray. Air bubbles can be removed from the gel by a pasteur pipette. The gel comb is added to the mixture and the agarose is allowed to solidify upon cooling.
8. After the agarose has solidified remove the gel comb and the sealing tape. The gel tray can be placed in the electrophoresis chamber. Add the minimum volume of 1× TBE to the electrophoresis chamber required to cover the gel.

9. DNA samples can be mixed with gel-loading buffer. 6X loading buffer 0.25% bromophenol blue, 0.25% xylene cyanol FF, 30% (v/v) glycerol is recommended but other equally appropriate loading buffers are available *(29)*. PCR amplification should generate enough DNA so that 5 μl of the amplicon mixed with 1 μl of 6× gel-loading solution should be sufficient to visualize the product. All of the mixture is loaded into the appropriate wells.
10. DNA size standards for agarose gel electrophoresis may be purchased from commercial vendors (e.g., Bio-Rad Laboratories, Hercules, CA or New England Biolabs, Ipswich, MA). These are available in a number of variations. PCR products to be examined using pyrosequencing are typically less than 500 bp in length. This makes the 100 bp ladder standards convenient for the purpose of evaluating the product.
11. After loading the size standard and samples onto the gel, the lid of the electrophoresis tank should be closed and the electrical leads connected. DNA is negatively charged and will migrate toward the anode (red lead). Apply a voltage of between 1 and 5 V/cm. For a 20 cm gel box this would be between 20 and 100 V.
12. Once the gel has run sufficiently the lead dye should have migrated roughly 3 cm into the agarose. Turn off the electric current and remove the gel tray from the tank. Be careful not to allow the agarose gel to slip off of the tray as it will likely tear. The DNA bands may be visualized by ultraviolet light and the data recorded by photography or captured by video imaging. Avoid excessive exposure of eyes and skin to the UV irradiation as it can cause retinal damage and burning of the skin.
13. Confirmation of successful PCR amplification occurs when a product of the expected length is observed. Products of other lengths or multiple products are considered evidence that the PCR amplification did not provide the proper material to prepare template for pyrosequencing.

3.9. Pyrosequencing of HLA Loci Using the PSQ 96MA Instrument

1. *HLA* pyrosequencing can be performed on a PSQ 96MA or PSQ HS 96A system (*see* **Note 7**). Pyrosequencing reactions are performed using reagents provided with the PSQ 96MA sample preparation kit and PSQ 96 SQA reagent kit. Reference DNA can also be obtained from human lymphoblastoid cell line of known *HLA* identity (*see* **Note 8**).
2. Materials to prepare prior to the pyrosequencing protocol:
   a. Prepare an 80°C heating block.
   b. Binding buffer: 10 mM Tris–HCl (pH 7.6), 2 M NaCl, 1 mM EDTA, 0.1% Tween 20.

c. Annealing buffer: 20 mM Tris–acetate (pH 7.6), 2 mM magnesium acetate.
d. Denaturation solution: 0.2 M NaOH.
e. Washing buffer: 10 mM Tris–acetate (pH 7.6).

3. Samples are prepared using 10 μl of amplified DNA from the PCR mixtures. These are mixed with 4 μl of streptavidin-coated beads and the volume adjusted to 50 μl with binding buffer.
4. Place the mixture of PCR product and streptavidin-coated beads onto an Eppendorf thermal mixer and vortex at room temperature using 1,400 rpm for 10 min.
5. Transfer the streptavidin-coated beads to a 96-well MultiScreen-HV membrane. Collect the biotinylated PCR product streptavidin-coated bead complex by vacuum filtration.
6. Beads are suspended in 150 μl denaturation buffer and incubated for 2 min at room temperature. Denaturation buffer is removed by vacuum filtration.
7. Immediately after removal of denaturation buffer, and while still under vacuum, the pH of the PCR product–bead complex is neutralized by addition of 200 μl washing buffer. Repeat this step for a total of two washings.
8. Resuspend beads in 50 μl of annealing buffer and transfer 45 μl to the pyrosequencing reaction tray.
9. Add the appropriate pyrosequencing primer to each well in a volume of 5 μl using 3 μM stock solution of primer. Recommended pyrosequencing primers are listed in **Table 8.9**. Annealing was performed by incubation at 80°C for 2 min prior to pyrosequencing. Single-stranded binding protein (SSBP) is added at a final concentration of 4 μg per pyrosequencing reaction (*see* **Notes 9** and **10**).

3.10. Pyrosequencing of HLA Loci Using the PSQ HS 96A Instrument

1. DNA profiling of *HLA* genotypes can also be performed using the PSQ HS96A system. This instrument differs from the PSQ 96MA in that it uses a smaller reaction volume and a corresponding low volume 96-well pyrosequencing reaction tray (*see* **Note 7**). Figure 8.3 illustrates a set of pyrograms obtained from a trio of family members consisting of father (panel A), mother (panel B), and offspring (panel C). Reference DNA can also be obtained from human lymphoblastoid cell line of known *HLA* identity (*see* **Note 8**).
2. Materials to prepare prior to the pyrosequencing protocol:
   a. Prepare a 90°C heating block.
   b. Binding buffer: 10 mM Tris–HCl (pH 7.6), 2 M NaCl, 1 mM EDTA, 0.1% Tween 20.

Table 8.9
DNA oligonucleotide primers for pyrosequencing of *HLA-DQB1* alleles

Name	Sequence (written 5′ to 3′)	Annealing site[a]	Dispensation order	Comment[b]
Forward strand pyrosequencing				
SR26	GACCCCGCAGAGGATTTCG	Residues 10–9	CTGTCATCACGTAGCAGT	-*DQB1* PS primer
SR30	GCCGAGTACTGGAACAGCC	Residues 159–177	CAGAGACGAGTCTGAGAG	-*DQB1* PS primer
PSQ1	ACGGAGCGCGTGCG	Residues 48–61	AGTCGTGTGTACAG	-*DQB1* PS primer
PSQ3	TCGACAGCGACGTG	Residues 106–119	CGAGTAGTAGC	-*DQB1* PS primer
YLPSQ4_05/06	GTGACGCCGCAGGGGCGGCCT	Residues 135–155	GACTGCTGCGCAGTA	-*DQB1*05/06* PS primer
YLPSQ4_02	GTGACGCTGCTGGGGCTGCCT	Residues 135–155	GACTGCTGCGCAGTA	-*DQB1*02* PS primer
YLPSQ4_03	GTGACGCCGCTGGGGCCGCCT	Residues 135–155	GACTGCTGCGCAGTA	-*DQB1*03* PS primer
YLPSQ4_04	GTGACGCCGCTGGGGCGGCTT	Residues 135–155	GACTGCTGCGCAGTA	-*DQB1*04* PS primer
Reverse strand pyrosequencing				
YLPSQ5	GTTCCAGTACTCGGC	Residues 159–173	AGCTACTATGCATGACT	-*DQB1* PS primer

[a]Numbering of nucleotides residues is based on the sequence of *HLA-DQB1* exon 2; [b]PS primer, pyrosequencing primer.

c. Annealing buffer: 20 mM Tris–acetate (pH 7.6), 2 mM magnesium acetate.
d. Denaturation solution: 0.2 M NaOH.
e. Washing buffer: 10 mM Tris–acetate (pH 7.6).

3. Prepare the binding reaction mixture by combining 38 μl binding buffer, 35 μl distilled water, and 2 μl streptavidin-coated beads for each well. Mix thoroughly by brief vortex.
4. Prepare the pyrosequencing primer mixture by combining 10.8 μl annealing buffer and 1.2 μl pyrosequencing primer solution (stock concentration of primer is 3 μM) for each well. Mix thoroughly by brief vortex.
5. Transfer 75 μl of the binding reaction mixture (prepared during step 3) and 5 μl of PCR product to each well of the pyrosequencing suction tray.
6. Seal the pyrosequencing suction tray and shake using an Eppendor thermal mixer at 1,400 rpm for at least 10 min at 23°C.
7. Transfer 12 μl of pyrosequencing primer mix (prepared during step 4) to each well of the pyrosequencing reaction tray. Set aside the reaction tray until step 14.
8. In order to prepare for the series of washing steps place pyrosequencing vacuum tool in the first distilled water trough. Turn on vacuum.
9. Remove the pyrosequencing suction tray from the thermal mixer. Hold the pyrosequencing vacuum tool upright until distilled water stops flowing through it.
10. Slowly place pyrosequencing vacuum tool into the pyrosequencing suction tray. Make sure that all liquid is removed.
11. Place the pyrosequencing vacuum tool in the 70% ethanol trough; after liquid starts visibly flowing wait for 5 s, then hold it upright until the flowing stops.
12. Repeat step 11 for denaturation solution (incubate for 5 s prior to vacuum filtration) and washing buffer (incubate for 10 s prior to vacuum filtration).
13. Turn off the pyrosequencing vacuum tool and unplug it to completely release the vacuum.
14. Slowly place pyrosequencing vacuum tool into the pyrosequencing reaction tray (prepared during step 7) and wiggle tool back and forth to dislodge beads into the pyrosequencing reaction tray.
15. Place pyrosequencing reaction tray on the 90°C heat block for 2 min. Remove from the heating block and allow it to cool for 15 min to room temperature.
16. Reconnect the pyrosequencing vacuum tool and place it in the last distilled water trough. Rinse the vacuum apparatus by allowing roughly 10 ml of distilled water to pass through.

17. Turn off the pyrosequencing vacuum tool, drain off all liquid from the vacuum apparatus, and place it in the storage trough.

4. Notes

1. Preparation of aqueous solutions: All solutions are prepared in deionized water (18.2 mΩ cm) filtered free of organic content. It is critical to keep reagents and laboratory equipment as clean as possible, this will help avoid strong background signals that can occur during the PCR or pyrosequencing stages.
2. When titrating solutions using concentrated acid care should be taken to avoid exposure to fumes or contact with liquid. Safe use of caustic chemicals is aided by wearing protective clothing (e.g., eye protection, gloves, and lab coat) and by proper use of a chemical fume hood.
3. Disposal of ethidium bromide: Ethidium bromide is a carcinogenic compound and should be handled with appropriate care. Dilute solutions of ethidium bromide should be decontaminated before disposal. For 0.5 μg/ml solutions of ethidium bromide the method described by Lunn and Sansone *(32)* and an alternate method described by Bensaude *(33)* are recommended. Treatment of dilute solution of ethidium bromide with bleach (hypochlorite) is not recommended. The latter method has been shown to convert the dye into a compound that is mutagenic *(31, 34)*.
4. Pyrosequencing provides distinct advantages for genetic typing. Samples can be assayed in 96-well trays making it compatible with common laboratory automation instruments. Pyrosequence-based typing strategies can be designed using a minimal number of nucleotide dispensation events. An entire 96-well tray can be assayed at a rate of roughly 1 min per base and approximately 20 min for a 20 nucleotide sequence. This is sufficient for analysis of most *HLA* alleles when primers are placed adjacent to hypervariable regions. Moreover, out-of-phase pyrosequencing of alleles *(18, 26)* allows unequivocal genotyping of allelic combinations that are ambiguous by conventional sequence-based typing techniques as well as those spaced too far apart to allow resolution by oligonucleotide hybridization-based approaches.
5. Since the study may involve the drawing of small volumes of blood there is minimal risk to the participants. The risks that are present are primarily those related to blood

collection, which may include bleeding, bruising, hematoma, or possible infection at the sight of phlebotomy as well as fainting.

6. Validation of PCR primers by gradient thermal cycler amplification: Optimization of PCR conditions is conveniently performed using a temperature gradient capable thermal cycler. This instrument can be purchased from the MJ Research division of Bio-Rad Laboratories Inc.. Hercules, CA. A range of PCR annealing temperatures can be examined and the relative yield of PCR product determined by comparing the intensity of DNA bands after ethidium bromide staining of bands separated by 1.5% agarose gel electrophoresis.
7. The pyrosequencing PSQ 96MA and PSQ HS 96A instruments can be used for DNA profiling of *HLA* alleles. The PSQ HS 96A instrument has roughly 2-fold greater sensitivity but is best used in pyrosequencing reactions requiring a read length of less than 30 nucleotides. For longer reads the 96MA system has been superior.
8. Recommended DNA standards for use during pyrosequencing of *HLA* samples: Human lymphoblastoid cell lines of known *HLA* identity can be obtained from the International Histocompatibility Working Group Cell and Gene Bank (Seattle, WA). These samples have been genotyped for *HLA* class I and class II loci *(35)*.
9. Validation of pyrosequencing primers: Pyrosequencing reactions can give background signals from self-priming of the 3′-end of the template strand or by self-annealing of the sequencing primer. The validation of the pyrosequencing step can be performed in order to examine the signals due to these events. A set of negative control reactions should be examined to estimate background pyrosequencing signal. A recommended set of negative control samples are listed in the following:

   a. Negative control 1: Sequencing primer alone in the pyrosequencing reaction.
   b. Negative control 2: Pyrosequencing of the no DNA template containing PCR negative control with sequencing primer.
   c. Negative control 3: Pyrosequencing of the no DNA template containing PCR negative control without sequencing primer.

10. Extending the pyrosequencing read length: Pyrosequencing has enabled DNA sequence reads of up to 150 residues *(18, 20, 26, 28)*. Optimization of the pyrosequencing reaction conditions has been reported to improve read length *(28, 36)* and focuses primarily on the use of enantiomer pure 2’-deoxyadenosine-5′-*O*’-(1-thiotriphosphate) and the use of single-stranded DNA binding protein *(26, 28)*.

Acknowledgments

We thank Alexis Styche for carefully critiquing the manuscript prior to publication. This work was supported by a grant from the Department of Defense (W81XWH-07-1-0619).

References

1. Mungall, A. J., Palmer, S. A., Sims, S. K., Edwards, C. A., Ashurst, J. L., Wilming, L., et al. (2003) The DNA sequence and analysis of human chromosome 6. *Nature* 425, 805–811.
2. Stewart, C. A., Horton, R., Allcock, R. J., Ashurst, J. L., Atrazhev, A. M., Coggill, P., et al. (2004) Complete MHC haplotype sequencing for common disease gene mapping. *Genome Res* 14, 1176–1187.
3. Campbell, R. D., Trowsdale, J. (1993) Map of the human MHC. *Immunology Today* 14, 349–352.
4. Robinson, J., Waller, M. J., Parham, P., de Groot, N., Bontrop, R., Kennedy, L. J., et al. (2003) IMGT/*HLA* and IMGT/MHC: sequence databases for the study of the major histocompatibility complex. *Nucl Acids Res* 31, 311–314.
5. Hurley, C. K., Baxter-Lowe, L. A., Begovich, A. B., Fernandez-Vina, M., Noreen, H., Schmeckpeper, B., et al. (2000) The extent of *HLA* class II allele level disparity in unrelated bone marrow transplantation: analysis of 1259 National Marrow Donor Program donor-recipient pairs. *Bone Marrow Transplant* 25, 385–393.
6. Cavalli-Sforza, L. L., Menozzi, P., Piazza, A. (1994) *The History and Geography of Human Genes.* Princeton University Press, Princeton, NJ.
7. de Groot, N. G., Otting, N., Doxiadis, G. G., Balla-Jhagjhoorsingh, S. S., Heeney, J. L., van Rood, J. J., et al. (2002) Evidence for an ancient selective sweep in the MHC class I gene repertoire of chimpanzees. *Proc Natl Acad Sci USA* 99, 11748–11753.
8. Kiepiela, P., Leslie, A. J., Honeyborne, I., Ramduth, D., Thobakgale, C., Chetty, S., et al. (2004) Dominant influence of *HLA-B* in mediating the potential co-evolution of HIV and *HLA*. *Nature* 432, 769–775.
9. Fellay, J., Shianna, K. V., Ge, D., Colombo, S., Ledergerber, B., Weale, M., et al. (2007) A whole-genome association study of major determinants for host control of HIV-1. *Science* 317, 944–947.
10. Jacob, S., McClintock, M. K., Zelano, B., Ober, C. (2002) Paternally inherited *HLA* alleles are associated with women's choice of male odor. *Nat Genet* 30, 175–179.
11. Boehm, T., Zufall, F. (2006) MHC peptides and the sensory evaluation of genotype. *Trends Neurosci* 29, 100–107.
12. Moore, C. B, John, M., James, I. R, Christiansen, F. T, Witt, C. S, Mallal, S. A. (2002) Evidence of HIV-1 adaptation to *HLA*-restricted immune responses at a population level. *Science* 296, 1439–1443.
13. Trachtenberg, E., Korber, B., Sollars, C., Kepler, T. B., Hraber, P. T., Hayes, E., et al. (2003) Advantage of rare *HLA* supertype in HIV disease progression. *Nat Med* 9, 928–935.
14. Ringquist, S., Nichol, L., Trucco, M. (2007) Transplantation genetics, in (Rimoin, D. L., Connor, J. M., Pyeritz, R. E., Korf, B. R., eds.), *Emery and Rimoin's Principles and Practice of Medical Genetics 5th Edition*, pp. 983–1010. Churchill Livingstone, Philadelphia, PA.
15. Cao, K., Hollenbach, J., Shi, X., Shi, W., Chopek, M., Fernandez-Vina, M. A. (2001) Analysis of the frequencies of *HLA-A, B*, and *C* alleles and haplotypes in the five major ethnic groups of the United States reveals high levels of diversity in these loci and contrasting distribution patterns in these populations. *Hum Immunol* 62, 1009–1030.
16. Klitz, W., Maiers, M., Spellman, S., Baxter-Lowe, L. A., Schmeckpeper, B., Williams, T. M., Fernandez-Vina, M. (2003) New *HLA* haplotype frequency reference standards: high-resolution and large sample typing of *HLA DR-DQ* haplotypes in a sample of European Americans. *Tissue Antigens* 62, 296–307.
17. Adams, S. D., Barracchini, K. C., Chen, D., Robbins, F., Wang, L., Larsen, P., et al. (2004) Ambiguous allele combinations in *HLA* Class I and Class II sequence-based typing: when precise nucleotide sequencing leads to imprecise allele identification. *J Transl Med* 2, 30–35.

18. Ringquist, S., Alexander, A. M., Styche, A., Pecoraro, C., Rudert, W. A., Trucco, M. (2004) *HLA* class II *DRB* high resolution genotyping by pyrosequencing: comparison of group specific PCR and pyrosequencing primers. *Hum Immunol* 65, 163–174.
19. Ringquist S, Styche A, Rudert WA, Trucco M. (2007) Pyrosequencing-based strategies for improved allele typing of human leukocyte antigen loci. *Methods Mol Biol* 373, 115–134.
20. Ringquist, S., Alexander, A. M., Rudert, W. A., Styche, A., Trucco, M. (2002) Pyrosequence based typing of alleles of the *HLA-DQB1* gene. *Biotechniques* 33, 166–175.
21. Ramon, D., Braden, M., Adams, S., Marincola, F. M., Wang, L. (2003) Pyrosequencing: a one-step method for high resolution *HLA* typing. *J Transl Med* 1, 9.
22. Entz, P., Toliat, M. R., Hampe, J., Valentonyte, R., Jenisch, S., Nurnberg, P., Nagy, M. (2005) New strategies for efficient typing of *HLA* class-II loci *DQB1* and *DRB1* by using pyrosequencing. *Tissue Antigens* 65, 67–80.
23. Marsh, S. (2007) Pyrosequencing applications. *Methods Mol Biol* 373, 15–24.
24. Pasquali, L., Bedeir, A., Ringquist, S., Styche, A., Bhargava, R., Trucco, G. (2007) Quantification of CpG island methylation in progressive breast lesions from normal to invasive carcinoma. *Cancer Lett* 257, 136–144.
25. Tost, J., Gut, I. G. (2007) Analysis of gene-specific DNA methylation patterns by pyrosequencing technology. *Methods Mol Biol* 373, 89–102.
26. Garcia, C. A., Ahmadian, A., Gharizadeh, B., Lundeberg, J., Ronaghi, M., Nyren, P. (2000) Mutation detection by pyrosequencing: sequencing of exons 5–8 of the p53 tumor suppressor gene. *Gene* 253, 249–257.
27. Ronaghi, M., Karamohamed, S., Pettersson, B., Uhlen, M., Nyren, P. (1996) Real-time DNA sequencing using detection of pyrophosphate release. *Anal Biochem* 242, 84–89.
28. Gharizadeh, B., Nordstrom, T., Ahmadian, A., Ronaghi, M., Nyren, P. (2002) Long-read pyrosequencing using pure 2'-deoxyadenosine-5′-O'-(1-thiotriphosphate) Sp-isomer. *Anal Biochem* 301, 82–90.
29. Sokal, R. R., Rohlf, F. J. (2003) *Biometry: The Principles and Practice of Statistics in Biological Research 3rd Edition.* W.H. Freeman and Company, New York, NY.
30. Snedecor, G. W., Cochran, W. G. (1989) *Statistical Methods 8th Edition.* Iowa State University Press, Ames, IA.
31. Sambrook, J., Fritsch, E. F., Maniatis, T. (1989) *Molecular Cloning: A Laboratory Manual, 2nd Edition.* Cold Spring Harbor Laboratory Press, Cold Spring Harbor, NY.
32. Lunn, G., Sansone, E. B. (1987) Ethidium bromide: destruction and decontamination of solutions. *Anal Biochem* 162, 453–458.
33. Bensaude, O. (1988) Ethidium bromide and safety-readers suggest alternative solutions. *Trends Genet* 4, 89–90.
34. Quillardet, P., Hofnung, M. (1988) Ethidium bromide and safety-readers suggest alternative solutions. *Trends Genet* 4, 89.
35. Faas, S. J., Menon, R., Braun, E. R., Rudert, W. A., Trucco, M. (1996) Sequence-specific priming and exonuclease-released fluorescence detection of *HLA-DQB1* alleles. *Tissue Antigens* 48, 97–112.
36. Ronaghi, M. (2000) Improved performance of pyrosequencing using single-stranded DNA-binding protein. *Anal Biochem* 286, 282–288.

Chapter 9

High-Throughput Multiplex HLA-Typing by Ligase Detection Reaction (LDR) and Universal Array (UA) Approach

Clarissa Consolandi

Abstract

One major goal of genetic research is to understand the role of genetic variation in living systems. In humans, by far the most common type of such variation involves differences in single DNA nucleotides, and is thus termed single nucleotide polymorphism (SNP). The need for improvement in throughput and reliability of traditional techniques makes it necessary to develop new technologies. Thus the past few years have witnessed an extraordinary surge of interest in DNA microarray technology. This new technology offers the first great hope for providing a systematic way to explore the genome. It permits a very rapid analysis of thousands genes for the purpose of gene discovery, sequencing, mapping, expression, and polymorphism detection. We generated a series of analytical tools to address the manufacturing, detection and data analysis components of a microarray experiment. In particular, we set up a universal array approach in combination with a PCR-LDR (polymerase chain reaction-ligation detection reaction) strategy for allele identification in the HLA gene.

Key words: Chitosan, amino-modified oligonucleotides, ligation detection reaction, universal array, human leukocyte antigen, single nucleotide polymorphisms.

1. Introduction

The completion of the large-scale sequencing of human genome has revealed a huge number of single nucleotide polymorphisms (SNPs), which in turn has opened new perspectives for the characterisation of disease-related molecular markers. A wide variety of different methods for scoring SNP markers have become available, but no consensus approach, so far, has been found *(1, 2)*. The molecular information that can be accessed by DNA analysis is generating the need for highly parallel and miniaturised techniques. Specifically, methods with high accuracy to

Peter Bugert (ed.), *DNA and RNA Profiling in Human Blood: Methods and Protocols, vol. 496*

DOI 10.1007/978-1-59745-553-4_9 Springerprotocols.com

be used in a high-throughput setting are needed for systematic surveys of the frequency and exact location of sequence variation and their influence on cellular behaviour. Among many others, DNA microarray technology is expected to fulfil the basic requirement for such a technology: flexibility, parallelisation, miniaturisation, and thus cost effectiveness. A key requirement for a scoring method for genomic SNPs is that it should be able to distinguish unequivocally between homozygous and heterozygous allelic variants in the diploid human genome. Thus the human leukocyte antigen (HLA) complex, which is characterised by the most extensive SNP patterns of its loci in the human genome, may represent a valid model system to develop a DNA microarray format for any polymorphisms detection, including multiplex genotyping of SNPs and large-scale mutations screening *(3)*. In the case of the traditional hybridisation reaction, the intensity of fluorescence bound to each oligonucleotide probe is intended to reveal which sequence is perfectly complementary to the query sequence. However, these approaches have not always been successful. During the hybridisation process, a disruption of secondary structure could occur: that is to say that perfectly matched sequence may assume a secondary structure that is eliminated in variant sequences. This structural change may lead to binding of a variant target to a perfectly matched probe with a higher binding affinity than true perfect-match target. Another consequence of direct hybridisation is the formation of stable duplexes by looping out of a non-complementary sequence during hybridisation. Either of these illegitimate hybridisations could produce false negative signals on an array. Alternative approaches, involving enzyme processing on the microarray have been developed, such as minisequencing *(4)*. In comparative studies it has been suggested that these enzyme-mediated allele recognition methods can show higher discrimination between competing alleles *(5)*. Ligation detection reaction (LDR) has been applied in combination to a universal DNA array to detect K-ras mutation in tumour and cell line DNA *(6, 7)* as well as on BRCA genes *(8)*. "Zip-code" arrays consist of selected oligonucleotides with similar hybridisation characteristics and minimal cross-hybridisation. Sequences with possible secondary structure are eliminated. This kind of array is "universal" because the sequences, complementary to immobilised probes, could be appended to any set of specific primers in the course of target-DNA processing. This methods combines a multiplex PCR and a multiplex LDR with zip-code hybridisation. A thermostable DNA–ligase links two adjacent oligonucleotides annealed to a complementary target if the nucleotides are perfectly base-paired at a junction. The assay allows for accurate detection of single-base mutations present at 1% or even less of the wild-type sequence.

This approach provides for an accurate, inexpensive, and high-throughput assay that does not exhibit false positive or false negative signals, thus making it highly suitable for gene-based testing in high-incidence, low-complexity diseases.

2. Materials

All chemicals and solvents are purchased from Sigma-Aldrich (Italy) and used without further purification.

All HPLC-purified oligonucleotides are purchased from Thermo Fisher Scientific (Ulm, Germany) and purity is checked by MALDI mass spectrometry. Lyophilised oligonucleotides are dissolved in bidistilled water to a final concentration of 100 μM and stored frozen at –20°C until use.

2.1. Sample Preparation

1. Extraction kit (QIAamp DNA Blood Mini Extraction Kit, Qiagen, Hilden, Germany) for the isolation of genomic DNA from whole blood of healthy persons (National Cancer Institute, IST, Advanced Biotechnology Center, Department of Oncology, Biology and Genetics, University of Genoa, Italy).
2. Primer3 (http://frodo.wi.mit.edu/cgi-bin/primer3/primer3_www.cgi), a free software for the PCR primer design.
3. PCR "upper mix": 5× Buffer I, 2.5 mM dNTPs, 1.5 U AmpliTaq PE DNA polymerase (Perkin Elmer Life Sciences, Boston, MA), 10 nM primer forward.
4. PCR "lower mix": 10 nM Primer reverse, 250 ng of genomic DNA, and ddH_2O up to 50 μl.
5. Ampliwax gem (Perkin Elmer Life Sciences, Boston, MA).
6. GeneAmp, PCR System 9700 (Applied Biosystem, Foster City, CA, USA).
7. GFX PCR DNA purification kit (Amersham Pharmacia Biotech Inc, Piscataway, NJ).
8. BioAnalyzer 2100 (Agilent Technologies, Palo Alto, CA, USA).

2.2. Universal Array (UA) Preparation

1. Slide washing solutions pre-silanisation: 1 M NaOH; 1 N HCl; 96% ethanol; acetone; methanol; ddH_2O.
2. Silanisation solution: 1% 3-Glycidoxypropylsilane (GOPS) in 95% ethanol.
3. Polymer solution: 10 g of chitosan (400 Kda) are dissolved in a liter of 50 mM HCl. 10% polymer (0,1% w/v) is added to 10% phosphate-buffered saline (PBS).
4. Activation solution: 0.2% solution of 1,4-phenylene diisothiocyanate (PDITC) in 90% *N,N*-dimethylformamide, and 10% pyridine.
5. Printing buffer: 150 mM sodium phosphate, pH 8.5.

6. Deactivation solution: 50 mM ethanolamine; 0.1 M Tris (pH 9); 0.1% sodium dodecyl sulphate (SDS).
7. Slide washing solution after spotting and surface deactivation: 4x saline sodium citrate (SSC) and 0.1% SDS.
8. MicroGrid II Compact Arrayer (BioRobotics Ltd, Cambridge, UK).
9. Home-made saturated NaCl humidification chamber.

2.3. Ligation Detection (LDR) and Hybridisation Reactions

1. ARB (www.arb-home.de), a publicly available software for the LDR probe design.
2. LDR mix: buffer (1×): 20 mM Tris–HCl (pH 7.5); 20 mM KCl; 10 mM $MgCl_2$; 0.1% Igepal; 0.01 mM rATP; 1 mM dithiothreitol (DTT) (Stratagene, La Jolla, CA, USA); 2 pmol of each discriminating oligo; 2 pmol of each common probe; 10 fmol of purified PCR products; 4 U of Pfu DNA ligase (Stratagene).
3. Hybridisation mix: 5× SSC and 0.1 mg/ml salmon sperm DNA.
4. Hybridisation washing solution: 1× SSC and 0.1% SDS.
5. Home-made multiple sample chambers: (8-well chamber) using Press-To-Seal silicone isolaters (Schleicher & Schuell, Germany).
6. Temperature controlled system (Shack'n'Stack, Hybaid, England).
7. Centrifuge 5415 D (Eppendorf, AG, Hamburg, Germany).

2.4. Signal Detection and Statistical Data Analysis

1. ScanArray® 4000 laser scanning system (Perkin Elmer Life Sciences, Boston, MA).
2. ScanArray 3.1 software (Perkin Elmer Life Sciences, Boston, MA).

3. Methods

The principle of the LDR combined to a UA format has been previously illustrated *(7)*. We used this molecular tool to accurately genotype a highly polymorphic region, such as the HLA gene. The overall process requires a careful probe design for targeting the selected polymorphisms, the fabrication of a UA, the preparation of PCR-specific products, the set up of the LDR reaction, the demultiplexing of LDR products onto UA by hybridisation, the measurement of the fluorescent signals and, finally, the statistical data analysis (**Fig. 9.1**).

3.1. Sample Preparation

1. Genomic DNA from the blood of healthy persons is provided by the National Cancer Institute, IST, Advanced

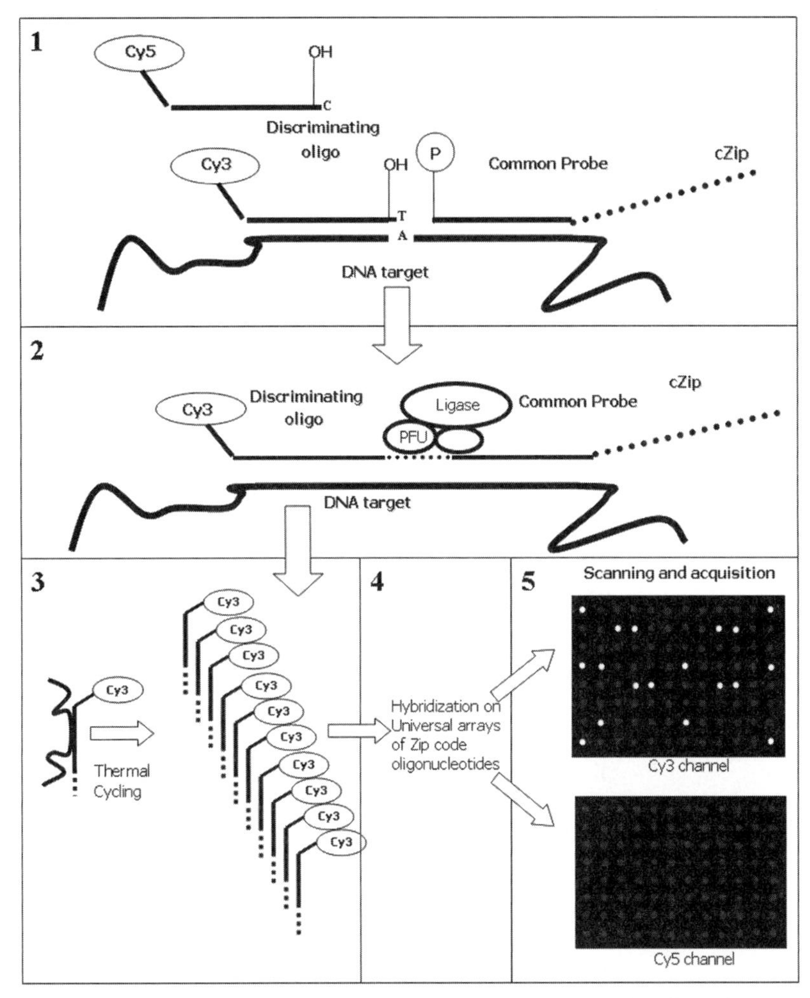

Fig. 9.1. Outline of the LDR method for detection of HLA polymorphisms. Cy3 or Cy5 fluorophores are employed to distinguish polymorphisms. The common probe is phosphorylated in 5′-terminal position and contains a sequence complementary to the zip-code (cZip code) in 3′ end. After hybridisation of the common probe and the discriminating oligonucleotide to the target sequence, ligation occurs only if there is perfect complementarity between the two oligonucleotides at the junction. The resulting reaction solution is hybridised onto a UA.

Biotechnology Center (Department of Oncology, Biology and Genetics, University of Genoa, Italy). DNA is isolated from whole blood by means of an extraction kit (QIAamp DNA blood mini extraction kit, Qiagen, Hilden, Germany), according to the manufacturer's protocol, and is used as template for PCR amplification of the targets.

2. In order to selectively amplify the HLA gene, it is fundamental to design primers, which are specific for the HLA loci of interest. PCR primer pairs to amplify SNP-containing regions are designed using Primer3 software (http://frodo.wi.mit.edu/cgi-bin/primer3/primer3_www.cgi) (*see* **Note 1**).

3. PCR reactions are performed by a "hot start" method using the "upper mix", the "lower mix", 250 ng of genomic DNA, and ddH_2O up to 50μl. PCR is performed by heating the lower mix containing the Ampliwax gem at 65°C for 2 min and then cooling at 4°C for a few minutes before adding the upper mix. PCR is carried out in the thermal cycler by the following thermal profile: eight cycles, each consisting of denaturation at 98°C for 5 s, annealing at 63°C for 30 s, and extension at 72°C for 2 min, followed by 32 cycles each consisting of denaturation at 98°C for 5 s, annealing at 58°C for 30 s, and extension at 72°C for 2 min *(9)*.
4. The PCR products are purified by using the GFX PCR DNA purification kit, according to the manufacturer's protocol (*see* **Note 2**). The purified HLA-PCR products of each individual are eluted in 50 μL of autoclaved water. They are always controlled and precisely quantified by capillary electrophoresis using the BioAnalyzer 2100 (Agilent Technologies) (*see* **Note 3**).

3.2. Universal Array Preparation

1. Glass slides are cleaned by soaking in 1M NaOH for 2 h on a shaker followed by rinsing with bidistilled water, immersed in 1N HCl solution overnight on a shaker and then rinsed again in bidistilled water (*see* **Note 4**). Slides are, then, immersed in 96% (v/v) ethanol for 10 min and then washed three times with bidistilled water. They are immersed in acetone for 10 min, removed and dried.
2. Slides are treated with the silanisation solution for 1 h. Excess silane is removed by dipping the slides in 95% (v/v) ethanol for 1 min. Finally, they are dried at 150°C in a temperature-controlled system for 20 min (*see* **Note 5**).
3. The polymer solution is prepared by dissolving 10 g of chitosan (400 Kda) in a liter of HCl 50 mM. The solution is heated at 70°C, stirred for 1 h and then filtered. Treated slides are left in a water solution containing 10% (v/v) PBS and 10% polymer (0.1% w/v) for 1 h on a shaker. Slides are washed repeatedly with bidistilled water (*see* **Note 6**), centrifuged at 800 rpm for 3 min and dried for 10 min at 45°C in an oven.
4. Slides are activated by immersion into a 0.2% (v/v) solution of PDITC in 90% (v/v) *N,N*-dimethylformamide and 10% (v/v) pyridine. The activation reaction is carried out at room temperature for 2 h (*see* **Note 7**). After washing with methanol and acetone (2 min each), the activated slides are dried and stored, until use, in a dark closed box under vacuum.
5. Zip-code oligonucleotides, synthesised with 5′-amino modified and a 10mer 5′-poly (A) linker (*see* **Note 8**), are diluted in printing buffer (150 mM sodium phosphate, pH 8.5) at a final concentration of 50 μM. They are spotted onto activated slides by contact printing using the MicroGrid II Compact

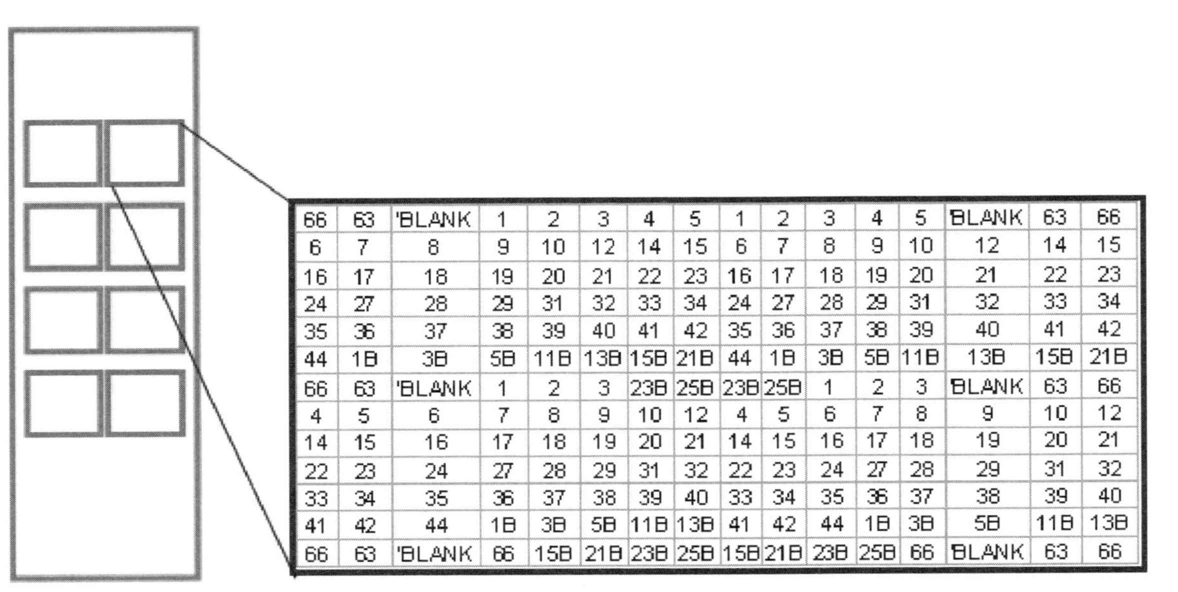

66	63	'BLANK	1	2	3	4	5	1	2	3	4	5	'BLANK	63	66
6	7	8	9	10	12	14	15	6	7	8	9	10	12	14	15
16	17	18	19	20	21	22	23	16	17	18	19	20	21	22	23
24	27	28	29	31	32	33	34	24	27	28	29	31	32	33	34
35	36	37	38	39	40	41	42	35	36	37	38	39	40	41	42
44	1B	3B	5B	11B	13B	15B	21B	44	1B	3B	5B	11B	13B	15B	21B
66	63	'BLANK	1	2	3	23B	25B	23B	25B	1	2	3	'BLANK	63	66
4	5	6	7	8	9	10	12	4	5	6	7	8	9	10	12
14	15	16	17	18	19	20	21	14	15	16	17	18	19	20	21
22	23	24	27	28	29	31	32	22	23	24	27	28	29	31	32
33	34	35	36	37	38	39	40	33	34	35	36	37	38	39	40
41	42	44	1B	3B	5B	11B	13B	41	42	44	1B	3B	5B	11B	13B
66	63	'BLANK	66	15B	21B	23B	25B	15B	21B	23B	25B	66	'BLANK	63	66

Fig. 9.2. Schematic illustration of the "array of arrays" format. Each microscope slide contains eight subarrays in a 2×4 conformation; each subarray contains 208 spots (each zip-code is replicated four times). Zip-code numbers 63 and 66 are associated, respectively, to ligation and hybridisation controls (with permission from Ref. *12*).

(BioRobotics Ltd, Cambridge, UK) and optimised protocols. The humidity during spotting is 80% and the temperature kept at ~ 25°C. Eight subarrays are positioned in an "array of arrays" conformation of four rows and two columns on the microscope slides. Zip-codes are printed in quadruplicate distributed within the printing area; two zip-codes, namely 63 and 66, are associated to ligation and hybridisation control, respectively; as negative control (blank) printing buffer is used. Each UA is a subarray consisting of 208 spots of 13 rows × 16 columns (**Fig. 9.2** and *see* **Note 10**).

6. The spotting is performed using a contact system (Micro Grid II Compact, Biorobotics, UK). Printed slides are placed, overnight at room temperature, in a saturated NaCl humidification chamber (*see* **Note 11**). The deactivation step is carried out by treatment with a solution consisting of 50 mM ethanolamine, 0.1M Tris (pH 9), 0.1% SDS at 50°C for 15 min in order to block residual reactive groups. After rinsing twice with distilled water, the spotted surfaces are immersed into a 4× SSC/0.1% SDS solution (pre-warmed to 50°C) for 15 min on a shaker. Finally, printed surfaces are washed twice with distilled water and spun at 800 rpm for 3 min *(10)*.

3.3. Ligation Probe Design

1. A preliminary phase implies the set up of a database consisting of the specific sequences retrieved from public repositories such as IMGT-HLA database (www.ebi.ac.uk/imgt/hla).
2. By means of a specific software, called ARB (www.arb-home.de) and working in a Linux environment, it is possible to organise a database by using the imported and aligned sequences and creating clusters. Using ARB-HLA SNPs clusters, allele-specific probes are designed surrounding the polymorphic sites (SNPs) of interest. Selection of polymorphic sites within the HLA genomic region are based on the method of Olerup for SSP primers designing *(11)*.
3. All probes for LDR (i.e., common probe and discriminating oligonucleotide) are designed to analyse the chosen polymorphic sites. Each allele specific probe is designed to have the 3′-position placed just over the polymorphic site, single based modification is discriminated by a labelled discriminating oligonucleotide and a common probe. For each SNP, two discriminating probes labelled with two distinct fluorochromes (Cyanine 3 and Cyanine 5) and a common probe are designed. Each zip-code is randomly assigned to a particular polymorphic site and the zip-code complement (cZip code) is attached to the 3′-terminal position of the related common probe. The possible hairpin loop formation in the common probe-cZip sequences is verified by proper software analysis.

4. A HLA control sequence is designed in a conserved region, representing an internal reference (HR) for the normalisation process and HLA genotyping calculations.

3.4. Ligation Detection Reaction

1. Ligation reaction is carried out in a final volume of 20 μL containing 20 pmol/μL Tris–HCl (pH 7.5), 20 pmol/μL KCl, 10 pmol/μL $MgCl_2$, 0.1% IgePal, 0.01 pmol/μL rATP, 1 pmol/μL DTT, 2 pmol of each discriminating oligo, 2 pmol of each common probe, and 10 fmol of purified PCR products. The reaction mixture is pre-heated for 2 min at 94°C, spun in a microcentrifuge for 1 min, then 1 μl of 4U/μl Pfu DNA ligase (Stratagene, La Jolla, CA) is added. The LDR is cycled for 30 rounds of 94°C for 30 s and 65°C for 4 min in a thermal cycler (GeneAmp, PCR System 9700, Applied Biosystem, Foster City, CA, USA).

3.5. LDR Capture by Hybridisation onto UA

1. The LDR mix (20 μl) is diluted to obtain 70 μl of hybridisation mixture containing 5× SSC buffer and 0.1 mg/ml salmon sperm DNA. The mix, after heating to 94°C for 2 min and chilling on ice (*see* **Note 12**), is applied onto the sub-array under a home-made multiple sample chamber (8-well chamber), using press-to-seal silicone isolaters (Schleicher & Schuell, Germany). Hybridisation is carried out in the dark at 65°C for 1.5 h, in a temperature-controlled system (Shack'n'Stack, Hybaid, England).
2. After removal of the chamber, the slide is washed for 15 min in pre-warmed (65°C) solution of 1× SSC buffer and 0.1% (v/v) SDS buffer on gentle shaking (Shack'n'Stack, Hybaid, England). Finally, the slide is spun at 800 rpm for 3 min.

3.6. Signal Detection and Statistical Data Analysis

1. Fluorescent signals are measured with a ScanArray® 4000 laser scanning system and ScanArray 3.1 software (Perkin Elmer Life Sciences, Boston, MA) using the green laser for Cy3 dye (λ_{ex} 543 nm/λ_{em} 570 nm) and the red laser for Cy5 dye (λ_{ex} 633 nm/λ_{em}670 nm). Both the laser and the photomultiplier (PMT) tube gain vary between fluorochromes and experiments. Typical settings for the laser and PMT gain are 75% for Cy3 and 65% for Cy5 (*see* **Note 13**).
2. Spot analysis of 5 μm resolution TIFF images are carried out by using QuantArray Quantitative Microarray Analysis software supplied with the scanner. The quantitation method chosen is the "fixed circle" method: the spot mask and background mask are constructed using the parameters of the spot diameter and the background inner and outer diameters. The spot intensities are calculated using the mean intensity option. The local backgrounds are subtracted from the intensity of each spot. This software generates an export file in Excel format: we operate a statistical analysis of these data.

3. Data analysis of each experiment is performed by the following calculations: (1) the average fluorescent intensity (IF) of quadruplicate spots background subtracted for each channel; (2) the coefficient of variation between the quadruplicate spots; (3) the IF ratios of averaged IF of Cy3 and Cy5 channels of HLA internal reference ($IF_{cy3}/IF_{Cy5} = R_{HR}$). Each IF signal derived from a given spot is subjected to two adjustments: a channel correction is carried out dividing Cy3 channel IF signal by R_{HR}; then, to compare results derived from different UA experiments, we normalise the corrected IF (cIF) of each spot using the HR (*see* **Note 14**).

3.7. Genotype Determination

To define the genotype for each allele HLA, we calculate the allelic fraction which estimates the relative amount of allele 1 Cy3-labelled in the target mixture by the relative proportion of the corrected intensities of fluorescent signal of each allele 1 Cy3 labelled

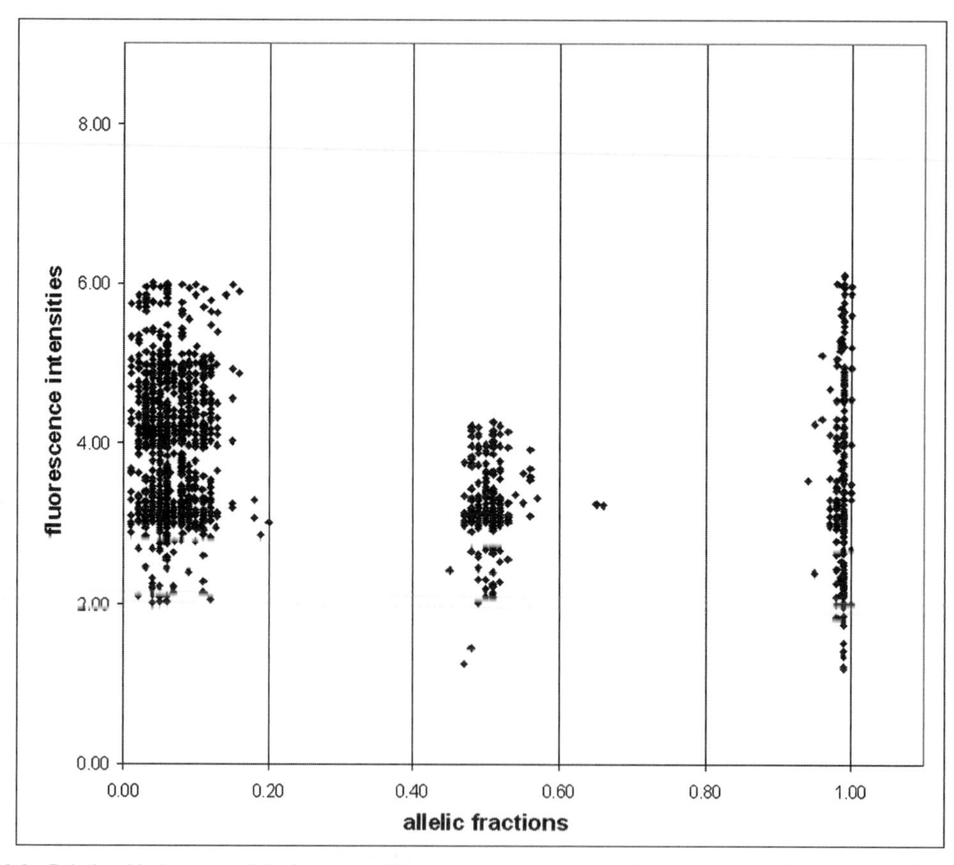

Fig. 9.3. Relationship between allelic fraction and fluorescent intensities for all probes. Cluster diagram shows the genotype assignment for 27 polymorphisms identified in 62 individuals. Allelic fractions $AF = cIF_{CY3}/(cIF_{cy3} + cIF_{CY5})$ versus the total normalised and corrected Cy3 and Cy5 fluorescent intensities (IFCy3+IFCy5) are plotted for each polymorphic site. Allelic fractions between 0.47 and 0.62 are scored as heterozygous while fractions ≤ 0.20 and ≥ 0.94 are scored as homozygous (with permission from Ref. *12*).

and allele 2 Cy5 labelled (allele1-cIF$_{CY3}$/(allele1-cIF$_{cy3}$+ allele2-cIF$_{CY5}$). Cluster analysis is performed on the entire collection of HLA single base modifications under study. The allelic fraction values are then plotted versus the normalised IF signal and clustering to three possible genotypes are computed. In **Fig. 9.3**, the very well-defined clusters, corresponding to the three possible genotypes, obtained from the analysis of 62 different human samples, are shown: allelic fractions ranging from 0.02 to 0.20 for allele 2 homozygous samples, 0.94–1.00 for allele 1 homozygous samples and 0.47–0.62 for heterozygous samples *(12)*.

4. Notes

1. To obtain good performances in the amplification reactions, it is important to verify that the oligonucleotide primers chosen have minimal free energy for primer–dimer or self-hairpin formation. Primer3 software has many different input parameters that tell primer3 exactly what characteristics make good primers for own goals.
2. It is possible to use any kit able to purify a PCR product obtaining sufficient yield (50 fmol) in order to perform the consequent LDR.
3. The Agilent Technology provides chips and reagents designed for sizing and analysis of DNA fragments with a quantitative range of 0.5–50 ng/μl. DNA chip contains an interconnected set of micro-channels that is used for separation of nucleic acid fragments based on their size as they are driven through it electrophoretically.
4. It is better to carry out the complete chemical treatment from beginning to end under an extractor fan.
5. All the solutions required for the chemical treatment of the glass slides are made fresh: re-used solutions can be invalidate the final outcome of the processing.
6. The slides are immersed in the bidistilled water about five times in order to eliminate the excess of the solution.
7. It is fundamental to conserve the PDITC in a dry place to avoid the humidity that could be damage the reagent.
8. Zip-codes sequences are chosen among those described by Chen (Chen et al., 2000) and Gerry (Gerry et al., 1999). It is important that all the oligonucleotides employed are purified by HPLC and that their quality is checked by MALDI mass spectrometry. These services can be required to the Thermo Fisher Scientific (http://www.thermo.com/). The zip-code oligonucleotides are synthesised with a 10mer 5′-poly (A) linker for two main reasons: this tail functions as spacer from the surface and permits a quality control of the oligonucleotide coupling onto the slides. It is possible

to perform a hybridisation on one of the eight subarrays on each slide with 1 μM 5′ Cy3 labelled p(dT)$_{10}$ in a solution containing 5× SSC buffer and 0.1 mg/mL DNA salmon sperm. The reaction is carried out at RT for 30 min. After incubation, slides are immersed for 5 min in 2× SSC buffer and 0.1% SDS buffer and fluorescent signals are controlled by laser scanning following the procedures described in Section 3.6.

9. To control the performance of each step of genotyping DNA microarray, it is better to include spot replicates and various controls.
10. The resulted spots are 160–180 μm in diameter and a 300 μm center-to-center distance between two adjacent spots. In this way, it is possible to identify until 47 SNPs. Using the same layout and maintaining the four replicates, it is possible to add other 20 zip-codes. Obviously, reducing the number of replicates (two instead of four for each zip-code), it is possible to double the number of SNPs identifiable.
11. To obtain a chamber with a relative humidity of approximately 75%, solid NaCl must be added to water as needed to form a 1 cm deep slurry in the bottom of a plastic container with an airtight lid. Overnight incubation has shown the best results. If possible incubate for a minimum of 4 h and a maximum of 72 h.
12. The denaturation step is crucial to permit the accessibility during the hybridisation reaction.
13. The processed arrays must be scanned taking care that the fluorescent signals are not saturated to permit the subsequent analysis.
14. A signal to background ratio >3 is taken as a threshold for including a spot within the calculation. A CV < 5% is taken as an index of good quality of the replicated spots (the overall success rate for the samples was 99.7%). The differences in performance of the entire process, consisting of PCR, LDR, and UA hybridisations, are normalised by adjustments of the signal intensity of each spot for each fluorochrome relative to the signal obtained from the HLA internal reference. Spurious signal (signals coming from spots expected to be negative that passed the $S/N>3$ threshold) are always below 1, thus preventing any miscall of the genotype.

Acknowledgments

The author would like to thank her supervisors Gianluca De Bellis and Cristina Battaglia for their continuous support, encouragement, extensive knowledge, and experience. Prof. GB Ferrara group (National Cancer Institute, IST, Advanced Biotechnology Center, Department of Oncology, Biology and Genetics, University of Genoa, Italy) is acknowledged for sharing their samples and their deep expertise in the immunology field.

References

1. Jenkins, S., Gibson, N. (2002) High throughput SNP genotyping. *Comp Funct Genomics* 3, 57–66.
2. Tsuchihashi, Z., Dracopoli, N. C. (2002) Progress in high throughput SNP genotyping methods. *Pharmacogenomics J* 2, 103–110.
3. Hacia, J. G., Fan, J. B., Ryder, O., Jin, L., Edgemon, K., Ghandour, G., et al. (1999) Determination of ancestral alleles for human single nucleotide polymorphisms using high-density oligonucleotide arrays. *Nat Genet* 22, 164–167.
4. Lindroos, K., Liljedahl, U., Raitio, M., Syvänen, A. C. (2001) Minisequencing on oligonucleotide microarrays: comparison of immobilisation chemistries. *Nucleic Acids Res* 29, E69-9.
5. Pastinen, T., Kurg, A., Metspalu, A., Peltonen, L., Syvänen, A. C. (1997) Minisequencing: a specific tool for DNA analysis and diagnostics on oligonucleotide arrays. *Genome Res* 7, 606–614.
6. Khanna, M., Park, P., Zirvi, M., Cao, W., Picon, A., Day, J., Paty, P., Barany, F. (1999) Multiplex PCR/LDR for detection of K-ras mutations in primary colon tumors. *Oncogene* 18, 27–38.
7. Gerry, N. P., Witowski, N. E., Day, J., Hammer, R. P., Barany, G., Barany, F. (1999) Universal DNA microarray method for multiplex detection of low abundance point mutations. *J Mol Biol* 292, 251–262
8. Favis, R., Day, J. P., Gerry, N. P., Phelan, C., Narod, S., Barany, F. (2000) Universal DNA array detection of small insertions and deletions in BRCA1 and BRCA2. *Nat Biotechnol* 18, 561–564.
9. Pera, C., Delfino, L., Morabito, A., Longo, A., Johnston-Dow, L., White, C. B., Colonna, M., Ferrara, G. B. (1997) HLA-A typing: comparison between serology, the amplification refractory mutation system with polymerase chain reaction and sequencing. *Tissue Antigens* 50, 372–379.
10. Consolandi, C., Severgnini, M., Castiglioni, B., Bordoni, R., Frosoni, A., Battaglia, C., Rossi Bernardi, L., De Bellis, G. (2006) A structured chitosan-based platform for biomolecule attachment to solid surfaces: application to DNA microarray preparation. *Bioconjug Chem* 17, 371–377.
11. Olerup, O., Zetterquist, H. (1992) HLA-DR typing by PCR amplification with sequence-specific primers (PCR-SSP) in 2 hours: an alternative to serological DR typing in clinical practice including donor-recipient matching in cadaveric transplantation. *Tissue Antigens* 39, 225–235.
12. Consolandi, C., Frosini, A., Pera, C., Ferrara, G. B., Bordoni, R., Castiglioni, B., Rizzi, E., Mezzelani, A., Rossi Bernardi, L., De Bellis, G., Battaglia, C. (2004) Polymorphism analysis within the HLA-A Locus by Universal Oligonucleotide Array. *Hum Mutation* 24, 428–434.

Chapter 10

Medium- to High-Throughput SNP Genotyping Using VeraCode Microbeads

Charles H. Lin, Joanne M. Yeakley, Timothy K. McDaniel, and Richard Shen

Abstract

Recent breakthroughs in multiplexed SNP (single nucleotide polymorphism) genotyping technology have enabled global mapping of the relationships between genetic variation and disease. Discoveries made by such whole-genome association studies often spur further interest in surveying more focused subsets of SNPs for validation or research purposes. Here we describe a new SNP genotyping platform that is flexible in assay content and multiplexing (up to 384 analytes), and can serve medium- to high-throughput applications. The Illumina BeadXpress platform supports the GoldenGate Genotyping Assay on digitally inscribed VeraCode microbeads to allow streamlined workflow, rapid detection, unparalleled data reproducibility and consistency. Thus, it is a highly valuable tool for biomarker research and validation, pharmaceutical development, as well as the development of molecular diagnostic tests.

Key words: High-throughput genotyping, biomarker validation, SNP, randomly assembled microarrays, allele-specific extension and ligation, universal capture oligonucleotides, universal arrays.

1. Introduction

Highly multiplexed SNP genotyping technologies have begun to transform the landscape of human genetics research by revealing the association between genetic variation and disease with unprecedented data quality and quantity *(1)*. Following this discovery-driven phase, more focused efforts to finely map or validate the newly identified biomarkers are often employed. The ideal technology tool for targeted genotyping, while requiring lower assay multiplexing, must be consistent in performance,

Peter Bugert (ed.), *DNA and RNA Profiling in Human Blood: Methods and Protocols, vol. 496*

DOI 10.1007/978-1-59745-553-4_10 Springerprotocols.com

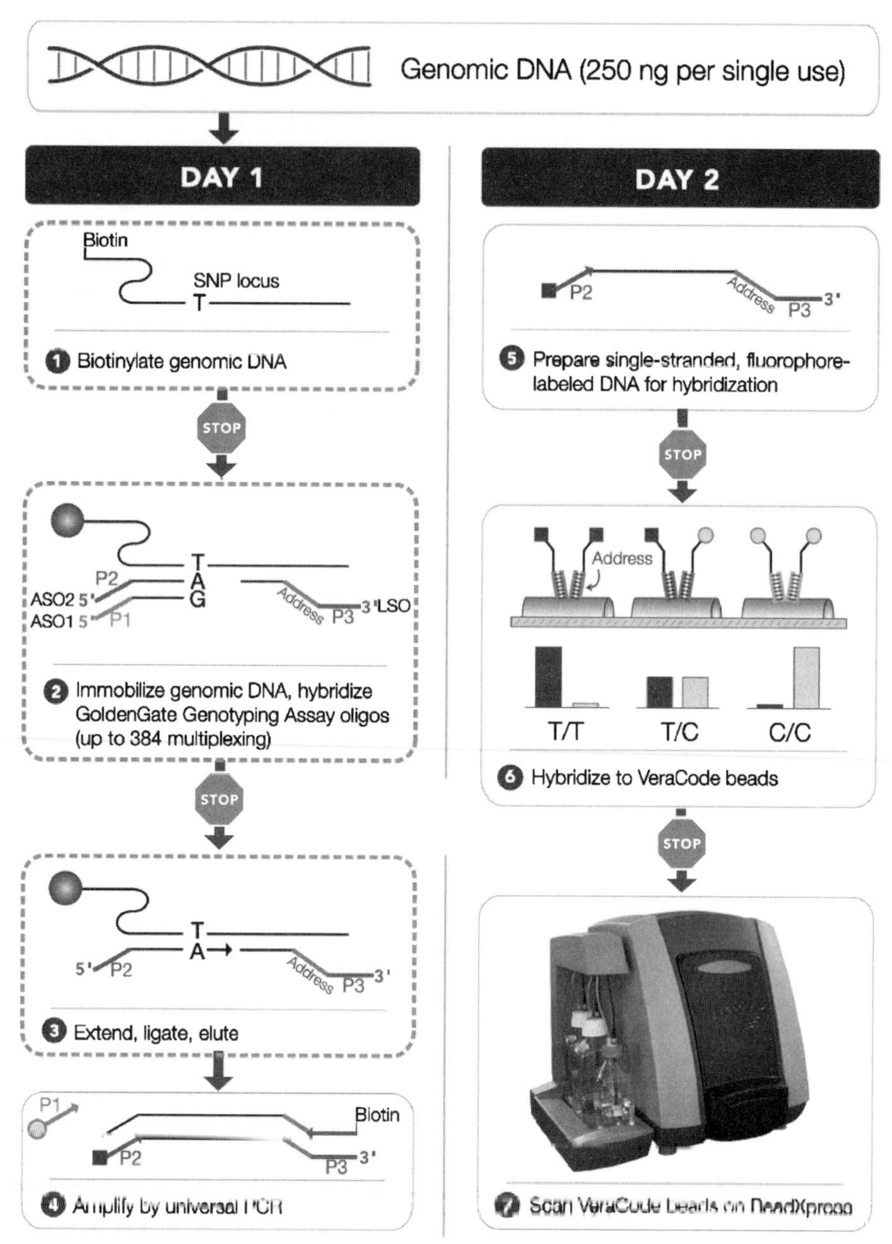

Fig. 10.1. Workflow of the GoldenGate Genotyping Assay on the VeraCode/BeadXpress platform. The schematic diagram illustrates the two-day assay workflow which can be broken down into several defined processes with optional stopping points, as indicated. Dashed boxes indicate the steps that must be performed in a dedicated pre-PCR area. The assay begins with chemical biotinylation of the input genomic DNA (step 1), followed by immobilization of the resulting biotinylated DNA to streptavidin-coated paramagnetic particles and simultaneous annealing of the assay oligonucleotides (step 2). The allele-specific oligonucleotides (ASO1 and ASO2) differ by the 3′ base and contain distinct universal priming sequences P1 and P2 as 5′ overhangs. The LSO contains an address sequence that identifies a specific SNP in a given assay, as well as a universal priming sequence P3 as a 3′ overhang. For robustness, the artificially created address sequences have been pre-screened bioinformatically to avoid cross-hybridization between themselves or with the human genome. The assay described here allows up to 384 multiplexing, meaning that 384 triplets of assay oligonucleotides can be annealed at once under the specified conditions. After unbound oligonucleotides are washed away, allele-specific extension and ligation of the hybridized oligonucleotides

flexible in content, fast in data turnaround, and affordable for studies involving large volumes of samples.

The GoldenGate Genotyping Assay with VeraCode microbeads was developed to fulfill those demands. Here, the same proven, robust GoldenGate Genotyping Assay biochemistry *(2, 3)*, which was used to generate nearly 70% of the Phase I International HapMap Project data, remains a central component of the new assay format. As illustrated in **Fig. 10.1**, allelic variants are captured as products of an extension–ligation reaction (steps 1–3), which in turn are labeled with respective fluorescent dyes and amplified at the same time in a universal polymerase chain reaction (PCR; step 4). The dye-labeled strand of the PCR product is hybridized to patented Illumina VeraCode microbeads (steps 5–6), instead of a BeadArray array matrix *(2,3)*. When analyzed by the high-throughput, two-color BeadXpress Reader (step 7), VeraCode beads allow immediate association of allele-specific fluorescence signals to the SNP loci being interrogated through the inscribed holographic barcodes (**Fig. 10.2**). Major advantages of this approach include: (a) bypassing the decoding process required during manufacturing that maps SNP loci to physical locations in a bead array *(2)*; (b) rapid reaction kinetics of solution-based assays; and (c) array customization as achieved simply by pooling together only the required number of VeraCode bead types.

Flexibility and ease of use of the VeraCode technology system certainly have broadened the potential of the GoldenGate Genotyping Assay into many areas including pharmaceutical development, pre-clinical, and clinical research. Using the method described in Section 3, up to 96 samples can be processed at a time in a typical PCR microplate by a single user without automation. For each DNA sample, a few to 384 SNP loci can be analyzed in parallel, translating into a maximum of 36,864 genotype calls. Most impressively, across such a wide multiplexing range, the VeraCode GoldenGate Genotyping Assay exhibits exquisite consistency, reproducibility and success rate (**Fig. 10.3**).

Fig. 10.1. (continued) is performed (step 3). The allelic variant that matches the genomic sequence is preferentially extended to produce a synthetic template which then gets amplified in the subsequent universal PCR, with universal primers P1, P2, and P3 (step 4). P1 and P2 are universal primers labeled with Cy3 (green, shown as open circles) and Cy5 (red, shown as solid squares) fluorescent dyes, respectively. In contrast, P3 is a universal primer biotinylated at the 5′ end. After PCR, the fluorescent strand is purified away from the non-fluorescent strand (step 5), and is hybridized to VeraCode beads through the address sequence (step 6). Finally, VeraCode beads are scanned by the high-throughput, two-color BeadXpress Reader (step 7). Scan data can be further analyzed using the BeadStudio software package (Illumina).

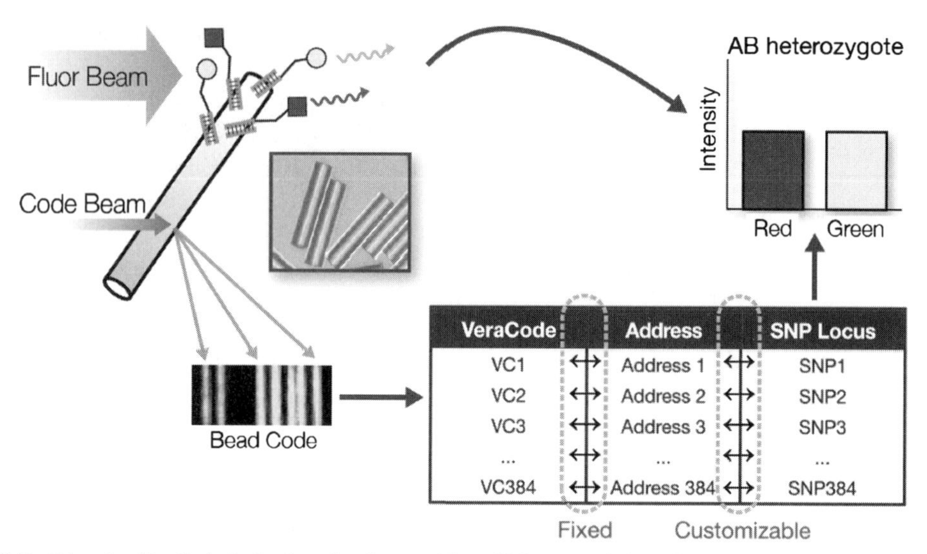

Fig. 10.2. Using the VeraCode technology to allow rapid, multiplex association of hybridization signals to SNP loci. The inset is a microscopic image of VeraCode beads (rod-shaped), each measuring 240 μm in length by 28 μm in diameter. Each VeraCode bead is inscribed with a specific barcode (VeraCode bead code) and its surface is available for the attachment of a single type of capture oligonucleotides complementary to a specific address sequence, introduced during the extension–ligation step (**Fig. 10.1**, step 3). Thus, the bead codes and the addresses are in a one-to-one relationship. The relationship between the addresses and SNP loci can vary from one test panel to another as needed (see below), but is known through the oligonucleotide manifest generated by the user in the assay design process. During the scan, each VeraCode bead is individually scanned for the barcode and green/red fluorescence, resulting in immediate association of the hybridization signal to a single SNP locus being analyzed in this assay. About 30 beads for each bead type are included in one assay and each bead is optically sampled up to a dozen times, leading to about 300 independent data points for each analyte. This strong bead and data redundancy has greatly enhanced both the precision and accuracy of the VeraCode GoldenGate Genotyping Assay. Custom test panels up to 384 loci each can therefore be made on demand simply by reassigning SNP contents to the address sequences. This "universal array" approach eliminates the requirement to create a new "custom array" for every new custom assay *(2)*.

2. Materials

2.1. Input DNA

Commercial kits are available from multiple vendors (e.g., QIAGEN) for extracting genomic DNA from blood or buccal samples. Successes in using such samples have been reported for the GoldenGate Genotyping Assay *(4, 5)*. High-purity DNA ($A_{260}/A_{280} = 1.7 - 1.9$) is generally desirable, whereas some degree of DNA degradation is tolerable by the assay *(2)*. In addition, highly accurate genotyping results have been obtained with whole-genome amplified DNA samples *(6)* and DNA samples derived from formalin-fixed, paraffin-embedded (FFPE) tissues (Fan et al., unpublished data). We recommend using PicoGreen reagent (Invitrogen) to quantitate genomic DNA samples, followed by normalization to 50 ng/μL final with TE buffer (see Section 3).

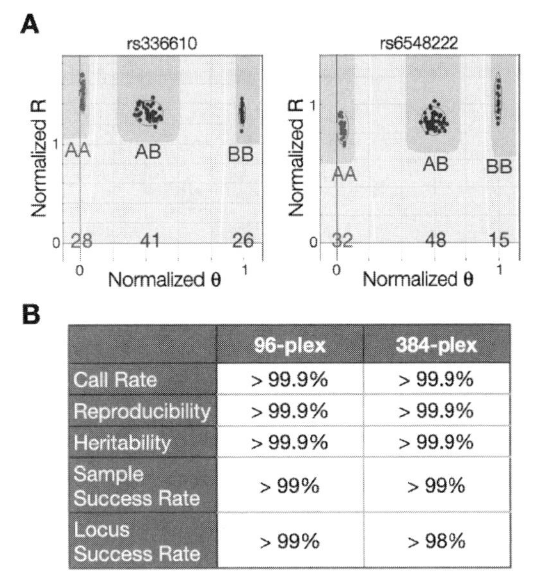

	96-plex	384-plex
Call Rate	> 99.9%	> 99.9%
Reproducibility	> 99.9%	> 99.9%
Heritability	> 99.9%	> 99.9%
Sample Success Rate	> 99%	> 99%
Locus Success Rate	> 99%	> 98%

Fig. 10.3. Data quality and assay design success rates of the VeraCode GoldenGate Genotyping Assay. (**A**) Sample clustering for two representative loci selected from 96-plex (left panel) and 384-plex (right panel) experiments, respectively. Each image consists of 95 data points corresponding to the 95 DNA samples analyzed. The x-axis represents a normalized value of the Cy5/Cy3 ratio (allele B vs allele A), which ideally is 0 if an AA homozygote and 1, if a BB homozygote. The y-axis represents a normalized value of the sum (R) of green and red fluorescence intensities. The AA, AB, and BB clusters were generated using BeadStudio (Illumina); and the images were enhanced for publication with Illustrator (Adobe). The number of samples belonging to each genotype cluster is also indicated. Dark shaded areas indicate the three call zones. (**B**) Typical performance of 96- and 384-plex assays. Metrics are defined as follows: call rate, (number of calls made) / (number of calls attempted, excluding bad samples or assignable errors); reproducibility, (number of calls agreeing between sample replicates) / (total number of calls made in each sample replicate); heritability, (number of child alleles consistent with parental alleles) / (total number of child alleles); sample success rate, (number of samples called) / (number of samples attempted); locus success rate, (number of SNP loci called) / (number of SNP loci attempted).

2.2. Assay Oligonucleotides

Two allele-specific oligonucleotides (ASO1 and ASO2) and one locus-specific oligonucleotide (LSO) are required for each SNP locus to be genotyped. The proprietary Assay Design Tool (ADT; Illumina), which has designed hundreds of thousands of successful GoldenGate assays, can assist researchers to design and optimize assay probes for custom contents. ADT evaluates assay probes based on T_{m}, self complementarity, as well as the sequences flanking the targeted SNP for neighboring polymorphisms, and any degree of repetitiveness or homology to the rest of the genome. Most ASOs designed by ADT have a T_{m} of 57–62°C; and the LSOs, 54–60°C. LSOs additionally must be synthesized with 5′-phosphate to allow ligation (**Fig. 10.1**, step 3).

2.3. VeraCode Bead Plates

VeraCode Bead Plates pre-loaded with 96 and 384 bead types in the 96-sample microplate format are available from Illumina. In addition, universal capture bead sets are available for researchers who would like to create their own test panels at any multiplexing level between 1 and 144.

2.4. Equipment

1. User-supplied: lab coats and gloves (especially for use in the pre-PCR area; *see* **Note 1**); two microtiter plate centrifuges with *g*-force range of 8–3,000 × *g* (one for pre-PCR use, the other for post-PCR use); sealing roller (for pre-PCR use; Bio-Rad); 96-well thermal cycler with heated lid (for post-PCR use; Bio-Rad); cap mat applicator (for post-PCR use; Corning); vacuum flask assembly and regulator (for post-PCR use; QIAGEN); multichannel pipettes (for both pre- and post-PCR use; 5–200 μL).
2. Illumina-supplied: raised bar magnet (for pre-PCR use); microplate vortexer (for pre-PCR use; *see* **Note 2**); heat sealer with 96-well adaptor base (for pre-PCR use); 96-well microplate heat block with heated lid (for pre-PCR use); vortex incubator for optimal hybridization to VeraCode beads (for post-PCR use); BeadXpress Reader and accessories.

2.5. Reagents and Consumables

1. User-supplied: 96-well 0.2 mL skirted microplates (Bio-Rad); microseal 'A' film (Bio-Rad); microplate clear adhesive film (Phenix Research); thermo-seal heat sealing foil sheets (ABgene); non-sterile solution basins (55 mL; Labcor Products); aerosol filter pipette tips (5–200 μL); serological pipettes (10, 25, and 50 mL); MultiScreen filter plates (0.45 μM, clear polystyrene; Millipore); 96-well polypropylene V-bottom plates (Corning Costar); 96-well cap mats (Corning).
2. User-supplied: Titanium *Taq* DNA polymerase (Clontech); TE buffer (10 mM Tris–HCl, pH 8.0, 1 mM EDTA); isopropanol; 0.1 N NaOH; deionized H_2O; 10% bleach (plain, unscented household bleach).
3. Illumina-supplied: all the proprietary VeraCode GoldenGate Genotyping Assay reagents used in the protocol described below contain sufficient volume for processing 96 samples at a time, using a multichannel pipette and solution basins. When processing smaller sample batches using a solution basin, dead volume, and pipetting error losses can increase. If that becomes an issue, single-pipette the reagents into each well. To store remaining reagents, we recommend freezing aliquots, rather than repeatedly freezing and thawing the supplied reagent tubes.

3. Methods

The GoldenGate Genotyping Assay with VeraCode microbeads can be broken down into several defined processes, offering the users potential stopping points should they wish to do so. Note that Sections 3.1–3.7 also match the steps shown in **Fig. 10.1**. Also note that the following protocol is based on a full plate (96 samples).

3.1. Biotinylation of Genomic DNA

Genomic DNA is first biotinylated in a heat-activated chemical reaction. The resulting biotinylated DNA is then cleaned up by isopropanol precipitation to remove excess biotinylation reagent (**Fig. 10.1**, step 1). Preheat the heat sealer and heat block (to 95°C).

1. Set up genomic DNA samples (5 μL of 50 ng/μL per well) in a 96-well 0.2-mL skirted microplate (plate name: SUD).
2. Use a multichannel pipette to add 5 μL biotinylation reagent (MS1; Illumina) to the DNA samples. Change pipette tips after each dispense.
3. Seal the SUD plate with a thermo-seal foil sheet using a heat sealer. Pulse-centrifuge at 250 × *g* to collect the contents.
4. Vortex at 2,300 rpm for 20 s. Pulse-centrifuge at 250 × *g*.
5. Incubate in the preheated 95°C heat block for exactly 30 min (*see* **Note 3**).
6. Pulse-centrifuge at 250 ×g. Meanwhile, reset the heat block to 70°C if planning on performing the steps in Section **3.2.** on the same day.
7. Remove the foil carefully. Add 5 μL precipitation reagent (PS1; Illumina) to each well (*see* **Note 4**). Vortex at 2,300 rpm for 20 s or until the solution is uniformly blue (*see* **Note 5**). Pulse-centrifuge at 250 ×*g*.
8. Add 15 μL isopropanol to each well. Vortex at 1,600 rpm for 20 s or until the solution is uniformly blue. Centrifuge at 3,000 ×*g* for 20 min. (DNA should appear as a faint blue pellet at the bottom of each well.)
9. Invert the plate and carefully smack it down a few times onto paper towels to blot off the supernatant. Air-dry the plate upside down at room temperature (RT) for another 15 min.
10. Resuspend the pellet in 10 μL resuspension reagent (RS1; Illumina) by vortexing at 2,300 rpm for 1 min. Pulse-centrifuge at 250 ×*g*. The SUD plate at this stage can be stored at 4°C overnight (*see* **Note 6**).

3.2. Immobilization of Biotinylated DNA and Annealing of Assay Oligonucleotides

In a single temperature ramp-down incubation, allele- and locus-specific GoldenGate Genotyping Assay oligonucleotides are annealed to the biotinylated genomic DNA template, which is simultaneously immobilized to streptavidin-coated paramagnetic beads (**Fig. 10.1**, step 2). Unbound oligonucleotides are removed by several washes afterwards.

1. Take a new 96-well 0.2-mL microplate (plate name: ASE), and add to each well 30 μL oligonucleotide annealing reagent (OB1, vortexed thoroughly after thawing; Illumina) and 10 μL of the oligonucleotide pool (OPA, 1.2 mL needed for 96 samples; Illumina).
2. Use a multichannel pipette to transfer the entire contents (10 μL) from the first column of the SUD plate to the corresponding wells in the ASE plate. Discard the pipette tips.
3. Repeat the previous step for columns 2–12. Change tips between columns.
4. Heat-seal the ASE plate with a thermoseal foil sheet. Vortex the ASE plate at 1,600 rpm for 1 min or until the beads are fully resuspended.
5. Place the plate in the preheated 70°C heat block. Switch the temperature setting immediately to 30°C, and allow the plate to cool down passively (*see* **Note 7**).
6. Remove the plate from the heat block. Pulse-centrifuge at 250 ×*g*. Meanwhile, reset the heat block to 45°C.
7. Place the plate on a raised bar magnet until beads are captured completely (about 2 min). Use a multichannel pipette to remove and discard the supernatant (*see* **Note 8**).
8. Add 50 μL AM1 (wash buffer; Illumina). Vortex at 1,600 rpm for 20 s or until beads are fully resuspended. Place the plate on the magnet until beads are captured completely. Remove and discard the supernatant.
9. Repeat AM1 wash once.
10. Add 50 μL UB1 (wash buffer; Illumina) (vortexing the plate is optional for UB1 washes). Place the plate on the magnet until beads are captured completely. Remove and discard the supernatant.
11. Repeat UB1 wash once. Immediately proceed to the steps in Section 3.3.

3.3. Extension–Ligation and Elution

Allele-specific extension and ligation reactions of the hybridized oligonucleotides take place simultaneously in a single incubation (**Fig. 10.1**, step 3). The extended-and-ligated product forms a synthetic template that is next eluted from the immobilized genomic DNA and transferred to a PCR reaction plate.

1. Add 37 μL extension–ligation reagent (MEL; Illumina) to each well. Vortex at 1,600 rpm for 1 min or until beads are fully resuspended. Incubate in the preheated 45°C heat block for exactly 15 min.
2. During this incubation, user should proceed to setting up the PCR plate: add 64 μL Titanium *Taq* DNA polymerase

(Clontech) and 50 μL uracil DNA glycosylase (UDG; Illumina) to one tube of fully thawed PCR master mix (MMP; Illumina). Vortex the tube briefly to mix the contents. Use a multichannel pipette to aliquot 30 μL of the resulting master mix to each well of a new 96-well 0.2-mL microplate (plate name: PCR). Set aside the plate until use (preferably in the dark (e.g., a drawer) to protect the fluorescent primers in the master mix).

3. Remove the ASE plate from the heat block. Reset the heat block to 95°C. Place the ASE plate on the magnet until beads are captured completely. Remove and discard the supernatant.
4. Add 50 μL UB1 (vortexing is optional). Place the plate on the magnet until beads are captured completely. Remove and discard the supernatant.
5. Add 35 μL IP1. Vortex at 1,800 rpm or until beads are fully resuspended.
6. Incubate the ASE plate in the preheated 95°C heat block for exactly 1 min. Place the plate back onto the magnet until beads are captured completely.
7. Use a multichannel pipette to carefully transfer 30 μL supernatant (*see* **Note 9**) from the first column of the ASE plate to the PCR plate. Mix the contents by gently pipetting up and down a few times. Discard the pipette tips.
8. Repeat the previous step for columns 2–12. Change tips between columns.
9. Use a microseal 'A' film (Bio-Rad) to seal the plate with the help of a sealing roller. The PCR plate can be transported to a thermal cycler in the post-PCR area, and must never be brought back to the pre-PCR area.

3.4. Universal PCR

During the PCR, each extended-and-ligated product is amplified with one of the two fluorophore-labeled universal primers (P1 or P2, depending on the allele variant present; **Fig. 10.1**, step 4) and a biotinylated universal primer (P3).

1. The settings below should be used for the thermal cycler: 37°C for 10 min (*see* **Note 10**); 95°C for 3 min; 34 thermal cycles (95°C for 35 s; 56°C for 35 s; 72°C for 2 min); 72°C for 10 min; 4°C for 5 min; 10°C for as long as needed.

3.5. Generation of Single-Stranded DNA for Hybridization to VeraCode Beads

The non-fluorescent strand of PCR product is removed through its 5′ biotin group (introduced by the universal primer P3; **Fig. 10.1**, step 5).

1. Pulse-centrifuge the PCR plate at 250 ×*g*. To each well, add 20 μL of paramagnetic bead mix (MPB, fully resuspended by vortexing and inverting several times; Illumina).
2. Use a multichannel pipette set to 85 μL, gently mix the samples in the first column by pipetting up and down a few times,

and then transfer the entire contents to the corresponding wells of a new filter unit assembly (*see* **Note 11**). Discard the pipette tips.

3. Repeat the previous step for columns 2–12. Change tips between columns.
4. Cover the filter unit assembly with the accompanying lid. Incubate at RT, protected from light, for 1 h.
5. Centrifuge at 1,000 ×*g* for 5 min. Add 50 μL UB2 (wash buffer; Illumina) to the upper plate of the assembly slowly and gently, without disturbing the beads.
6. Centrifuge at 1,000 ×*g* for 5 min. Meanwhile, take a new 96-well V-bottom plate and dispense 30 μL of hybridization reagent (MH2; Illumina) into each well (plate name: INT).
7. When centrifugation is finished, discard the bottom part of the filter unit assembly, and replace with the INT plate. Confirm that the A1 position of the filter plate matches the A1 position of the INT plate.
8. Add 30 μL 0.1N NaOH to the upper plate of the assembly slowly and gently, without disturbing the beads. Immediately centrifuge the assembly at 1,000 ×*g* for 5 min (*see* **Note 12**).
9. The INT plate contains the single-stranded DNA ready to hybridize to VeraCode beads (see below). Alternatively, the INT plate can be stored at –20°C until use. The lid and upper plate of the filter unit assembly can be discarded.

3.6. Hybridization to VeraCode Beads

The address sequence attached to the extended-and-ligated product by the LSO (**Fig. 10.1**, steps 2 and 3) directs hybridization to a specific type of VeraCode bead. Each VeraCode bead type is manufactured with a high-density coat of oligonucleotides; the sequence of which is the reverse complement of a unique address sequence (**Fig. 10.1**, step 6).

1. Preheat the vortex incubator to 45°C. Remove the VeraCode Bead Plate (plate name: VBP) from 4°C. Pulse-centrifuge at 250 ×*g* to ensure that all the bead pellets remain at the bottom of the plate.
2. Neutralize 3 mL MH2 with 3 mL 0.1N NaOH in a solution basin. Use a multichannel pipette to add 50 μL of neutralized MH2 into each well of the INT plate.
3. Remove the cap mat from the VBP plate, and save the cap mat for subsequent use. Use a multichannel pipette set to 100 μL, gently mix the first column of samples in the INT plate by pipetting up and down, and then transfer 100 μL to the corresponding wells of the VBP plate. Discard the pipette tips.
4. Repeat the previous step for columns 2–12. Change tips between columns.
5. Reapply the cap mat to the VBP plate with a cap mat applicator. Secure the plate in the preheated vortex incubator. Use another 96-well plate as a balance.

6. With the incubator lid closed, shake the VBP plate at 850 rpm for 3 h at 45°C.
7. Pulse-centrifuge at 250 ×*g* to collect any condensation.
8. Add 200 μL VeraCode bead wash buffer (VW1; Illumina), and make sure to agitate the bead pellets. Gently swirl the plate in a circular motion on the bench top. Aspirate off the supernatant with the vacuum manifold (at 50 mbar).
9. Repeat VW1 wash once. The VBP plate is ready to be scanned by a BeadXpress Reader. Alternatively, the VBP plate can be stored at RT, protected from light, for up to 72 h.

3.7. Scanning VeraCode Bead Plate

The VeraCode Bead Plate is scanned in the BeadXpress Reader (**Fig. 10.1**, step 7).

1. Connect and initialize the BeadXpress Reader from the computer. Prime the fluidics.
2. Load the VBP plate onto the tray. Make sure that the plate is correctly oriented with A1 of the plate matching the engraved A1 on the reader tray.
3. Load VeraCode assay scan settings (*see* **Note 13**). Start the scan (*see* **Note 14** for example scan times).
4. GoldenGate Genotyping Assay data consists of green and red fluorescence intensity values for each VeraCode bead type (corresponding to a specific SNP locus), and can be analyzed by the BeadStudio software package (Illumina; *see* **Note 15**).

4. Notes

1. Due to the serious concern of retrograde contamination by amplicons generated in the universal PCR (**Fig. 10.1**, step 4), the GoldenGate Genotyping Assay requires the steps before the universal PCR (steps 1–3) to be performed in a physically separate, dedicated lab area. In addition, the pre-PCR area should be routinely cleaned and monitored for contaminant levels. Every lab must establish procedures for preventing PCR amplicon contamination and train new users properly. The procedures should include, but may not be limited to, the following guidelines: (a) store all area-specific supplies separately, and do not share supplies between areas; (b) when working in the pre-PCR area, users should wear gloves and lab coats (which are meant to stay in the pre-PCR area) at all times; (c) users should decontaminate the bench top with 10% bleach before and after use; (d) wipe contamination hot spots (e.g., door/refrigerator/freezer handles, computer mouse, and keyboards) often with 10% bleach; (e) mop floors with freshly diluted 10% bleach weekly; and (f) anything that falls to the floor is treated as contaminated, and non-disposable items that fall to the floor

should be decontaminated with 10% bleach immediately. Care should be taken, however, not to overwhelm the area with bleach fumes because they can degrade the fluorescent dyes on the PCR primers. As an added precaution, oligonucleotide pools received from Illumina contain internal controls which can serve as an indicator when significant contamination has occurred.

2. The displayed speed of the pre-PCR microplate vortexer may differ from the actual speed. It is important to use a digital stroboscope to calibrate the vortexer periodically for the following speeds: 1,800, 2,000, and 2,300 rpm. Each lab may want to establish its own lookup table for displayed and actual speeds. Using an uncalibrated vortexer may cause the samples to splash and increase the likelihood of cross-contamination.
3. Use the heated lid for all incubations done in the heat block to minimize condensation.
4. From this step on, unless noted otherwise, dispensing reagent into the sample plate can be done for all columns using the same set of eight pipette tips if the tips touch only the top edge of the wells. However, if it is suspected that the tips may have become contaminated with the contents of the well, they should be discarded and replaced immediately.
5. Unless noted otherwise, always use a clear adhesive film with the help of a sealing roller to seal the plate before vortexing.
6. For long-term storage –20°C is recommended. However, if stored frozen, the contents of the plate need to be fully resuspended prior to use after thawing.
7. Cooling from 70 to 30°C takes approximately 2.5 h. Alternatively, the plate may be left in the heat block at 30°C overnight for up to 16 h.
8. If an Illumina-recommended raised bar magnet is used, the beads in odd-numbered columns will be pulled to the right wall of the wells, and the beads in even-numbered columns, to the left. When removing unwanted supernatant from a plate using a multichannel pipette, the pipette should be set to a volume slightly greater than the volume of supernatant to be removed (e.g., 60 μL if removing 50 μL of supernatant). Next, push the tips almost down to the bottom of the wells, gently touching the side opposite where the beads are captured (left if odd-numbered columns, and right if even-numbered columns) but not touching the bead pellets. By doing so, it is not necessary to change pipette tips until the entire plate has been processed. Remember to visually inspect the pipette tips upon drawing the liquid. If beads are visible (brownish-turbid), return the liquid into the same wells, change pipette tips and allow the magnet to re-capture the beads.

9. The supernatant now contains the extended-and-ligated product, while leaving the biotinylated genomic DNA template behind on the paramagnetic beads.
10. The PCR master mix (MMP) contains dUTP instead of dTTP. The incorporation of dUTP does not affect downstream analysis, but offers an effective measure for controlling PCR contamination that would otherwise affect the accuracy of the assay. An initial incubation of 37°C for 10 min in the PCR cycling conditions allows UDG to digest away any dUTP-incorporated PCR product existing prior to the current PCR (= contaminant).
11. The filter unit should be assembled in this order, from bottom up: a 96-well V-bottom plate (Corning); filter plate adaptor (Illumina); MultiScreen filter plate; and lastly the lid (Millipore).
12. Due to the sensitivity of the fluorescent dyes to 0.1 N NaOH, proceed with this step quickly. Prolonged incubation with NaOH is unnecessary. Almost instantly, the fluorescent strand is dissociated from the complementary strand that gets retained by the streptavidin-coated paramagnetic beads. Centrifugation must proceed within 5 min to prevent damage to the assay products.
13. Instructions for scan settings and a template file are provided with the BeadXpress Reader.
14. For 96-plex assays, the BeadXpress Reader scans about 80 samples per hour (= 128 genotype calls per minute). For 384-plex assays, the BeadXpress Reader scans about 24 samples per hour (= 154 genotype calls per minute).
15. The genotyping module of the BeadStudio software package consists of several features that allow automatic sample clustering and genotype calling for custom SNP genotyping assays.

Acknowledgments

We thank M. Chen, I. Lewis, and S. Oeser for their crucial contributions to the development of the VeraCode Golden Genotyping Assay. We also thank J. Stuelpnagel, M. Henshall, C. Allred and S. Valentini for critical reading of this chapter. Illumina, GoldenGate, BeadArray, BeadXpress, and VeraCode are registered trademarks or trademarks of Illumina, Inc. All other brands and names contained herein are the property of their respective owners.

References

1. Steemers, F. J., Gunderson, K. L. (2007) Whole genome genotyping technologies on the BeadArray platform. *Biotechnol J* 2, 41–49.
2. Fan, J. B., Gunderson, K. L., Bibikova, M., Yeakley, J. M., Chen, J., Wickham Garcia, E., Lebruska, L. L., Laurent, M., Shen, R., Barker, D. (2006) Illumina universal bead arrays. *Methods Enzymol* 410, 57–73.
3. Shen, R., Fan, J. B., Campbell, D., Chang, W., Chen, J., Doucet, D., Yeakley, J., Bibikova, M., Wickham Garcia, E., McBride, C., Steemers, F., Garcia, F., Kermani, B. G., Gunderson, K., Oliphant, A. (2005) High-throughput SNP genotyping on universal bead arrays. *Mutat Res* 573, 70–82.
4. Paynter, R. A., Skibola, D. R., Skibola, C. F., Buffler, P. A., Wiemels, J. L., Smith, M. T. (2006) Accuracy of multiplexed Illumina platform-based single-nucleotide polymorphism genotyping compared between genomic and whole genome amplified DNA collected from multiple sources. *Cancer Epidemiol Biomarkers Prev* 15, 2533–2536.
5. García-Closas, M., Malats, N., Real, F. X., Yeager, M., Welch, R., Silverman, D., Kogevinas, M., Dosemeci, M., Figueroa, J., Chatterjee, N., Tardón, A., Serra, C., Carrato, A., García-Closas, R., Murta-Nascimento, C., Rothman, N., Chanock, S. J. (2007) Large-scale evaluation of candidate genes identifies associations between VEGF polymorphisms and bladder cancer risk. *PLoS Genet* 3, e29.
6. Barker, D. L., Hansen, M. S., Faruqi, A. F., Giannola, D., Irsula, O.R., Lasken, R. S., Latterich, M., Makarov, V., Oliphant, A., et al. (2004) Two methods of whole-genome amplification enable accurate genotyping across a 2320-SNP linkage panel. *Genome Res* 14, 901–907.

Chapter 11

The Use of Maternal Plasma for Prenatal RhD Blood Group Genotyping

Kirstin Finning, Pete Martin, and Geoff Daniels

Abstract

Alloimmunization to the blood group antibody anti-RhD (anti-D) is the most common cause of hemolytic disease of the fetus and newborn. Knowledge of fetal D type in women with anti-D makes management of the pregnancy much easier and avoids unnecessary procedures in those women with a D-negative fetus. Fetal D typing can be performed by detection of an *RHD* gene in cell-free DNA in the plasma of D-negative pregnant women. The technology involves real-time quantitative polymerase chain reactions targeting exons 4, 5, and 10 of *RHD*, with the exons 4 and 10 tests performed as a multiplex. Testing for *SRY* in multiplex with the *RHD* exon 5 test provides an internal control for the presence of fetal DNA when the fetus is male. Fetal D typing has become the standard of care in England in pregnant women with a significant level of anti-D.

Key words: Blood groups, Rh, D, fetal testing, free fetal DNA, hemolytic disease of the fetus and newborn.

1. Introduction

1.1. Fetal RHD Genotyping for Assessing Risk of Hemolytic Disease of the Fetus and Newborn

Hemolytic disease of the fetus and newborn (HDFN) results from the immune destruction of fetal erythroid cells by maternal IgG antibodies, following their transfer across the placental barrier *(1)*. Alloimmunization against the RhD (D or RH1) red cell surface antigen was the commonest cause of fetal and neonatal morbidity and mortality prior to introduction of prophylaxis with RhD immunoglobulin (Ig) in the 1960s. Despite the success of anti-D immunoglobulin prophylaxis in substantially reducing the prevalence of HDFN caused by anti-D, some D-negative women

Peter Bugert (ed.), *DNA and RNA Profiling in Human Blood: Methods and Protocols, vol. 496*

DOI 10.1007/978-1-59745-553-4_11 Springerprotocols.com

still produce anti-D. In the UK alone about 500 fetuses develop HDFN each year, leading to 25–30 deaths *(2)*.

The most common use of molecular blood group genotyping is for predicting the D type of the fetus of a woman with anti-D, to assist in assessing the risk of HDFN *(3–5)*. If the fetus is D-positive, it is at risk from HDFN and the appropriate management of the pregnancy can be arranged; if it is D-negative, there is no risk and unnecessary invasive procedures can be avoided. The usual source of fetal DNA is now cell-free fetal (cff) DNA in maternal plasma. This fetal DNA represents 3–6% of free DNA in the mother's blood *(6)* and cannot be separated from the maternal DNA, creating complications in testing and providing suitable controls. The use of cff-DNA avoids the invasive procedures of amniocentesis and chorionic villus sampling. At the International Blood Group Reference Laboratory in England, we have developed a method for testing for *RHD* in cff-DNA and use this to provide a service for fetal blood group genotyping in alloimmunized pregnant women *(5, 7)*.

After anti-D, the next most common causes of HDFN are anti-K (KEL1) and anti-c (RH4), and anti-E (RH3) and anti-C (RH2) are occasionally implicated *(1)*. We have also developed methods for predicting K, c, C, and E phenotypes from testing cff-DNA in maternal plasma *(8)*, but these will not be described here.

1.2. Genetics of RhD

Rh phenotypes are controlled by *RHD* and *RHCE* a pair of homologous genes on chromosome 1, which have ten exons each and share about 94% identity *(9)*. In people of European origin, the D-negative phenotype almost always results from homozygosity for a complete deletion of *RHD (10)*. Most D-positive people are either homozygous or hemizygous for *RHD*. Variant haplotypes exist in which all or part of a *RHD* is present, but no D antigen is expressed on the red cells, and in which some or most of *RHD* is missing, yet some D epitopes are expressed *(9)*. Fortunately, most of these *RHD* variants are rare, but there is one, *RHDΨ*, that is particularly important because it is common in people of African origin *(11)*. *RHDΨ* contains all ten exons, but is inactive because of two inactivating mutations: a 37 bp duplication in exon 4 and a nonsense mutation in exon 6. There are also four characteristic single nucleotide changes in exons 4 and 5. Another abnormal gene that is relatively common in Africans and produces no D antigen, despite the presence of some *RHD* exons, is *RHD–CE–D*^{S}. This hybrid gene comprises exons 1 and 2, the 5′ end of exon 3, and exons 9 and 10 from *RHD*, and the 3′ end of exon 3 and exons 4–8 from *RHCE (12)*.

2. Materials

2.1. Maternal Blood Processing

1. Conical tubes, 15 mL (BD Biosciences, B9320 Erembodegem, Belgium, 352096).
2. Screwcap microtubes, 1.6 mL (BioQuote, York, UK 14201).
3. Aerosol-resistant disposable filter pipette tips (Anachem Ltd, Luton, Bedfordshire, UK, RT-1000F, RT-100F).

2.2. Extraction of DNA from Maternal Plasma

1. QIAamp DNA Blood Mini Kit (Qiagen, Hilden, Germany). Contains buffer AL, wash buffers AW1, AW2, QIAamp spin columns, lysis tubes and elution tubes (*see* **Note 1**). Store at room temperature.
2. Qiagen protease (Qiagen). Store at 4°C.
3. Collection tubes, 2 mL (Qiagen).
4. Ethanol, 100% (Sigma, Steinheim, Germany). Store at room temperature.
5. Aerosol-resistant disposable filter pipette tips (Anachem RT-1000F, RT-200F, RT-100F).
6. Microtubes, 1.5 mL (Anachem).

2.3. Real-Time Quantitative PCR

1. Primers and probes are listed in **Table 11.1** (also *see* **Note 2**). Reconstitute primers and probes with nuclease-free water to 1m*M* and store at –30°C. Make 5 μ*M* working solutions of each primer and probe and store at –30°C. Probes should be protected from light during storage.
2. TaqMan Universal PCR Master Mix 2× (Applied Biosystems, Foster City, CA, USA). Store at 4°C.
3. Nuclease-free water (Sigma). Aliquot and store at room temperature for up to 12 months.
4. Microtubes, 1.5 mL (Anachem).
5. Human male genomic DNA (Promega, Madison, USA). Dilute in nuclease-free water to 20 ng/μL and store at –30°C in 20 μL aliquots. This is DNA prepared from pooled male blood. It will provide the DNA for standard curve generation and will also be the "positive control" (D POS) for the *RHD* exons and *SRY*.
6. Control DNAs (prepared in house from whole blood). Dilute to and store at the following concentrations: female D-negative (2 ng/μL), the "negative control" (D NEG) for all *RHD* exons and *SRY*; *RHDΨ*-positive DNA (0.1 ng/μL), a negative control (*RHDΨ*) for *RHD* exon 4 and 5.
7. Optical reaction PCR plates, 96-well (Applied Biosystems) and MicroAmp Optical Caps (Applied Biosystems). Store at room temperature in sealed plastic bags.
8. Aerosol-resistant disposable filter pipette tips (Anachem RT-1000F, RT-200F, RT-100F, RT-10GF).

Table 11.1
Primers and probes for real-time PCR detection of DNA in maternal plasma

Name	Sequence (5′ to 3′)	Specificity
RHDEX4_F	CTG CCA AAG CCT CTA CAC G	*RHD*
RHDEX4_R	ATG GCA GAC AAA CTG GGT GTC	*RHD*
RHDEX4_P	(FAM)-TTG CTG TCT GAT CTT TAT CCT CCG TTC CCT-(TAMRA)	*RHD*
RHDEX5_F	CGC CCT CTT CTT GTG GAT G	*RHD*
RHDEX5_R	GAA CAC GGC ATT CTT CCT TTC	*RHD*
RHDEX5_P	(FAM)-TCT GGC CAA GTT TCA ACT CTG CTC TGC T-(TAMRA)	*RHD*
RHDEX10_F	CCT CTC ACT GTT GCC TGC ATT	*RHD, RHDΨ*
RHDEX10_R	AGT GCC TGC GCG AAC ATT	*RHD, RHDΨ*
RHDEX10_P	(VIC)-TAC GTG AGA AAC GCT CAT GAC AGC AAA GTC T-(TAMRA)	*RHD, RHDΨ*
SRY_F	TGG CGA TTA AGT CAA ATT CGC	*SRY*
SRY_R	CCC CCT AGT ACC CTG ACA ATG TAT T	*SRY*
SRY_P	(VIC)-AGC AGT AGA GCA GTC AGG GAG GCA GA-(TAMRA)	*SRY*
CCR5_F	TAC CTG CTC AAC CTG GCC AT	*CCR5*
CCR5_R	TTC CAA AGT CCC ACT GGG C	*CCR5*
CCR5_P	(FAM)-TTT CCT TCT TAC TGT CCC CTT CTG GGC TC-(TAMRA)	*CCR5*

3. Methods

DNA is extracted from maternal plasma and amplified by real-time PCR to detect the presence of fetal *RHD* and/or *SRY* genes if the baby is D-positive and/or male. Three exons of the *RHD* gene are targeted: the primers for exons 4 and 5 are designed to amplify only the *RHD* gene and not *RHDΨ*; the primers for exon 10 will amplify both genes (*see* **Note 3**). The amplification of a ubiquitous gene, *CCR5*, is used to confirm the successful extraction of maternal DNA, to quantify the amount of maternal DNA in the sample and to highlight the potential presence of maternal silent *RHD* genes. The fetus is predicted to be D-positive or D-negative depending on the cycle threshold (C_t) values obtained for the *RHD* assays (*see* **Note 4**). Control wells are run concurrently to ensure the assays are performing optimally.

Gloves should be worn at all times when handling samples and reagents and should be changed frequently between steps and if they become contaminated. Pipette tips should be changed between liquid transfers to avoid cross-contamination. For the same reason, open only one tube at a time. Briefly pulse spin microtubes after mixing steps to remove droplets from the top of the tube and the tube lid.

3.1. Maternal Blood Processing (see Note 5)

1. Two 6 mL EDTA blood tubes are pooled into a 15 mL tube and centrifuged at 4,000 × *g* for 10 min.
2. The plasma layer is carefully removed to a new 15 mL tube leaving at least 1 mL plasma above the buffy coat. Mix the transferred plasma by inverting the tube several times, then transfer 1 mL aliquots of plasma into labeled 1.6 mL microtubes.
3. Centrifuge microtubes at 4,600 × *g* for 10 min in a microfuge.
4. Remove 800 μL aliquots of plasma into fresh microtubes without disturbing the cell pellet. Aliquots of plasma may either be used immediately or stored at −30°C until required.

3.2. DNA Extraction from Maternal Plasma

1. Use one 800 μL plasma aliquot per patient and allow to thaw at room temperature if frozen (*see* **Note 6**).
2. Add 60 μL Qiagen protease to the plasma and mix by vortexing.
3. Add 400 μL buffer AL to each of two 1.5 mL (lysis) tubes, then add 430 μL of the plasma/protease mix to each tube. Mix by vortexing.
4. Incubate tubes in a waterbath at 56°C for 15 min.
5. Add 400 μL ethanol to each tube and mix by vortexing. Leave for 5 min at room temperature.
6. Using one spin column per patient, load 600 μL lysate to the spin column and microcentrifuge for 1 min at ≥6,000 × *g*.
7. Remove spin column to new 2 mL collection tube and repeat step 6 until all the lysate has been spun through the column (four centrifugation spins in total). Remove spin column to a new collection tube.
8. Add 500 μL wash buffer AW1 to the column and centrifuge at full speed (15,000–20,800 × *g*) for 1 min.
9. Remove spin column to a new collection tube and add 500 μL wash buffer AW2. Centrifuge at full speed (15,000–20,800 × *g*) for 1 min.
10. Remove spin column to a new collection tube and centrifuge at full speed (15,000–20,800 × *g*) for 3 min to dry the column completely.
11. Remove spin column to a new 1.5 mL (elution) tube. Add 55 μL nuclease-free water directly to the center of the spin column membrane. Close the lid and incubate for 1 min

at room temperature. Centrifuge at full speed (15,000–20,800 × *g*) for 1 min to elute the DNA. Discard spin column and close lid of elution tube. The DNA can be used directly in PCR or stored for several hours at 4°C or on ice. Freezing the DNA once extracted is not recommended as this may result in loss of fetal DNA.

3.3. Real-Time PCR

1. Prepare DNA standards for standard curve generation (*see* **Note 7**).
   a. Thaw an aliquot of male human DNA (20 ng/μL; standard 1) at room temperature.
   b. Make a standard curve DNA dilution series. Add 45 μL nuclease-free water to each of three 1.5 mL tubes.
   c. Add 5 μL standard 1 to one tube containing 45 μL water to make standard 2 (2 ng/μL). Mix well.
   d. Add 5 μL standard 2 to one tube containing 45 μL water to make standard 3 (200 pg/μL). Mix well.
   e. Add 5 μL standard 3 to one tube containing 45 μL water to make standard 4 (20 pg/μL). Mix well.
2. Thaw aliquots of the negative controls (D NEG and *RHD*Ψ) at room temperature.
3. Prepare PCR reaction mixes:
   a. Prepare the *RHD* exon 4 + 10 reaction mix. Four × 20 μL of each reaction mix is needed per patient sample, plus 14 × 20 μL for controls. The final reaction volume should be increased by 10% to allow for loss during pipetting. A bulk mix should be made by multiplying up for the required volume (e.g., for one patient sample, make a 20 × 20 μL bulk mix). For 1 × 20 μL *RHD* exon 4 + 10 reaction mix the following reagents should be combined in a clean 1.5 mL tube and mixed by vortexing: 12.5 μL TaqMan Universal PCR Master Mix, 1.5 μL *RHD* exon 4 forward primer (5 μ*M*), 1.5 μL *RHD* exon 4 reverse primer (5 μ*M*), 1.5 μL *RHD* exon 10 forward primer (5 μ*M*), 1.5 μL *RHD* exon 10 reverse primer (5 μ*M*), 0.25 μL *RHD* exon 4 probe (5 μ*M*), 0.25 μL *RHD* exon 10 probe (5 μ*M*), 1.0 μL nuclease-free water.
   b. Prepare the *RHD* exon 5 + *SRY* reaction mix, making a bulk mix as described in step 3a. For 1 × 20 μL *RHD* exon 5 + *SRY* reaction mix, the following reagents should be combined in a clean 1.5 mL tube and mixed by vortexing: 12.5 μL TaqMan Universal PCR Master Mix, 1.0 μL *RHD* exon 5 forward primer (5 μ*M*), 1.0 μL *RHD* exon 5 reverse primer (5 μ*M*), 1.5 μL *SRY* forward primer (5 μ*M*), 1.5 μL *SRY* reverse primer (5 μ*M*), 0.25 μL *RHD* exon 5 probe (5 μ*M*), 0.5 μL *SRY* probe (5 μ*M*), 1.75 μL nuclease-free water.

c. Prepare the *CCR5* reaction mix. Reaction mix of 2 × 20 μL is required for each patient sample. In addition, 12 × 20 μL mix is needed for the DNA standard curve. The final reaction volume should be increased by 10% to allow for loss during pipetting. A bulk mix should be made by multiplying up for the required volume (e.g., for one patient sample, make a 16 × 20 μL bulk mix). For 1 × 20 μL *CCR5* reaction mix the following reagents should be combined in a clean 1.5 mL tube and mixed by vortexing: 12.5 μL TaqMan Universal PCR Master Mix, 1.0 μL *CCR5* forward primer (5 μ*M*), 1.0 μL *CCR5* reverse primer (5 μ*M*), 0.5 μL *CCR5* probe (5 μ*M*), 5.0 μL nuclease-free water.

4. Prepare the 96-well reaction plate: Load 20 μL of the relevant PCR reaction mixes into the wells of a 96-well PCR plate; a suggested plate plan showing allocation of wells for each patient sample, the controls, and standards for each of the assays is indicated in Table 11.2 (*see* **Note 8**).
   a. Load 5 μL of maternal plasma DNA to the wells assigned to patient test samples.
   b. Load 5 μL control DNA and standard DNA to the appropriate wells.
   c. Load 5 μL nuclease-free water to each no-template control (NTC) well.
5. Seal plate and centrifuge at 1,000 rpm for 1 min to remove air-bubbles and to ensure all liquid is pooled in bottom of the wells.
6. Run the PCR using the following reaction conditions: 50°C for 2 min; 95°C for 10 min; 45 cycles with 95°C for 15 s and 60°C for 1 min.

3.4. Interpretation of Results

1. Analyze the reaction (*see* **Note 9**): A negative result (no PCR amplification) is indicated by a C_t value of 45. A positive result (PCR product produced) is indicated by a C_t value <42.
2. Control and quality validation:
   a. Negative control and NTC: not more than one replicate should have a C_t value <42. If more than one negative control or more than one NTC control has a C_t <42, this could indicate the PCR is contaminated; the assay has failed and results are unreliable (*see* **Note 10**).
   b. Positive controls: all replicates should have C_ts <42 for all *RHD* exons and for *SRY*, otherwise this could indicate that the reaction is not performing optimally and results will be unreliable.
   c. The slope of the standard curve should be between –3.5 and –4.5 for accurate DNA quantitation. The predicted mean amount of maternal plasma DNA per well should be <10 ng. If it is higher than this, the plasma contains

Table 11.2
PCR plate (96-well) layout showing the suggested allocation of wells for patient samples, controls, and DNA quantitation standards. Each filled well contains 20 μL PCR reaction mix, plus 5 μL test, control, or standard DNA, or water

	1	2	3	4	5	6	7	8	9	10	11	12
A	*RHD* ex4/10 Patient 1	*RHD* ex4/10 Patient 1	*RHD* ex4/10 Patient 1	*RHD* ex4/10 Patient 1	*RHD*ex5/ SRY Patient 1	*RHD*ex5/ SRY Patient 1	*RHD*ex5/ SRY Patient 1	*RHD*ex5/ SRY Patient 1		CCR5 Patient 1	CCR5 Patient 1	
B	*RHD* ex4/10 Patient 2	*RHD* ex4/10 Patient 2	*RHD* ex4/10 Patient 2	*RHD* ex4/10 Patient 2	*RHD*ex5/ SRY Patient 2	*RHD*ex5/ SRY Patient 2	*RHD*ex5/ SRY Patient 2	*RHD*ex5/ SRY Patient 2		CCR5 Patient 2	CCR5 Patient 2	
C	*RHD* ex4/10 Patient 3	*RHD* ex4/10 Patient 3	*RHD* ex4/10 Patient 3	*RHD* ex4/10 Patient 3	*RHD*ex5/ SRY Patient 3	*RHD*ex5/ SRY Patient 3	*RHD*ex5/ SRY Patient 3	*RHD*ex5/ SRY Patient 3		CCR5 Patient 3	CCR5 Patient 3	
D	*RHD* ex4/10 Patient 4	*RHD* ex4/10 Patient 4	*RHD* ex4/10 Patient 4	*RHD* ex4/10 Patient 4	*RHD*ex5/ SRY Patient 4	*RHD*ex5/ SRY Patient 4	*RHD*ex5/ SRY Patient 4	*RHD*ex5/ SRY Patient 4		CCR5 Patient 4	CCR5 Patient 4	
E	*RHD* ex4/10 D NEG	*RHD* ex4/10 D NEG	*RHD* ex4/10 D NEG	*RHD* ex4/10 D NEG	*RHD*ex5/ SRY D NEG	*RHD*ex5/ SRY D NEG	*RHD*ex5/ SRY D NEG	*RHD*ex5/ SRY D NEG				
F	*RHD* ex4/10 RHDΨ	*RHD* ex4/10 RHDΨ	*RHD* ex4/10 RHDΨ	*RHD* ex4/10 RHDΨ	*RHD*ex5/ SRY RHDΨ	*RHD*ex5/ SRY RHDΨ	*RHD*ex5/ SRY RHDΨ	*RHD*ex5/ SRY RHDΨ				
G	*RHD* ex4/10 D POS	*RHD* ex4/10 D POS	*RHD* ex4/10 D POS	*RHD* ex4/10 NTC	*RHD* ex4/10 NTC	*RHD* ex4/10 NTC	*RHD*ex5/ SRY D POS	*RHD*ex5/ SRY D POS	*RHD*ex5/ SRY D POS	*RHD*ex5/ SRY NTC	*RHD*ex5/ SRY NTC	*RHD*ex5/ SRY NTC
H	CCR5 Standard 1	CCR5 Standard 1	CCR5 Standard 2	CCR5 Standard 2	CCR5 Standard 3	CCR5 Standard 3	CCR5 Standard 4	CCR5 Standard 4	CCR5 Standard 4	CCR5 NTC	CCR5 NTC	CCR5 NTC

excessive amounts of maternal DNA, which could interfere with detection of fetal DNA (*see* **Note 11**).

3. Check the amplification plots for each *RHD* exon, *SRY*, and *CCR5* for patient samples. The fetus is predicted to be D-positive if at least 2/4 replicates for *RHD* exons 4, 5, and 10 are positive (i.e., C_t value <42), with an additional three positive replicates from any of the *RHD* exons (meaning that a total of at least 9/12 replicates must be positive) (**Fig. 11.1** and *see* **Note 12**).

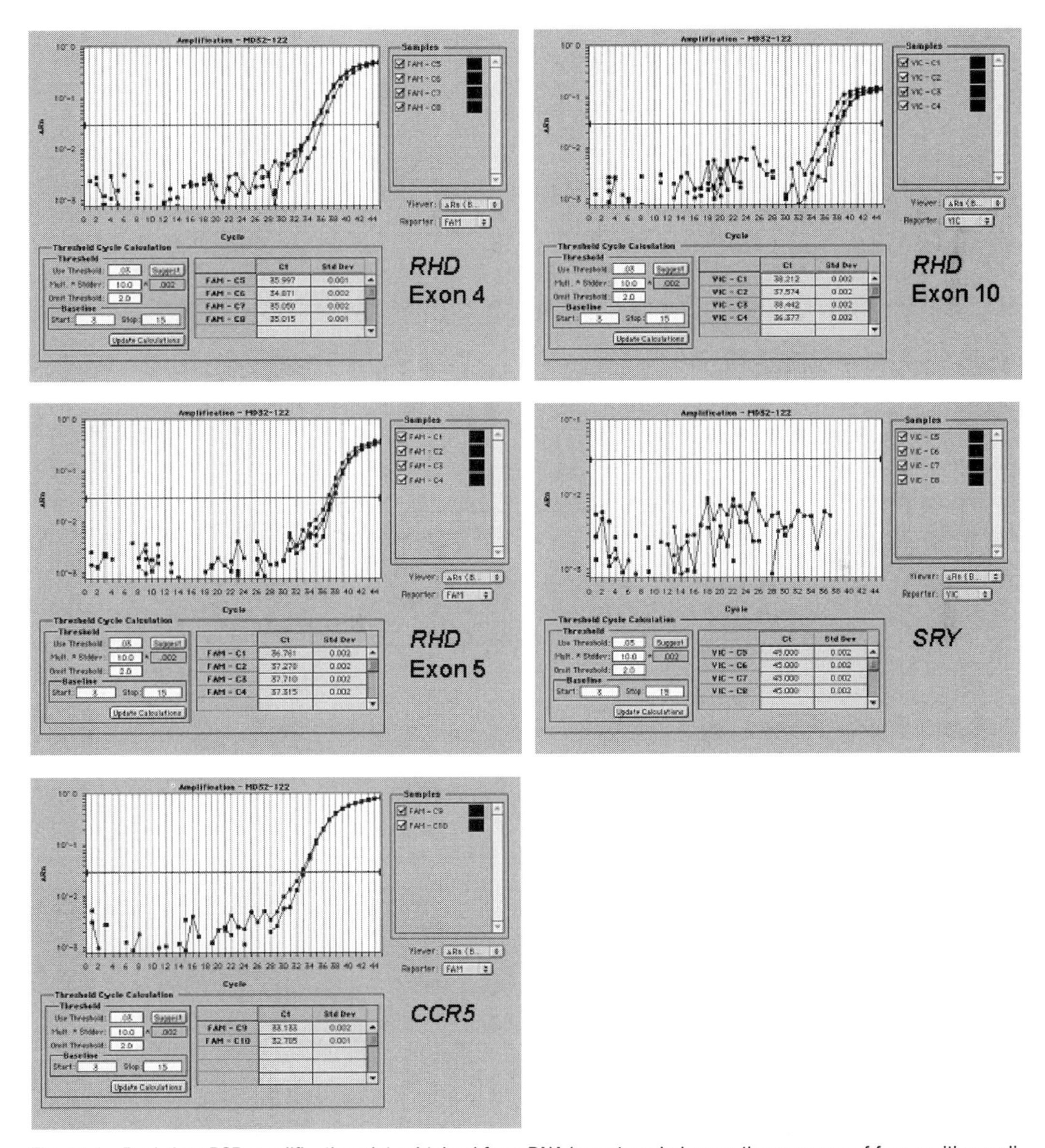

Fig. 11.1. Real-time PCR amplification plots obtained from DNA in maternal plasma; the presence of four positive replicates for each of *RHD* exons 4, 5, and 10, and the lack of positive replicates for *SRY*, indicate that the fetus is D-positive (and female).

4. If, however, the difference between any C_t value for *RHD* exon 10 and *CCR5* (C_t exon 10 – C_t *CCR5*) is less than 2, this could indicate that the mother has an unexpressed *RHD* gene and the test results may be unreliable (**Fig. 11.2**). Further investigation of the maternal genome should be carried out using the frozen buffy coat (*see* **Note 13**).

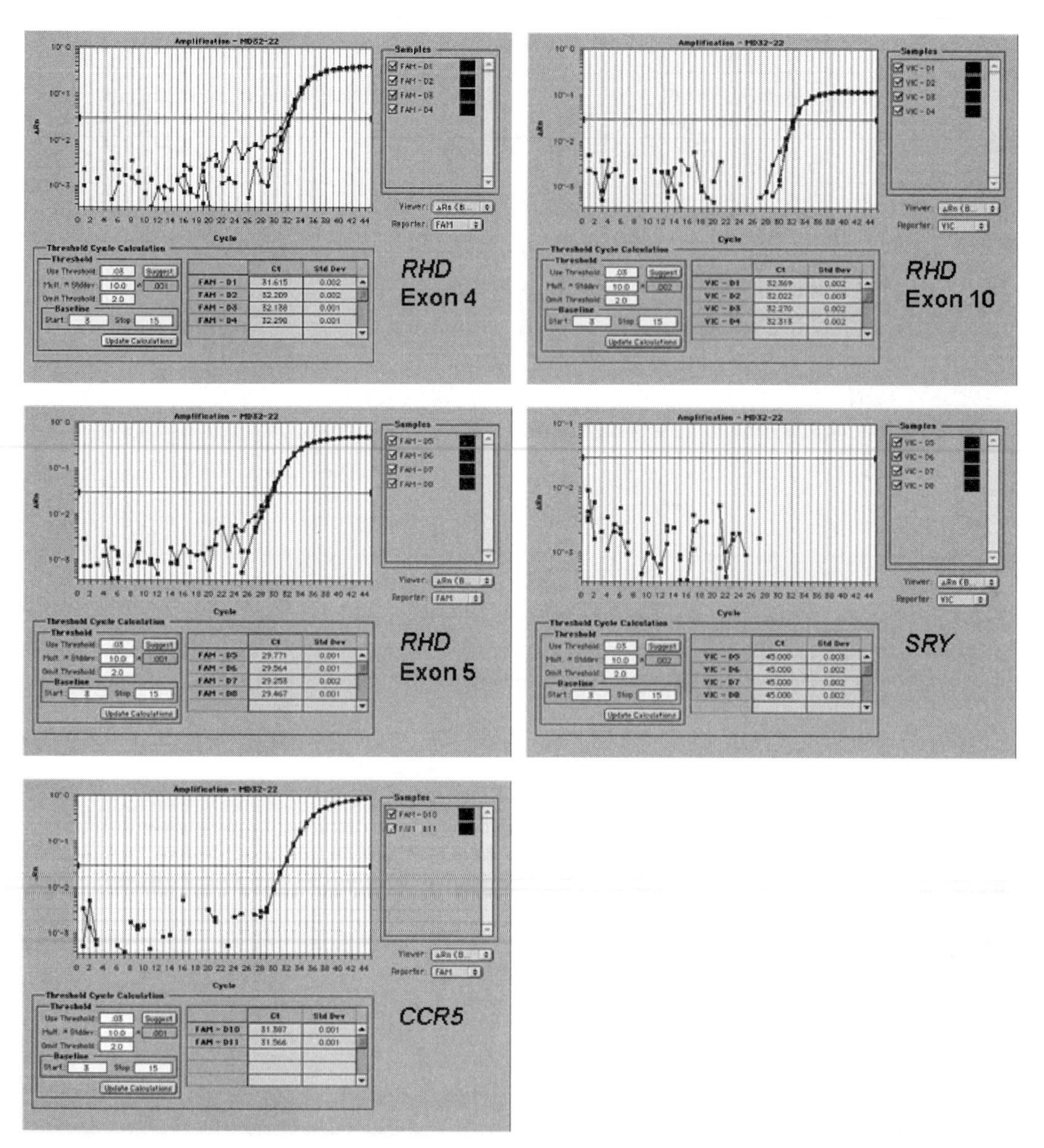

Fig. 11.2. Real-time PCR amplification plots obtained when the mother has an unexpressed *RHD* gene. If the difference in C_t values between *RHD* exon 10 and *CCR5* (C_t exon 10 – C_t *CCR5*) is less than 2, this could identify the presence of a maternal silent gene (either *RHD*Ψ, *RHD-CE-D*^{S}, or a D-variant *RHD* gene). In the example shown, low C_t values were obtained for all *RHD* exons suggesting that the mother has a largely intact *RHD* gene; further investigation showed her to have a weak D phenotype and therefore it was not possible to determine the fetal *RHD* genotype from maternal plasma.

5. The fetus can be predicted to be D-negative if no replicates, or only 1/12 replicates, are positive. The remaining 11/12 replicates should be negative (i.e., C_t values of 45) (**Fig. 11.3** and *see* **Note 14**). If the fetus is predicted to be D-negative, the *SRY* result may be used to confirm the presence of fetal DNA if the fetus is male. We recommend that at least 2/4 replicates of *SRY* should be positive (C_t values <42) to confirm the presence of fetal DNA (*see* **Note 15**).

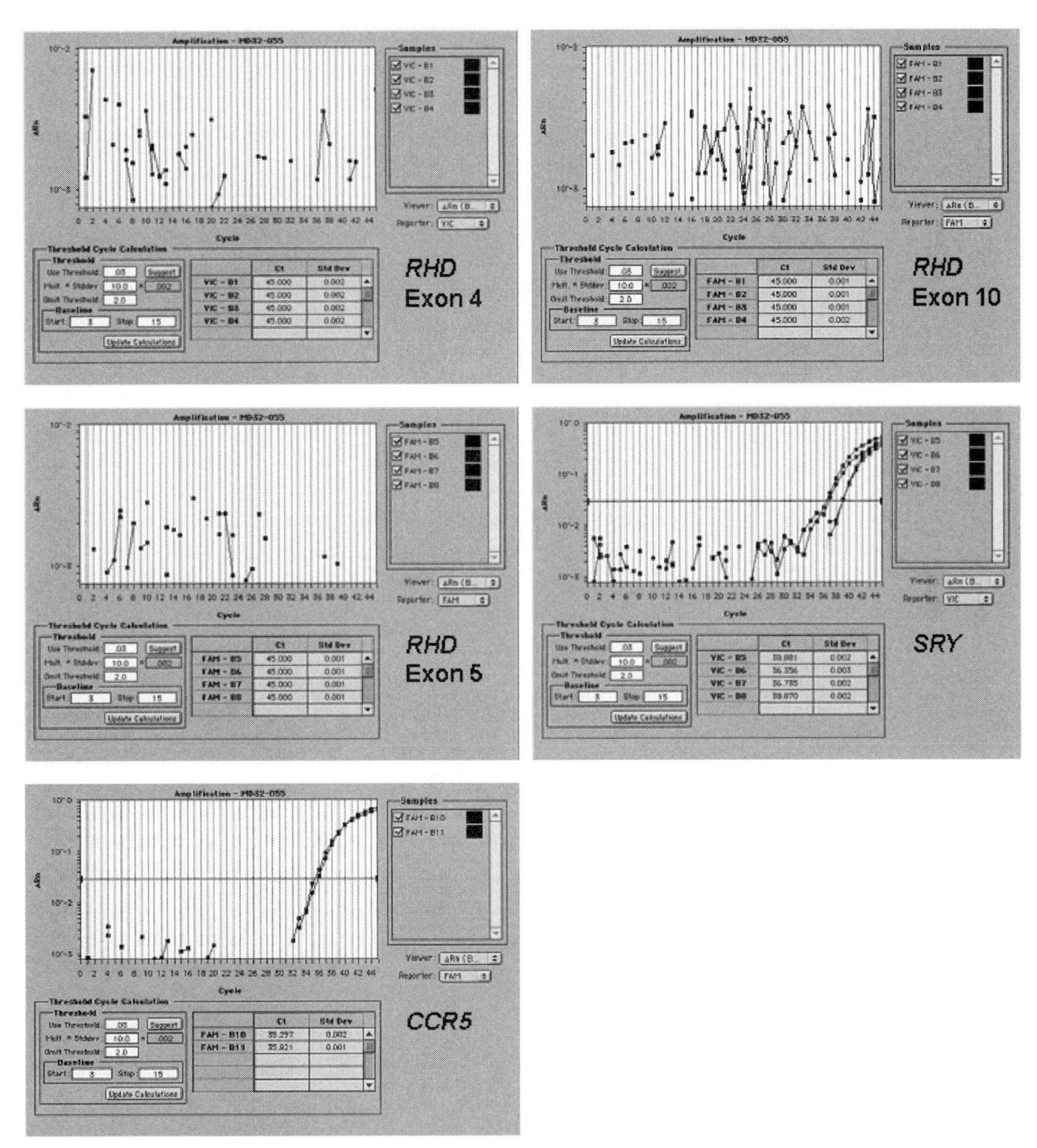

Fig. 11.3. Real-time PCR amplification plots obtained from DNA in maternal plasma. The lack of any positive replicates for each of *RHD* exons 4, 5, and 10, but the positive replicates for *SRY*, indicate that the fetus is D-negative (and male).

6. If *RHD* exon 10 replicates are positive, but exons 4 and 5 are negative, this could indicate that the fetus has a D-variant *RHD* gene, has *RHD*Ψ, or has the hybrid *RHD-CE-D*^{S} gene.

4. Notes

1. The kit contains lysis, elution, and collection tubes. However, extra amounts of these are required to deal with the larger volume of plasma than that intended in the manufacturer's protocol. Sterile nuclease-free 1.5 mL microtubes can be used in place of the lysis and elution tubes. Additional collection tubes are available from Qiagen. It is not advisable to reuse the collection tubes as eluate may be transferred to the bottom of the spin column, which can result in a reduction in DNA yield.
2. We suggest using HPLC or PAGE purified primers, as they give an increased batch-to-batch consistency of performance. FAM-labeled probes can be purchased from a variety of sources (e.g., Sigma-Genosys); VIC-labeled probes are only available from Applied Biosystems. In our experience, the performance of FAM and VIC labeled probes in multiplex reactions is superior to that of probes labeled with other dyes.
3. Fetal *RHD* genotyping may also be accomplished by targeting alternative exons, exon 7 being a popular choice *(4)*. However, if the test population were likely to contain individuals of black African descent, we would strongly recommend that an assay, such as that for exon 5 as described here, is also used to prevent false-positive results arising from the presence of *RHD*Ψ.
4. TaqMan real-time PCR relies upon PCR primers to define the specificity of the reaction and a probe with reporter and quencher dyes attached. If the target DNA sequence is present, the increase in PCR product formation is monitored by the increase in probe reporter dye fluorescence throughout each cycle and converted by the software into an amplification plot (increase in reporter dye fluorescence versus PCR cycle). The cycle at which the reporter dye reaches a threshold level of fluorescence (C_t) is dependent on the starting amount of target DNA present. The more target DNA is present in a sample at the start of PCR, the lower the C_t value.
5. Maternal blood should be processed within 48 h of venipuncture to reduce the amount of maternal DNA in the plasma resulting from breakdown of maternal white cells. At each stage, the centrifuge settings should be set to minimize physical disruption to the blood, i.e., slow speed increase and no brake on slowdown. After the initial centrifugation, any sign

of hemolysis should be noted as this may reduce the quality of the DNA. Care should be taken not to disturb the buffy coat layer when transferring tubes from the centrifuge and when withdrawing plasma. It is advisable to take and store at –30°C a small amount of maternal buffy coat (e.g., 50 μL buffy coat mixed with 150 μL water) for later analysis of maternal genome if necessary.

6. The plasma should be completely thawed before use. If a cryoprecipitate is seen in the plasma, centrifuge the tube at 4,600 × *g* for 1 min to pellet the precipitate and remove the plasma to a fresh tube.
7. The standard DNA should be made up just prior to setting up the PCR and discarded after each assay set up. Great care should be taken not to contaminate reagents, test DNA, or PCR plate wells with control DNA or standard DNA.
8. Four patient samples are comfortably accommodated on one 96-well plate; five may be fitted on if non-consecutive wells are used.
9. Using the ABI 7700 or ABI 7900 sequence detector the baseline should be set between 3 and 15 cycles and the threshold at 0.03 (ABI 7700) or 0.06 (ABI 7900) for both FAM and VIC dyes. These settings may not be suitable for other real-time PCR machines; the correct baseline and threshold settings need to be established in-house and measurements must be taken during the exponential phase of the PCR. The settings will affect the C_t values of the test results and possible interpretation of the results.
10. Theoretically all negative control and NTC wells should have a C_t of 45 (i.e., no PCR amplification) because the target DNA sequence was not present In practice, however, it is almost impossible to prevent fluorescent signal from being detected in these wells every time (possibly due either to contamination of very small amounts of airborne DNA or artifacts causing non-specific amplification or fluorescence). If contamination of the reaction is suspected, working stocks of reagents should be discarded and fresh stocks made. If necessary, purchase new reagents, primers, probes, etc.
11. We use a maximum cut-off value of 10ng DNA per well (equivalent to 125 ng DNA per mL maternal plasma). Although the assay may still be viable if the maternal DNA concentration is greater than this, if the blood is received within 48 h of venipunture the DNA concentration rarely exceeds this.
12. For positive results, it is important to ensure that all exons show some amplification; this will help prevent mis-typing of rare *RHD* variant genes which have some *RHD* exons replaced by the corresponding exon of the homologous *RHCE* gene.

13. DNA could be extracted from the maternal buffy coat and various PCRs (and possibly DNA sequencing) carried out to elucidate the nature of the maternal *RHD* gene. If the mother is thought to carry *RHDΨ* or an *RHDVI* variant gene, it may still be possible to genotype the fetus. In our laboratory, when this situation occurs, we extract DNA from a further aliquot of plasma and perform real-time PCR for *RHD* exon 5 and *SRY* (as well as *CCR5* as described) using eight replicates per patient. The fetus is predicted to be D-positive if at least 5/8 replicates are positive for *RHD* exon 5, and D-negative if zero or one replicate is positive.
14. In practice, if between two and four replicates (not more than two from any exon) are positive, we repeat the assay using freshly prepared plasma DNA. If after the repeat assay, the total combined number of positive replicates from the two assays is ≤ 4, the fetus can be predicted to be D-negative.
15. Detection of *SRY* is not an ideal control to confirm the presence of fetal DNA in maternal plasma, as it is only applicable for male fetuses. Detection of paternally inherited biallelic insertion/deletion polymorphisms which are not present in the maternal genome may be used to indicate presence of fetal DNA *(13)*. However, unless the assay for the paternally inherited marker and for the *RHD* are multiplexed and tested in the same tube, there is still the possibility that a paternally inherited marker may be detected but *RHD* remains undetected. At present there is no universal marker to indicate the presence of fetal DNA. It is possible that in the future difference in the methylation status of maternal and fetal genes may be used to provide a fetal control *(14)*.

References

1. Klein, H. G., Anstee, D. J. (2005) *Blood Transfusion in Clinical Medicine, 11th edition.* Blackwell Publishing, Oxford, UK.
2. National Institute for Clinical Excellence (2002) *Technology Appraisal Guidance 41. Guidance on the Use of Routine Antenatal Anti-D Prophylaxis for RhD-Negative Women.* NICE, London, UK.
3. van der Schoot, C. E., Tax, G. H. M., Rijnders, R. J. P., de Haas, M., Christiaens, G. C. M. L. (2003) Prenatal typing of Rh and Kell blood group system antigens: the edge of a watershed. *Transfus Med Rev* 17, 31–44.
4. Daniels, G., Finning, K., Martin, P., Soothill, P. (2004) Fetal blood group genotyping from DNA from maternal plasma: an important advance in the management and prevention of haemolytic disease of the fetus and newborn. *Vox Sang* 87, 225–232.
5. Daniels, G., Finning, K., Martin, P., Summers, J. (2006) Fetal blood group genotyping. Present and future. *Ann NY Acad Sci* 1075, 88–95.
6. Lo, Y. M., Tein, M. S. C., Lau, T. K., Haines, C. J., Leung, T. N., Poon, P. M. K., et al. (1998) Quantitative analysis of fetal DNA in maternal plasma and serum: implications for noninvasive prenatal diagnosis. *Am J Hum Genet* 62, 768–775.
7. Finning, K. M., Martin, P. G., Soothill, P. W., Avent, N. D. (2002) Prediction of fetal D status from maternal plasma: introduction of a new noninvasive fetal *RHD* genotyping service. *Transfusion* 42, 1079–1085.
8. Finning, K., Martin, P., Summers, J., Daniels, G. (2007) Fetal genotyping for the K (Kell) and Rh C, c, and E blood groups on cell-free

fetal DNA from maternal plasma. *Transfusion* 47, 2126–2133.
9. Daniels, G. (2002) *Human Blood Groups, 2nd edition*. Oxford: *Blackwell Science*.
10. Colin, Y., Chérif-Zahar, B., Le Van Kim, C., Raynal, V., Van Huffel, V., Cartron, J-P. (1991) Genetic basis of the RhD-positive and RhD-negative blood group polymorphism as determined by Southern analysis. *Blood* 78, 2747–2752.
11. Singleton, B. K., Green, C. A., Avent, N. D., Martin, P. G., Smart, E., Daka, A., et al. (2000) The presence of an *RHD* pseudogene containing a 37 base pair duplication and a nonsense mutation in most Africans with the Rh D-negative blood group phenotype. *Blood* 95, 12–18.
12. Faas, B. H. W., Beckers, E. A. M., Wildoer, P., Ligthart, P. C., Overbeeke, M. A., Zondervan, H. A., et al. (1997) Molecular background of VS and weak C expression in blacks. *Transfusion* 37, 38–44.
13. Page-Christiaens, G. C., Bossers, B., van der Schoot, C. E., De Haas, M. (2006) Use of bi-allelic insertion/deletion polymorphisms as a positive control for fetal genotyping in maternal blood: first clinical experience. *Ann NY Acad Sci* 1075, 123–129.
14. Chan, K. C., Ding, C., Gerovassili, A., Yeung, S. W., Chiu, R. W., Leung, T. N., et al. (2006) Hypermethylated *RASSF1A* in maternal plasma: A universal fetal DNA marker that improves the reliability of noninvasive prenatal diagnosis. *Clin Chem* 52, 2211–2218.

Part II

RNA Profiling in Blood Cells

Chapter 12

Nanoliter High-Throughput PCR for DNA and RNA Profiling

Colin J. H. Brenan, Douglas Roberts, and James Hurley

Abstract

The increasing emphasis in life science research on utilization of genetic and genomic information underlies the need for high-throughput technologies capable of analyzing the expression of multiple genes or the presence of informative single nucleotide polymorphisms (SNPs) in large-scale, population-based applications. Human disease research, disease diagnosis, personalized therapeutics, environmental monitoring, blood testing, and identification of genetic traits impacting agricultural practices, both in terms of food quality and production efficiency, are a few areas where such systems are in demand. This has stimulated the need for PCR technologies that preserves the intrinsic analytical benefits of PCR yet enables higher throughputs without increasing the time to answer, labor and reagent expenses and workflow complexity. An example of such a system based on a high-density array of nanoliter PCR assays is described here. Functionally equivalent to a microtiter plate, the nanoplate system makes possible up to 3,072 simultaneous end-point or real-time PCR measurements in a device, the size of a standard microscope slide. Methods for SNP genotyping with end-point TaqMan PCR assays and quantitative measurement of gene expression with SYBR Green I real-time PCR are outlined and illustrative data showing system performance is provided.

Key words: 5′-exonuclease assay, TaqMan PCR, SNP genotyping, nanofluidic, high-throughput genotyping, SYBR Green I real-time PCR, nanoliter PCR, quantitative PCR.

1. Introduction

The shift in life science research from *de novo* discovery to utilization of genetic and genomic information underlies the need for high throughput technologies capable of analyzing the expression of multiple genes or the presence of informative single nucleotide polymorphisms (SNPs) across large numbers of specimens. Human disease research *(1–3)*, disease diagnosis *(4, 5)*, environmental monitoring and testing *(6)*, blood testing *(7, 8)*, and identification of genetic traits impacting agricultural practices

Peter Bugert (ed.), *DNA and RNA Profiling in Human Blood: Methods and Protocols, vol. 496*

DOI 10.1007/978-1-59745-553-4_12 Springerprotocols.com

(9–11), both in terms of food quality and production efficiency, are a few areas where such systems are in demand.

The polymerase chain reaction (PCR) has become the standard for detection and quantification of targeted nucleic acid sequences in a sample. Although strategies for increasing PCR throughput are varied *(12–15)*, common to each is the goal of preserving the intrinsic analytical benefits of PCR yet at the same time enabling more PCR analyses to be performed without an increase in labor and reagent expenses, analysis time, and workflow complexity.

Described here is an approach based on the simultaneously processing of PCR assays in a high density, rectilinear array of etched 3,072 through-holes in a thin stainless steel platen (25 mm × 75 mm x 0.3 mm) with polymer coatings engineered to make the inside surface of each hole hydrophilic and PCR-compatible and the exterior surface of the plate hydrophobic *(16–18)*. The through-holes are grouped in 48 subarrays of 64 holes each and spaced on a 4.5 mm pitch equal to that of wells in a 384-well microplate (**Fig. 12.1**). Each 33 nanoliter hole in the nanotiter plate is a separate and fluidically isolated container into which primer or primer-probe pairs are deposited in such a way that each through-hole in the plate can support a different PCR assay. Once a platen is fully populated with primer or primer–probe pairs, the solvent is evaporated in a controlled manner leaving the primers immobilized on the inside surface of each through-hole. The primer/primer-probe deposition process and quality control is performed at BioTrove Inc., the nanotiter plate supplier, rather than at the user's site. The company provides plates to the user pre-loaded with primers for their application.

The nanoplate system supports TaqMan or SYBR Green I PCR chemistry for end-point analyses and SYBR Green I PCR chemistry for real-time quantitative PCR (qPCR). A key advantage of this system is its compatibility with commercially available DNA or RNA preparation methods developed for PCR in microplates. The analysis workflow (**Fig. 12.2**) begins with the samples arrayed in a microplate, mixed with off-the-shelf PCR reagents and then transferred into each nanoplate subarray (or a portion of a subarray) by a block of automated 48 pipette tips. The high accuracy and precision of fluid dispensing with this method is based on the combination of capillary action and the precise dimensions of the through-holes etched in the stainless steel platen. A great deal of flexibility in experimental design is inherent to this fluid transfer scheme, allowing a large number of assay–sample combinations to be analyzed starting with one sample per nanoplate and 3,072 PCR assays interrogating that sample or 144 samples per nanoplate and 16 PCR assays against each sample.

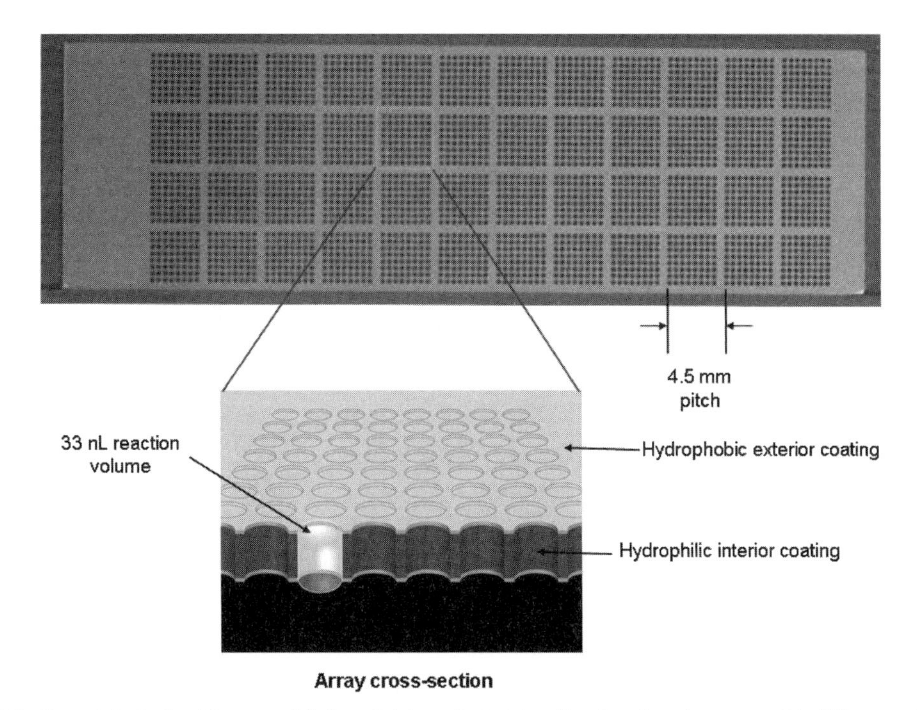

Fig. 12.1. Nanoplate design. The nanoplate is a stainless steel platen, the size of a microscope slide (25 mm × 75 mm × 0.3 mm) with a rectilinear array of 3,072 micro-machined, 320 m diameter holes of 33 nL each. The 48 groups of 64 holes are spaced at 4.5 mm to match the pitch of the wells in a 384-well microplate. A PCR compatible poly(ethylene glycol) hydrophilic layer is covalently linked to the interior surface of each hole and a hydrophobic fluoroalkyl layer is bonded to the exterior surface of the platen. The surface tension differential between the interior and exterior surfaces results in the isolated retention in each through-hole of PCR reagents and sample introduced into the nanotiter plate.

Once loaded with sample and PCR reagents, the platen is inserted into a glass-walled cassette containing an immiscible perfluorinated liquid (Fluorinert^{TM}) to prevent evaporation during thermal cycling. The cassette is hermetically sealed with UV curable epoxy and is ready for thermal cycling. The workflow at this point bifurcates depending on whether the application requires end-point or real-time PCR. Blood typing by SNP genotyping of adult human blood samples with 64 SNPs by TaqMan end-point PCR is an example of DNA profiling. DNA is isolated and purified from each sample using a standard DNA prep kit for whole blood. Measurement of 64 SNPs (one subarray) in one sample requires a starting DNA concentration of >75 ng/uL to achieve >0.5 ng of DNA in each TaqMan reaction after dilution of the sample by PCR reagents and dispensing into the nanoplate. After the cassette is sealed, it can be thermal cycled in one of several commercially available flat block thermal cyclers to an end-point and then imaged in a computer-controlled imager (NT Imager) to record the VIC and FAM dye fluorescent signals that distinguish the homozygote and heterozygote SNP genotypes. This

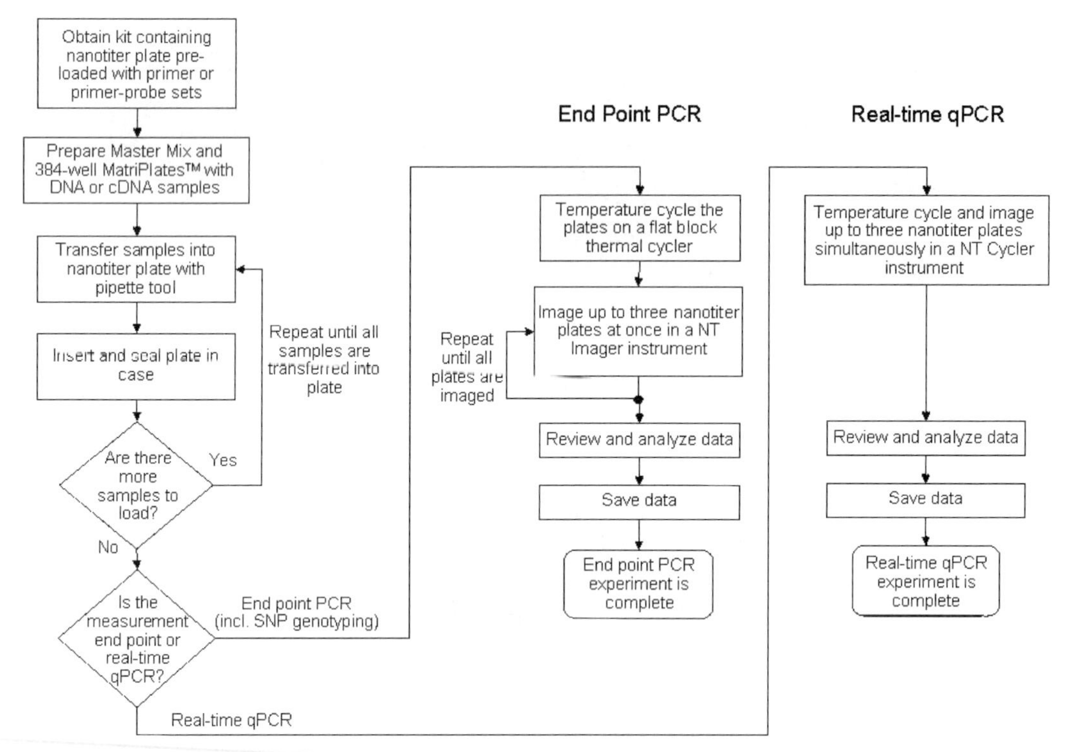

Fig. 12.2. Nanoplate workflow. The nanoplate workflow is similar for both end-point and real-time PCR applications. Preparation of DNA and RNA samples for nanoplate analysis use commercially available preparation kits and loading the sample and PCR master mix into the nanoplate is the same for either application. The nanoplate is then thermally cycled to an end point and imaged or imaged during thermal cycling for real-time PCR.

workflow is readily scalable for batch processing several tens of nanotiterplates in an 8-h day for genotyping many hundreds of samples with a panel comprising tens of SNPs, reaching a sustainable throughput of up to 98,000 SNP analyses per person per 8-hr shift.

The workflow for RNA profiling of a sample in the nanoplate with SYBR Green I real-time PCR is similar. The example to illustrate this application is measurement of the expression of 56 kinase genes in healthy and diseased adult human heart total RNA samples. Including eight no assay controls, the total number of real-time PCR assays interrogating each sample is 64 or one subarray in the nanoplate for each sample. The total RNA concentration for each assay should be >180 ng/uL or >2.5 ng per reaction prior to reverse transcription (RT) into cDNA. Similar to the SNP genotyping protocol, each cDNA sample is dispensed into a different well of a 384-well microplate, mixed with PCR reagents, and transferred into an individual subarray of the nanoplate. The loaded nanotiterplate is inserted into the cassette, sealed and placed in a computer-controlled imaging thermal cycler instrument (NT Cycler) that implements a real-time PCR method resulting in 9,216 PCR amplifications and dissociation curves in less than 3.5 h. Post-acquisition data processing generates fluorescence amplification and melt curves for each through-hole in the array, from which cycle threshold (C_T) and melt temperature (T_m) are computed for each reaction. For both SNP genotype and real-time PCR applications, all data is stored in a flat file (*.csv) format for ready export to a database or third party software for further analysis.

2. Materials

2.1. SNP Genotyping Application

1. Whole fresh or frozen human blood samples (48 × 200 μL) treated with citrate, heparin or EDTA.
2. QIAamp DNA Blood Mini Kit (Qiagen, Valencia, CA).
3. PCR master mix with final concentrations of 1x Universal Mastermix (Applied Biosystems, Foster City, CA, USA), 0.5 % glycerol (Sigma-Aldrich, St Louis, MO), 1% Pluronic F38 (BASF), 50 ng/μl BSA (Sigma-Aldrich).
4. Sixty-four TaqMan SNP PCR assays (Applied BioSystems, Foster City, CA) at a 1 × concentration.
5. OpenArray™ SNP Genotyping Kit (BioTrove Inc., Woburn, MA)
   - OpenArray™ plates (nanotiter plates) pre-loaded with TaqMan SNP assays
   - OpenArray™ Genotyping Cases.
   - Immersion Fluid.
   - Case Sealing Glue.
   - Plate-File CDROM specifying location of the SNP assays in the nanotiter plate.
6. NT Imager (BioTrove Inc.)
7. Flat block thermal cycler (MJ Tower PTC-200 recommended).
8. Optional: GenomiPhi™ V2 DNA Amplification Kit (GE Healthcare, Piscataway, NJ).

2.2. Real-Time PCR Application

1. Human heart, diseased total RNA (Clontech, Mountain View, CA), human heart, total RNA; human fetal heart, total RNA.
2. High-capacity cDNA archive kit (Applied BioSystems, Foster City, CA).
3. Exonuclease I (10 U/μL) (GE Healthcare, Piscataway, NJ).
4. PCR master mix with final concentrations of 1× LightCycler™ FastStart DNA Master SYBR Green I (Roche Applied Science, Indianapolis, IN), 0.2 % (w/v) Pluronic F-68 (Invitrogen, Carlsbad, CA), 1 mg/ml BSA (Sigma-Aldrich), 2.5 mM $MgCl_2$ (Roche Applied Science), 1:24,000 SYBR Green I (Sigma-Aldrich), 0.5 % (v/v) glycerol (Sigma-Aldrich), 8% (v/v) formamide (Applied Biosystems).

5. User supplied primer pairs
   - Fifty-six primer pairs (400 nM concentration) targeting 56 kinase genes in this example.
6. OpenArray™ Real-Time qPCR Kit (BioTrove Inc., Woburn, MA)
   - OpenArray™ plates (nanotiter plates) pre-loaded with user supplied primer pairs.
   - OpenArray™ Real-Time qPCR Cases.
   - OpenArray™ Real-Time qPCR Plate Frame and Frame Adhesive.
   - Immersion Fluid.
   - Case Sealing Glue.
   - Plate-File CDROM specifying location of each primer pairs in the nanotiter plate.
7. NT Cycler (BioTrove Inc.)

2.3. Materials Common to Both Applications

1. Finnpipette pipettor (16-channel), 5–50 μL (VWR).
2. MatriPlate plates, 384-well, black, low volume, polypropylene (MatriCal).
3. Vertical slide holder (Ted Pella Inc.).
4. AutoLoader (BioTrove Inc.)
5. Case sealing station
6. Autoloader supplies (BioTrove Inc.)
7. Plate holder
8. Plate guide set
9. Tip block
10. Box loader tips (384 surface-treated 20 μL Finn tips).

3. Methods

3.1. TaqMan PCR for SNP Genotyping

3.1.1. TaqMan SNP Assays

The 64 TaqMan SNP assays are loaded into the nanoplates and dried down at BioTrove Inc. at a 1x concentration. Store plates at –20°C.

3.1.2. DNA Samples

Isolate and purify DNA from the 48 whole blood samples per Qiagen kit instructions. Either freeze at –20°C or keep the microplate on ice until ready for analysis (*see* **Notes 1–4**).

3.1.3. Preparing Source Microplates

1. Label each microplate with a unique numeric or alphanumeric code for entry into the SNP genotyping software for tracking of samples transferred from the microplate into the nanoplate.
2. Thaw plate containing DNA samples at room temperature, centrifuge the plates at 1,000 rpm for 1 min. Place samples on ice to keep cool.

3. Prepare 211 μL of PCR master mix (*see* **Note 5**).
4. Pipette 4 μL aliquots of master mix into 48 adjacent wells of the MatriPlate.
5. Pipette 1 μL of DNA from each sample such that there is a different sample in the 48 wells of the source plate; gently aspirate and dispense to mix.
6. Microplate layout: Dispense sample and master mix into wells A1–A12, B1–B12, C1–C12, and D1–D12 where each well corresponds to a subarray in the nanoplate.
7. Cover the sample plate with aluminum foil sealing tape and centrifuge the plate for 1 min at 1,000 rpm to eliminate bubbles. Place the plate on ice to keep the samples cold.

3.1.4. OpenArray Loading

1. With forceps, peel off the foil exposing the 48 samples to be loaded onto the nanoplate.
2. Place the plate guide exposing the wells from which samples will be transferred into the nanoplate. To load the first nanoplate, the mask exposes only wells A1–A12, B1–B12, C1–C12, and D1–D12.
3. A tip block is prepared by placing 20 μL Finn pipette tips into each hole of the tip block with a 16-channel pipettor. Inserting 12 pipette tips at a time fills all 48 pipette tip positions in four insertion operations.
4. The tip block is inserted into the plate guide so that each individual pipette tip draws up fluid from each of 48 contiguous wells in the MatriPlate. One minute of gentle vertical movement of the tip block relative to the microplate ensures at least 4 μL of liquid is drawn into each pipette tip.
5. Insert the nanotiterplate into the Plate Holder and place into the Autoloader.
6. After liquid is loaded into each pipette tip, insert the tip block into the Autoloader. The Autoloader brings the pipette tip block into intimate contact with the nanoplate and moves the plate relative to the tip block to dispense liquid from each pipette tip into the through-holes of the nanoplate (*see* **Note 6**).
7. Fill the case three fourths with immersion fluid. This is to prevent thermal evaporation of the assay during thermal cycling.
8. Remove the plate from the Autoloader, insert into the cassette, and pipette UV curable epoxy into the case to seal. Insert the cassette into the case sealer where the glue is exposed to UV light to cure the epoxy forming a hermetic seal.

3.1.5. Thermal Cycler Protocol and Imaging

1. The case is thermal cycled on a MJ Research PTC-200 peltier thermal cycler equipped with a dual tower for simultaneously thermal cycling of up to 32 nanoplates. The thermal cycle protocol starts with a 10 min, 91°C polymerase activation step followed by 50 cycles of 23 s at 51°C; 30 s at 53.5°C; 13 s

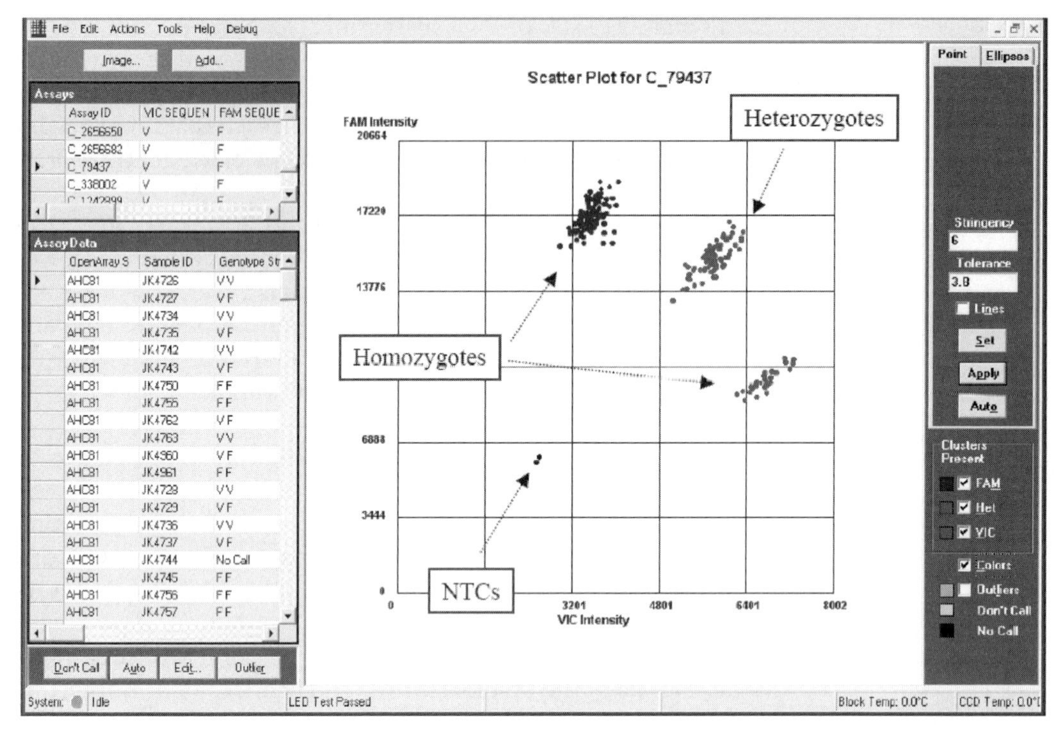

Fig. 12.3. Example cluster plot for one SNP TaqMan assay. Individuals in the population analyzed are categorized based on whether they are homozygote or heterozygote for the particular SNP genotype. The SNP genotype accuracy of TaqMan PCR assays in the nanoplate system is >99.5% with a call rate >95%.

at 54.5°C; 22 s at 97°C, and 7 s at 92°C. After temperature cycling, the plates are held at 20°C until they are removed for imaging.

2. Plates are imaged in groups of three in the NT Imager at the FAM and VIC fluorescent dye wavelengths. Instrument software captures and processes fluorescent images of the plates to measure the fluorescent signal from each through-hole and display the data as cluster plots for manual or semi-automated genotype calling by the user (**Fig. 12.3**).
3. SNP genotype call accuracy for TaqMan assays in the nanoplate is >99.5 % and call rates are >95%; similar in performance to the same SNP assays in a microplate (*see* **Notes 7** and **8**).

3.2. SYBR Green I Real-Time PCR

3.2.1. Primer Design

1. Primers to detect expression of 56 kinase genes are selected based on their classification in Gene Ontology and their presence in the RefSeq database (*see* **Notes 9–11**).
2. The primers are loaded at a concentration of 400 nM and dried down in the nanoplates by BioTrove.

3.2.2. cDNA Preparation from Total RNA

1. Thaw reagents in the high-capacity cDNA reverse transcription kit, total RNA and exonuclease I.

2. Prepare 2× RT mix from the high-capacity cDNA reverse transcription kit with a final volume adjusted to the number of cDNA samples to be converted plus one tenth for pipetting errors. With 10 μL of RT mix for one reaction, the final volume of RT mix would be 330 μL if 30 samples are to be converted. For 10 μL of 2× RT solution, mix the following reagents: 2 μL 10× RT buffer, 0.8 μL 25× dNTP mixture, 2 μL 10× random primers, 1 μL 5 U/μL MultiScribe reverse transcriptase and 4.2 μL RNase-free water.
3. Combine equal volumes of 250 ng/μL total RNA and 2× RT mix.
4. Incubate at room temperature for 10 min.
5. Incubate at 37°C for 2 h
6. Put on ice for 5 min and spin.
7. Incubate at 75°C for 10 min to inactivate the reverse transcriptase.
8. Put on ice for 5 min and spin.
9. Add 3 μL of 10 U/μL exonuclease I.
10. Incubate at 37°C for 1 h.
11. Incubate at 85°C for 10 min to inactive the exonuclease I.
12. Put on ice and spin down.
13. The sample is now ready for addition of PCR reagents and loading into the nanoplate or can be stored at –20°C.

3.2.3. Preparing Source Microplates

1. Label each microplate with a unique numeric or alphanumeric code for entry into the real-time PCR software for tracking of samples transferred from the microplate into the nanoplate.
2. Thaw plate containing cDNA samples at room temperature; centrifuge the plates at 1,000 rpm for 1 min. Place samples on ice to keep cool.
3. Prepare 185 μL of PCR master mix.
4. Pipette 3.5 μL of master mix into 48 adjacent wells of the MatriPlate.
5. Pipette 1.5 μL of cDNA from each sample such that there is a different sample in the 48 wells of the source plate; gently aspirate and dispense to mix.
6. Microplate layout: Dispense sample and master mix into wells A1–A12, B1–B12, C1–C12, and D1–D12 where each well corresponds to a subarray in the nanoplate.
7. Cover the sample plate with aluminum foil sealing tape and centrifuge the plate for 1 min at 1,000 rpm to eliminate bubbles. Place the plate on ice to keep the samples cold.

3.2.4. OpenArray Loading

1. Follow the same procedure as described for loading nanoplates for SNP genotyping.
2. Attach the plate frames with the adhesive to the case prior to thermal cycling. The frame keeps the case in intimate contact

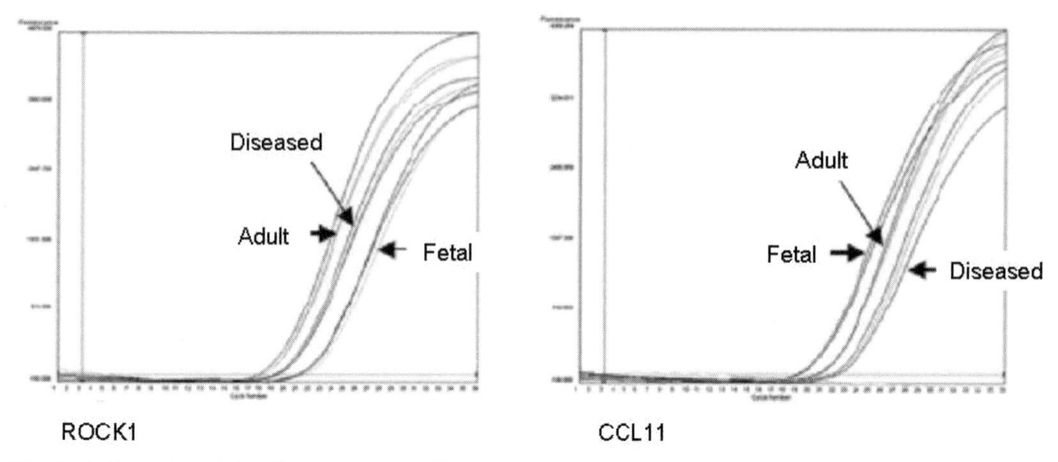

Fig. 12.4. Example real-time fluorescence amplification curves. Real-time fluorescence amplification curves showing expression two of 56 kinase genes in adult normal heart tissue, diseased heart tissue, and fetal heart tissue. Note the difference in transcript abundance in ROCK1 (Rho-associated, coiled-coil containing protein kinase 1), and CCL11 ((C–C motif) ligand 1) between the different tissues.

with the thermal block to ensure thermal uniformity during temperature cycling.

3.2.5. Thermal Cycler Protocol and Imaging

1. The OpenArray thermal cycling protocol consisted of 10 min, 92°C polymerase activation step followed by 35 cycles of 15 s at 92°C, 1 min at 55°C and 1 min at 72°C (imaging step).
2. Following amplification, amplicon dissociation is measured by cooling the PCR array to 70°C then slowly heated to 92°C at 1°C/min, with images collected for every 0.25°C.
3. **Figure 12.4** shows fluorescence amplification curves indicating the expression differences of two kinase genes in the three different heart tissues: control (adult), diseased, and fetal.

4. Notes

1. Greater than 1 ng of DNA per TaqMan assay in the nanoplate is recommended, consistent with Applied Biosystem's recommendation of 1–20 ng of DNA for the same assay in a microplate. To achieve this goal, a starting concentration of 150 ng/ul is required. Adding 1 μl of the concentrated DNA to 4 μl of PCR reagents in a microplate results in a 30.3 ng/ul final DNA concentration. This leads to the required 1 ng of DNA per reaction when dispensed into the plate.
2. If the DNA concentration is <150 ng/uL but >75 ng/uL, the amount of DNA per reaction can be kept at 1 ng by adding up to 2 μL of DNA sample per reaction and decreasing the water volume in the master mix so each well has a total volume of 5 μL (including the DNA sample).

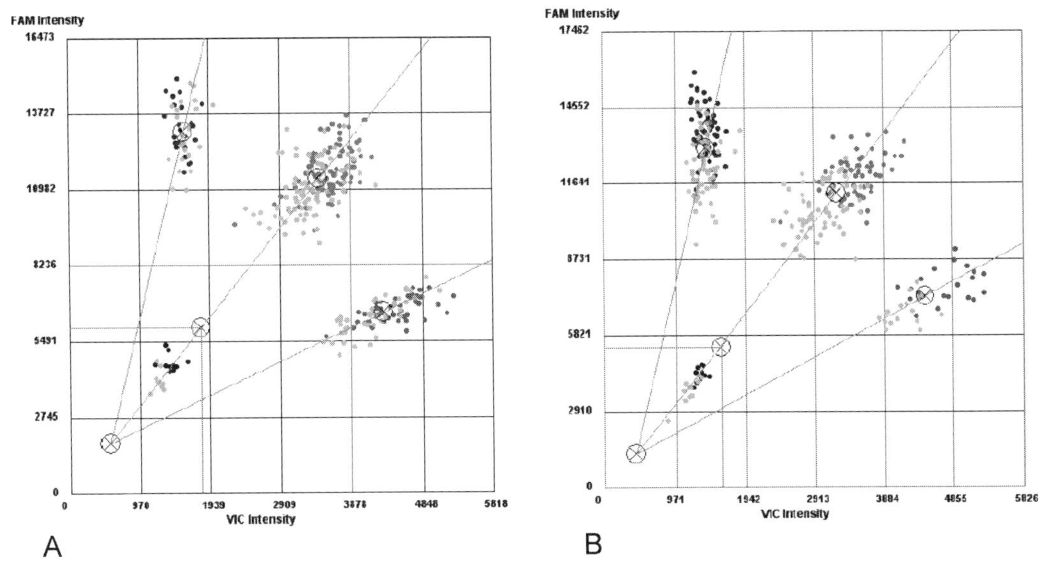

Fig. 12.5. SNP genotyping after whole genome amplification. Results for two different TaqMan SNP assays in A and B with whole-genome amplified (WGA) samples in light grey. The starting DNA concentration in the sample is 30 ng/uL and after WGA it is 448 ng/uL. In all cases the SNP call rate was 99% and accuracy 100%.

3. There are several options available to increase the DNA concentration if it is <75 ng/μl. The sample concentration can be increased by first evaporative drying and reconstituting with a smaller volume of buffer. Formulating the modified master mix with a higher concentration (2× for example) of PCR master mix allows a larger volume of sample to be added up to a total volume of 5 μL.
4. Another option is to amplify the sample DNA with a whole-genome amplification kit such as GenomiPhi™ V2 DNA amplification kit. Diluting the DNA sample by 100-fold, whole genome amplifying the diluted sample and comparing the results with the unamplified DNA sample (**Fig. 12.5**) shows the results are similar.
5. Volume of PCR master mix is sufficient to load one nanoplate plus an additional 10 % volume to account for pipetting dead volume.
6. The tips ride on a layer of liquid sandwiched between the end of the tip and the nanoplate. Fluid does not wet the nanoplate surface because of the hydrophobic exterior surface but is quickly, rapidly, and precisely wicked into each through-hole of the array because of the hydrophilic polymer coating the inside surfaces of the through-holes themselves.
7. DNA quality and amount per reaction can affect SNP assay performance. TaqMan assay accuracy with amount of DNA per reaction at concentrations ranging 0.01–1 ng per through-hole shows the accuracy of each assay exhibits a dif-

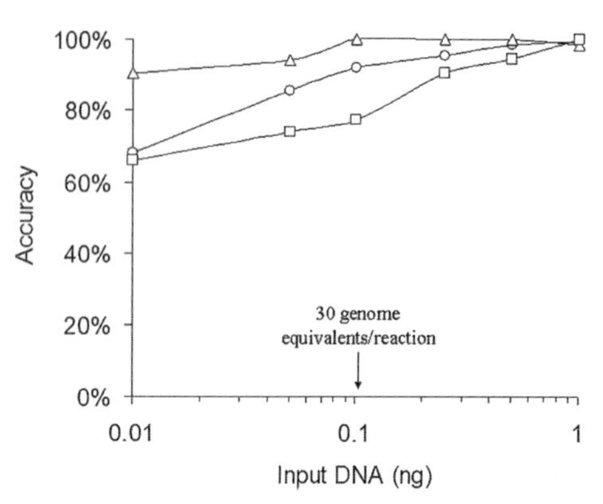

Fig. 12.6. Effect of DNA concentration on PCR Array TaqMan SNP assay accuracy. The plot indicates average accuracy for assays BT1000026 (square), BT1000094 (triangle), or combined results from 24 different assays (circle) at indicated starting quantities of human DNA.

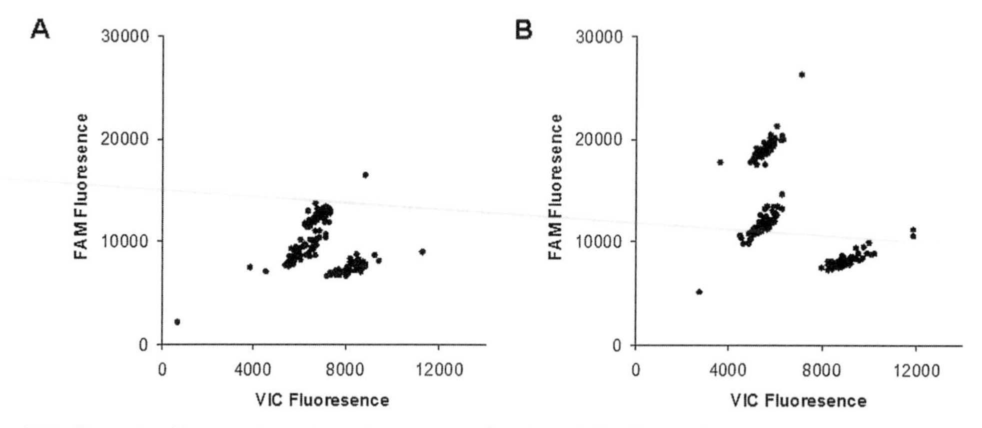

Fig. 12.7. Example of improved genotype cluster separation through TaqMan probe redesign. Plots indicate clusters from original TaqMan assay-by-design (A) and an assay differing by a single additional nucleotide to 5′-prime end of the TaqMan probe (B).

ferent degree of sensitivity (**Fig. 12.6**). Some assays maintain high accuracy essentially independent of the amount of DNA per reaction to as low as 0.1 ng genomic human DNA (33 genomic equivalence); while others begin to lose accuracy rapidly at <1 ng genomic DNA (<330 genomic equivalence). Commensurate with this loss of accuracy is an increase in heterozygote cluster variance that makes it more difficult to distinguish a heterozygote from a homozygote.

8. Redesign of the TaqMan assay can also improve performance in the nanoplate system. Shown in **Fig. 12.7** are cluster plots for a TaqMan SNP assay before (**Fig. 12.7A**) and after (**Fig. 12.7B**) the addition of a single base to the 5′ probe end. This modification improves the FAM probe performance and leads to good cluster separation.

9. Primers are designed using Primer3 or selected from PrimerBank, the primers have annealing temperatures between 60 and 65°C using nearest neighbor thermodynamics. The primer sequences have a length of 18–25 nucleotides and GC content of 30–60%.
10. Each primer pair is supplied phosphorylated prior to deposition in the nanotiter plate. An ionic bond between the primer pairs and the interior surface of the through-hole prevents carry-over until the thermal cycling process begins.
11. The primers are synthesized by a commercial supplier (Integrated DNA Technologies, www.idtdna.com) and validated according to the following protocol.
    a. A product greater than 0.2 Rn for each primer set run against a 37 tissue cDNA library (BD Quick-Clone II Human Universal cDNA, BD BioSciences, Franklin Lakes, NJ) in an ABI 7900 real-time thermal cycler (Applied BioSystems Inc., Foster City, CA).
    b. Amplicon product within 10% of the predicted length measured with 4% agarose E-gels (Invitrogen, Carlsbad, CA) from images captured with a gel imager (AlphaImager, AlphaInnotech, San Leandro, CA) and processed by software (Quantity One 1-D Analysis Software; BioRad, Hercules, CA).
    c. Amplicons generated from primer pairs passing the above criteria are pooled and gel purified to remove fragments <80 and >400 base pair in size. This pool is the source of template in subsequent primer pair validation experiments.
    d. Amplicon sequence where ~ 70% matched all or part of the expected target sequence. Although the remainder could not be confirmed by sequencing, these amplicons nonetheless matched the expected size of the predicted amplified product.
    e. PCR primer efficiencies measured between 96%–110%.

References

1. Murray, S., Oliphant, A., Shen, R., McBride, C., Steeke, R., Shannon, S., Rubano, T., Kermani, B., Fan, J-B., Chee, M., Hansen, M. (2004) A highly informative SNP linkage panel for human genetic studies. *Nature Meth* 1, 113–117.
2. Saito, K., Nakayama T., Sato N., Morita A., Takahashi T., Soma M., Usami, R. (2006) Haplotypes of the plasminogen activator gene associated with ischemic stroke. *Thromb Haemost* 96, 331–336.
3. Shi, M. (2001) Enabling large-scale pharmacogenetic studies by high-throughput mutation detection and genotyping technologies. *Clin Chem* 47, 164–172.
4. Beuselinck K., van Ranst M., van Eldere, J. (2005) Automated extraction of viral-pathogen RNA and DNA for high-throughput quantitative real-time PCR. *J Clin Microbiol* 43, 5541–5546.
5. Balashov S. V., Gardiner R., Park S., Perlin D. S. (2005) Rapid, high-throughput, multiplex, real-time PCR for identification of mutations in the cyp51A gene of *Aspergillus fumigatus* that confer resistance to itraconazaole. *J Clin Microbiol* 43, 214–222.

6. Stephens, A. J., Huygens F., Inman-Bamber J., Price E. P., Nimmo G. R., Schooneveldt J., Munckhof W., Giffard P. M. (2006) Methicillin-resistant *Staphylococcus aureus* gentoytping using a small set of polymorphisms. *J Med Microbiol* 55, 43–51.
7. Denomme G. A., Van Oene, M. (2005) High-throughput multiplex single-nucleotide polymorphism analysis for red cell and platelet antigen genotypes. *Transfusion* 45, 660–666.
8. Higgins, M., Hughes, A., Buzzacott, N., Lown J. (2004) High-throughput genotyping of human platelet antigens using the 5'-nuclease assay and minor groove binder probe technology. *Vox Sang* 87, 114–117.
9. Smith, C. T., Baker, J., Park, L., Seeb, W., Elfstrom, C., Abe, S., Seeb, J. E. (2005) Characterization of 13 single nucleotide polymorphism markers for chum salmon. *Molecular Ecology Notes* 5, 259–262.
10. Smith, C. T., Templin, W. D., Seeb, J. E., Seeb, L. W. (2005) Single nucleotide polymorphisms provide rapid and accurate estimates of the proportions of U.S. and Canadian chinook salmon caught in Yukon river fisheries. *North Am J Fisheries Management* 25, 944–953.
11. Giancola, S., McKhann, H., Berard, A., Camilleri, C., Durand, S., Libeau, P., Roux, F., Reboud, X., Gut, I., Brunel, D. (2006) Utilization of the three high-throughput SNP genotyping methods, the GOOD assay, Amplifluor and TaqMan, in diploid and polyploidy plants. *Theoretical Appl Genetics* 112, 1115–1124.
12. Liu, R. H. et al. (2004) Self-contained, fully integrated biochip for sample preparation, polymerase chain reaction amplification, and DNA microarray detection. *Anal Chem* 76, 1824–1831.
13. Lee, D. -S. et al. (2004) Bulk-micromachined submicroliter-volume PCR chip with very rapid thermal response and low power consumption. *Lab Chip* 4, 401–407.
14. Zhang, C., Xing, D. (2007) Miniaturized PCR chips for nucleic acid amplification and analysis: latest advances and future trends *Nucleic Acids Research* 35, 4223–4237.
15. Dahl, A. et al. (2007) Quantitative PCR based expression analysis on a nanoliter scale using polymer nano-well chips *Biomed Microdevices* 9, 307–314.
16. Kanigan T. K., Brenan C. J. H., Lafontaine S., et al. (2000) Living chips for drug discovery. *SPIE Proceedings* 3929, 172–180.
17. Brenan C. J, H., Morrison T., Stone K., et al. (2002) A massively parallel microfluidics platform for storage and ultra high throughput screening. *SPIE Proceedings* 4626, 560–569.
18. Morrison T., Hurley J., Garcia J., et al. (2006) Nanoliter high throughput quantitative PCR. *Nucleic Acids Res* **34**, e123.

Chapter 13

RNA Stabilization of Peripheral Blood and Profiling by Bead Chip Analysis

Svenja Debey-Pascher, Daniela Eggle, and Joachim L. Schultze

Abstract

Gene expression profiling of peripheral blood is a very attractive approach for the development of new diagnostic markers of blood-borne but also systemic diseases as well as the development of biomarkers for drug development. Since most cellular components of peripheral blood are specialized to quickly respond to exogenous stimuli, sample procurement approaches are required that reduce the overall impact of ex vivo changes in gene expression due to technical issues such as prolonged sample handling or temperature changes during transportation of the blood prior to genome-wide analysis. To address these needs, a whole blood RNA stabilization technology was combined with a bead-based oligonucleotide microarray technology for genome-wide transcriptome analysis. Cells, and thereby also RNA is immediately stabilized after the blood draw by a commercially available device (PAXgene). Total RNA is then extracted from PAXgene-stabilized blood and subjected to microarray analysis. In our hands, the Illumina BeadChip array platform outperformed other microarray platforms. Combining RNA stabilization of peripheral blood with bead-based oligonucleotide microarray technology is not only applicable to small single-center studies with optimized infrastructure but also to large scale multi-center trials that are mandatory for the development of predictive markers for disease and treatment outcome.

Key words: Transcriptome, gene expression profiling, peripheral blood, RNA stabilization, bead chip arrays.

1. Introduction

Genome-wide gene expression profiling is the first mature genome-wide technology to be considered for routine clinical use *(1, 2)*. Landmark studies in cancer research, have highlighted the power of this technology to improve molecular diagnostics *(3–5)*, prediction of prognosis at diagnosis *(6–10)*, and even prediction of drug efficacy *(11–13)*.

Peter Bugert (ed.), *DNA and RNA Profiling in Human Blood: Methods and Protocols, vol. 496*

DOI 10.1007/978-1-59745-553-4_13 Springerprotocols.com

During the last years, gene expression profiling has also been applied to peripheral blood. Peripheral blood mononuclear cells (PBMC) were used to study gene expression in acute myeloid leukaemia *(9)*, renal cell carcinoma (RCC) *(14)*, systemic lupus erythematosus (SLE) *(15–17)*, bladder cancer *(18)*, and acute infections *(19)*. The MIAME consortium has established guidelines for reporting, annotation, and data analysis of microarray data *(20)* and recent reports have focused on the impact of different microarray platforms and lab-to-lab variability in reproducibility and comparability of microarray results, suggesting the use of highly standardized protocols from RNA amplification to data analysis *(21–28)*. We have recently demonstrated that sample procurement, handling and transportation also have significant impact on data quality and therefore need to be thoroughly standardized to optimize results from gene expression profiling studies, particularly when analyzing peripheral blood *(29–31)*. In fact, significant effort will be necessary to achieve a high grade of standardization leading to good quality gene expression data when using PBMC in multi-center trials *(32)*. Here we describe a methodology that overcomes previous hurdles of gene expression profiling of peripheral blood. The PAXgene™ Blood RNA system, which allows standardized blood collection, is one of the most promising techniques for whole blood sample handling in clinical trials. PAXgene™ derived whole blood RNA samples are applied to a very cost-effective bead-based oligonucleotide microarray (Illumina BeadChip) platform. Overall, the methodology is reliable, robust, and cost-effective and can be easily applied to genome-wide expression profiling of peripheral blood in large clinical trials.

2. Materials

2.1. Blood Collection

1. PAXgene blood RNA system
2. BD Safety-Lok™ blood collection set
3. BD Vacutainer™ holder

2.2. RNA Isolation

1. PAXgene blood RNA kit (Qiagen) including following components (*see* **Note 1**):

   Buffer BR1 (resuspension buffer) 20 ml, store at room temperature.
   Buffer BR2 (binding buffer) 18 ml, store at room temperature.
   Buffer BR3 (wash buffer) 45 ml, store at room temperature.
   Buffer BR4 (wash buffer) 11 ml, add four volumes of absolute ethanol before use, store at room temperature.
   Buffer BR5 (elution buffer) 5 ml, store at room temperature.

RNase-free water (bottle) 2 × 125 ml, store at any temperature.
Proteinase K (green lid) 2 × 1.4 ml, store at room temperature.
PAXgene RNA spin columns (red) 5 × 10, store at room temperature.
Processing tubes (2 ml) 6 × 50, store at room temperature.
Secondary BD Hemogard^{TM} Closures 50, store at room temperature.
Microcentrifuge tubes (1.5 ml), store at room temperature.
RNase-free water (tube, red lid) 2 ml, store at 4°C.

2. RNase-free DNase set (Qiagen) including following components:

DNase I, RNase-free (lyophilized) 1500 Kunitz units, store at 4°C. Buffer RDD (white lid) 2 × 2 ml, store at 4°C.
RNAse-free water, 1.5 ml.

3. Ethanol absolute (molecular biology grade).
4. Microcentrifuge tubes (1.5 or 2 ml).
5. Sterile, RNase-free pipet tips.

2.3. cDNA and cRNA Synthesis for Bead Chip Analysis

1. Illumina TotalPrep RNA amplification kit including following components:

T7 Oligo(dT) primer, 26 μl, store at –20°C.
First strand buffer (10×), 51 μl, store at –20°C.
dNTP mix, 205 μl, store at –20°C.
RNase inhibitor, 26 μl, store at –20°C.
ArrayScript^{TM}, 27 μl, store at –20°C.
Second strand buffer (10×), 252 μl, store at –20°C.
DNA polymerase, 51 μl, store at –20°C.
RNase H, 26 μl, store at –20°C.
T7 Reaction buffer (10×), 64 μl, store at –20°C.
T7 enzyme mix, 64 μl, store at –20°C.
Biotin–NTP mix, 64 μl, store at –20°C.
RNase-free water, 1.75 ml, store at any temperature.
Control RNA (1 mg/ml HeLa total RNA), 10 μl, store at –20°C.
RNase-free water, 10 ml, store at any temperature.
Wash buffer (30 ml), add 24 ml absolute ethanol before use, room temperature.
cDNA binding buffer, 7 ml, store at room temperature. (*see* **Note 2**).
cRNA binding buffer, 9 ml, store at room temperature.
cRNA filter cartridges, 24, store at room temperature.
cRNA collection tubes, 48, store at room temperature.
cDNA filter cartridges, 24 + tubes, store at room temperature.
cDNA elution tubes, 24, store at room temperature.

2. Ethanol (96–100%), molecular biology grade.
3. Microcentrifuge tubes (1.5 or 2 ml).
4. Sterile, RNase-free pipet tips.

2.4. Hybridization, Washing, and Scanning of Bead Chips

1. Illumina Sentrix® BeadChip and Accompanying CD
2. Beadchip Hyb chamber (Illumina)
3. Hyb chamber inserts (Illumina)
4. Hyb chamber gaskets
5. Hybridization oven
6. Illumina bead station 500x
7. Beadchip wash tray
8. Tray covers
9. Hybex heat block with water bath insert
10. Rocker and orbital shaker
11. Stain dishes and rack, 3 × 250 ml
12. Glass dish, 190 × 100 mm
13. GEX-HYB-buffer, store at –20°C (supplied with arrays).
14. GEX-HCB-buffer store at –20°C (supplied with arrays).
15. High-temp wash buffer, 10 x, store at room temperature (supplied with array).
16. Wash E1BC buffer, store at room temperature (supplied with arrays).
17. Block E1 buffer, store at 4 °C (supplied with arrays).
18. Streptavidin Cy3, FluoroLink™ Cy™3 (Amersham Biosciences) diluted to 1 μg/μL with RNase-free water (*see* **Note 3**).
19. Ethanol (100%) (=99.8% p.a.).
20. RNase-free water, store at any temperature.

2.5. RNA Preparation Alternatives

2.5.1. Globin Reduction by RNase H Digestion

1. Globin reduction oligos, HPLC-purified, store at –20°C
   α1: 5′- TGC AGG AAG GGG AGG AGG GGC TG-3 (nt 512 – 534)
   α2: 5′- TGC AAG GAG GGG AGG AGG GCC CG-3′ (nt 512 – 534)
   β: 5′- CCC CAG TTT AGT AGT TGG ACT TAG GG-3′ (nt 539 – 564)
2. SUPERase•In™, 2,500 U (Ambion), store at –20°C
3. RNase H, *Escherichia coli*, 10 U/μL, 200 U, store at –20°C.
4. EDTA, 0.5 M RNase-free.
5. GeneChip® sample cleanup module (Affymetrix).
6. RNase-free water.
7. Oligo Hyb buffer (10×): 100 mM Tris–HCI, pH 7.6, 200 mM KCI, store at –20
8. RNase H buffer (10×): 100 mM Tris–HCI, pH 7.6, 10 mM DTT, 20 mM $MgCl_2$, store at –20

2.5.2. Globin Reduction by Magnetic Beads

1. GlobinClear Kit (Ambion) including the following components:

Capture oligo mix, 20 μl, store at –20°C.
Streptavidin magnetic beads, 600 μl, store at 4°C, do not freeze.
Hybridization buffer (2×), 1.5 ml, store at room temperature.
Streptavidin bead buffer, 1.5 ml, store at room temperature.
RNase-free water, 1.75 ml, store at any temperature.
RNA binding beads, 200 μl, store at 4°C, do not freeze.
RNA wash solution concentrate, 5 ml; add 4 ml of 100% ethanol before use, store at room temperature or 4°C.
Elution buffer, 1 ml, store at 4°C or room temperature.
RNA bead buffer, 80 μl, store at room temperature.
RNA binding buffer concentrate, 4 ml, add 2 ml of 100% isopropanol before use, store at room temperature.
Non-stick tubes (1.5 ml) 100, store at room temperature.

2. Magnetic stand for 1.5 ml tubes, e.g. Ambion's six tube magnetic stand.
3. Ethanol (100%), ACS grade or higher quality.
4. Isopropanol (100%), ACS grade or higher quality.
5. RNase-free water.

2.6. Microarray Procedure Alternatives Using Affymetrix Arrays

1. Poly-A RNA control kit with: Poly-A control stock and poly-A control dilution buffer, both stored at –20°C.
2. GeneChip® one-cycle cDNA synthesis kit including following components:

   7-Oligo(dT) primer, 50 μM,120 μl, store at–20°C.
   First strand reaction mix (5×), 120 μl, store at –20°C.
   DTT, 0.1 M, 60 μl, store at –20°C.
   dNTP, 10 mM, 120 μl, store at –20°C.
   SuperScript™ II 200 U/μl, 60 μl, store at –20°C.
   Second strand reaction mix (5×), 900 μl, store at –20°C.
   E.coli DNA ligase 10 U/μl, 30 μl, store at –20°C.
   E.coli DNA polymerase I 10 U/μl, 120 μl, store at –20°C.
   RNase H 2 U/μl, 30 μl, store at –20°C.
   T4 DNA polymerase 5 U/μl, 60 μl, store at –20°C.
   EDTA, 0.5 M, 300 μl, store at –20°C.
   RNase-free water, 3.1 ml, store at any temperature.

3. IVT labeling kit including following components:

   IVT labeling buffer (10×), 120 μl, store at –20°C.
   IVT labeling enzyme mix, 120 μl, store at –20°C.
   IVT labeling NTP mix, 360 μl, store at –20°C.
   3′-Labeling control (0.5 μg/μl),10 μl, store at –20°C.
   RNase-free water, 825 μl, store at any temperature.

4. Sample cleanup module including following components:

   cDNA cleanup spin columns, 30, store at 4°C.
   cDNA binding buffer, 20 ml, store at room temperature.

cDNA wash buffer concentrate, 6 ml, add 24 ml of ethanol (96–100%) before use, store at room temperature.
cDNA elution buffer, 15 ml, store at room temperature.
30 IVT cRNA cleanup spin columns, store at 4°C.
IVT cRNA binding buffer, 18 ml, store at room temperature.
IVT cRNA wash buffer concentrate, 5 ml, add 20 ml of ethanol (96–100%) before use, store at room temperature.
RNAse-free water 3 × 1.2 ml, store at any temperature.
Fragmentation buffer (5×), 1 ml, store at room temperature.
Collection tubes (1.6 ml), 60 (for elution), store at room temperature.
Collection tubes, (2 ml) store at room temperature.

5. Hybridization control kit including following components:

   Hybridization control mix (20×), 450 μl, store –20°C.
   Control Oligo B2, 3 nM, store –20°C.
   Reagents should be aliquoted and stored at –20°C. Performance may be comprised after three freeze/thaw cycles.

6. Acetylated bovine serum albumin (BSA) solution (50 mg/ml) (Invitrogen).
7. Herring sperm DNA (Promega Corp.)
8. Micropure separator (Millipore), optional
9. GeneChip® eukaryotic hybridization control Kit (Affymetrix), contains control cRNA and control Oligo B2.
10. Control Oligo B2, 3 nM (Affymetrix), can be ordered separately.
11. NaCl (5M), RNase-free, DNase-free (Cambrex).
12. MES free acid monohydrate (Sigma-Aldrich).
13. MES sodium salt (Sigma-Aldrich).
14. EDTA disodium salt, 0.5 M solution (Sigma-Aldrich).
15. Surfact-Amps 20 (Tween-20), 10% (Pierce Chemical).
16. Sterile, RNase-free, microcentrifuge tubes, 1.5 ml
17. Sterile-barrier pipet tips and non-barrier pipet tips.
18. MES stock (12×): 1.22 M MES, 0.89 M [Na^+]. For 1,000 ml: 70.4 g MES free acid monohydrate, 193.3 g MES sodium salt, dissolve in Molecular Biology Grade water and adjust volume to 1,000 ml (pH should be between 6.5 and 6.7). Filter through a 0.2 μm filter. Do not autoclave. Store at 2–8°C and protect from light. Discard solution if yellow.
19. Hybridization buffer (2×): 1× concentration is 100 mM MES, 1 M [Na^+], 20 mM EDTA, 0.01% Tween 20. For 50 ml: 8.3 ml of 12× MES stock, 17.7 ml of 5 M NaCl, 4.0 ml of 0.5 M EDTA, 0.1 ml of 10% Tween 20, 19.9 ml of Molecular Biology Grade water. Store at 2–8°C, and shield from light. Do not autoclave. Discard solution if yellow.

20. Ethanol absolute, molecular biology grade
21. Water, molecular biology grade
22. Water, HPLC-grade.
23. Ampuwa, distilled water.
24. SSPE, 20 × (Cambrex).
25. R-Phycoerythrin streptavidin (Molecular Probes).
26. SSPE, 20 ×: 3 M NaCl, 0.2 M NaH_2PO_4, 0.02 M EDTA (Accugene).
27. Goat IgG, reagent grade (Sigma-Aldrich). 10 mg/ml goat IgG stock: resuspend 50 mg IgG in 5 ml PBS. Store at 4°C.
28. Anti-streptavidin antibody (goat), biotinylated (Vector Laboratories).
29. Stringent wash buffer: 100 mM MES, 0.1 M [Na^+], 0.01% Tween 20. For 1,000 ml: 83.3 ml of 12× MES stock buffer, 5.2 ml of 5 M NaCl, 1.0 ml of 10% Tween 20, 910.5 ml of water (HPLC-grade). Filter through a 0.2 μm filter. Store at 2–8°C and protect from light.
30. Non-stringent wash buffer: 6× SSPE, 0.01% Tween 20. For 1,000 ml: 300 ml of 20× SSPE, 1.0 ml of 10% Tween 20, 699 ml of water (HPLC-grade). Filter through a 0.2 μm filter.
31. Stain buffer (2×): 1× concentration is 100 mM MES, 1 M [Na^+], 0.05% Tween 20. For 250 ml: 41.7 ml 12× MES stock buffer, 92.5 ml 5 M NaCl, 2.5 ml 10% Tween 20, 112.8 ml Molecular Biology Grade water. Filter through a 0.2 μm filter. Store at 2–8°C and protect from light.

3. Methods

3.1. Blood Collection

The PAXgene system allows standardized and easy blood collection based on convenient BD Vacutainer™ technology. The PAXgene tube contains a proprietary reagent that immediately stabilizes intracellular RNA for 3 days at room temperature (18–25°C) and 5 days at 2–8°C *(31)*.

1. Ensure that the PAXgene blood RNA tube is at room temperature (18–25°C) prior to use and properly labeled with patient identification.
2. If the PAXgene blood RNA tube is the only tube to be drawn, a small amount of blood should be drawn into a "discard tube" prior to drawing blood into the PAXgene blood RNA tube. Otherwise, the PAXgene blood RNA tube should be the last tube drawn in the phlebotomy procedure.
3. Using a BD Vacutainer® Safety-Lok™ blood collection set, collect blood into the PAXgene blood RNA tube holding the PAXgene blood RNA tube vertically, below the blood donor's arm, during blood collection.

4. Allow at least 10 s for a complete blood draw to take place. Ensure that the blood has stopped flowing into the tube before removing the tube from the holder.
5. Gently invert the PAXgene blood RNA tube 8–10 times.
6. Store the PAXgene blood RNA tube upright at room temperature (18–25°C) for at least 2 h. After 24 h proceed to RNA isolation or store at –20 or –80°C for long-term storage.
7. To freeze PAXgene™ blood RNA tubes stand them upright in a wire rack. Do not freeze tubes upright in a Styrofoam tray as this may cause the tubes to crack. Tubes can be stored at –20°C and below. If tubes are to be frozen at temperatures below –20°C, freeze the tubes first at –20°C for 24 h, then transfer them to –70 or –80°C.
8. To thaw PAXgene™ blood RNA tubes, place them in a wire rack at ambient temperature (18–22°C) for approximately 2 hours. Do not thaw PAXgene™ Blood RNA Tubes at temperatures above 22°C. One suggestion is to remove the tubes from the freezer the afternoon before processing, thaw overnight, and process tubes the next day. After thawing, carefully invert the tubes ten times before starting RNA preparation.

3.2. RNA Preparation

1. Centrifuge the PAXgene blood RNA tube for 10 min at 3,000–5,000 × *g* at room temperature using a swing out rotor. The rotor must contain tube adapters for round-bottom tubes. If other types of tube adapter are used, the tubes may break during centrifugation.
2. Remove the supernatant by decanting or pipetting. Add 5 ml RNase-free water to the pellet, and close the tube using a fresh secondary Hemogard closure. If the supernatant is decanted, take care not to disturb the pellet, and dry the rim of the tube with a clean paper towel.
3. Vortex until the pellet is visibly dissolved, and centrifuge for 10 min at 3,000–5,000 × *g* using a swing-out rotor. Remove and discard the entire supernatant. Small debris remaining in the supernatant after vortexing but before centrifugation will not affect the procedure. Important. Incomplete removal of the supernatant will inhibit lysis and dilute the lysate, and therefore affect the conditions for binding RNA to the PAXgene membrane.
4. Thoroughly resuspend the pellet in 360 μl buffer BR1 by vortexing.
5. Pipet the sample into a 1.5 ml microcentrifuge tube. Add 300 μl buffer BR2 and 40 μl proteinase K. Mix by vortexing for 5 s, and incubate for 15 min at 55°C using a shaker–incubator at 1,400 rpm. After incubation, set the temperature of the shaker–incubator to 65 °C (for step 20). Important: Do not mix buffer BR2 and proteinase K together before adding them to the sample.

6. Centrifuge for 5 min at maximum speed in a microcentrifuge. Transfer the supernatant to a fresh 2 ml microcentrifuge tube. A minimum g-force of $10,000 \times g$ is required. Transfer of small debris remaining in the supernatant after centrifugation at full speed will not affect the procedure.
7. Add 350 μl absolute ethanol. Mix by vortexing, and centrifuge briefly (1–2 s at 500–1,000 × g) to remove drops from the inside of the tube lid. Important: The length of the centrifugation must not exceed 1–2 s, as this may result in pelleting of nucleic acids and reduced yields of total RNA.
8. Pipet 700 μl sample into the PAXgene RNA spin column (red) placed in a 2 ml processing tube, and centrifuge for 1 min at 8,000–20,000 × g. Place the spin column in a new 2 ml processing tube, and discard the old processing tube containing flow-through.
9. Pipet the remaining sample into the PAXgene RNA spin column, and centrifuge for 1 min at 8,000–20,000 × g. Place the spin column in a new 2 ml processing tube, and discard the old processing tube containing flow-through. Important: Carefully pipet the sample into the spin column and visually check that the sample is completely transferred to the spin column.
10. Pipet 350 μl buffer BR3 into the PAXgene RNA spin column. Centrifuge for 1 min at 8,000–20,000 × g. Place the spin column in a new 2 ml processing tube, and discard the old processing tube containing flow-through.
11. Add 10 μl DNase I stock solution to 70 μl buffer RDD in a 1.5 ml microcentrifuge tube. Mix by gently flicking the tube, and centrifuge briefly to collect residual liquid from the sides of the tube. If processing, for example, ten samples, add 110 μl DNase I stock solution to 770 μl buffer RDD. Important: DNase I is especially sensitive to physical denaturation. Mixing should only be carried out by gently flicking the tube. Do not vortex.
12. Pipet the DNase I incubation mix (80 μl) directly onto the PAXgene RNA spin column membrane, and place on the benchtop (20–30°C) for 15 min. Important: Ensure that the DNase I incubation mix is placed directly onto the membrane. DNase digestion will be incomplete if part of the mix is applied to and remains on the walls or the O-ring of the spin column.
13. Pipet 350 μl buffer BR3 into the PAXgene RNA spin column, and centrifuge for 1 min at 8,000–20,000 × g. Place the spin column in a new 2 ml processing tube, and discard the old processing tube containing flow-through.
14. Pipet 500 μl buffer BR4 to the PAXgene RNA spin column, and centrifuge for 1 min at 8,000–20,000 × g. Place the spin column in a new 2 ml processing tube, and discard the old processing tube containing flow-through. Important: Buffer

BR4 is supplied as a concentrate. Ensure that ethanol is added to buffer BR4 before use.

15. Add another 500 μl buffer BR4 to the PAXgene RNA spin column. Centrifuge for 3 min at 8,000–20,000 × *g*.
16. Discard the tube containing the flow-through, and place the PAXgene RNA spin column in a new 2 ml processing tube. Centrifuge for 1 min at 8,000–20,000 × *g*.
17. Discard the tube containing the flow-through. Place the PAXgene RNA spin column in a 1.5 ml microcentrifuge tube, and pipet 40 μl RNAse-free water directly onto the PAXgene RNA spin column membrane. Centrifuge for 1 min at 8,000–20,000 × *g* to elute the RNA. It is important to wet the entire membrane with buffer H_2O in order to achieve maximum elution efficiency.
18. Repeat the elution step as described, using the same 40 μl RNAse-free water and the same microcentrifuge tube.
19. Incubate the eluate for 5 min at 65°C in the shaker–incubator (from step 5) without shaking. After incubation, chill immediately on ice. Important: This incubation at 65°C denatures the RNA for downstream applications. Do not exceed the incubation time or temperature.
20. If the RNA samples will not be used immediately, store at –20 or –70°C. Since the RNA remains denatured after repeated freezing and thawing, it is not necessary to repeat the incubation at 65°C.

3.3. Quantification and Determination of Quality of Total RNA

1. The concentration of RNA should be determined by measuring the absorbance at 260 nm (A_{260}) in a spectrophotometer. To ensure significance, readings should be greater than 0.15. An absorbance of 1 unit at 260 nm corresponds to 44 μg of RNA per ml ($A_{260} = 1$, $c = 44\mu g/ml$). This relation is valid only for measurements in 10 mM Tris–HCl, pH 7.5. Therefore, if it is necessary to dilute the RNA sample, this should be done in 10 mM Tris–HCl.
2. The ratio of the readings at 260 and 280 nm (A_{260}/A_{280}) provides an estimate of the purity of RNA with respect to contaminants that absorb in the UV, such as protein. However, the A_{260}/A_{280} ratio is influenced considerably by pH. Lower pH results in a lower A_{260}/A_{280} ratio and reduced sensitivity to protein contamination. For accurate values, we recommend measuring absorbance in 10 mM Tris–HCl, pH 7.5. Pure RNA has an A_{260}/A_{280} ratio of 1.9–2.1 in 10 mM Tris–HCl, pH 7.5. Always calibrate the spectrophotometer with the same solution.
3. The integrity and size distribution of total RNA purified using the PAXgene Blood RNA System can be checked by denaturing agarose gel electrophoresis (1% w/v) and ethidium bromide staining. The respective ribosomal bands should appear

sharp on the stained gel. Ribosomal RNA, 28S (5 kb) bands should be present with an intensity approximately twice that of the 18S RNA (1.9 kb) band. If the ribosomal bands in a given lane are not sharp, but appear as a smear of smaller sized RNAs, it is likely that the RNA sample suffered major degradation during preparation. Only RNA samples of good purity and integrity should be used for further microarray analysis.

3.4. cDNA and cRNA Synthesis Using Illumina TotalPrep RNA Amplification Kit

RNA amplification has become the standard method for preparing RNA samples for array analysis *(33, 34)*. The Illumina® TotalPrep RNA amplification kit is a complete system for generating biotinylated, amplified RNA for hybridization with Illumina Sentrix® arrays. It is based on the RNA amplification protocol developed in the laboratory of James Eberwine *(35)*. The procedure consists of reverse transcription with an oligo(dT) primer bearing a T7 promoter using Array-Script™, a reverse transcriptase (RT) engineered to produce higher yields of first strand cDNA than wild-type enzymes. ArrayScript catalyzes the synthesis of virtually full-length cDNA. The cDNA then undergoes second strand synthesis and cleanup to become a template for in vitro transcription with T7 RNA polymerase where hundreds to thousands of biotinylated, antisense RNA copies are synthesized of each mRNA in a sample. The labeled cRNA produced with the kit was developed for hybridization with Illumina arrays. The here described protocol is according to the manufacturer´s instructions.

3.4.1. Reverse Transcription to Synthesize First Strand cDNA

1. Place a maximum volume of 11 μl of total RNA (100 ng) into a non-stick, sterile, RNase-free, 0.2 ml microcentrifuge tube.
2. Add RNase-free water as necessary to bring all samples to 11 μl.
3. At room temperature, prepare reverse transcription master mix in a nuclease-free tube: for a single RNA sample mix 1 μl T7 oligo(dT) primer, 2 μl 10× first strand buffer, 4 μl dNTP mix, 1 μl RNase inhibitor, and 1 μl ArrayScript.
4. Mix well by gently vortexing. Centrifuge briefly (~5 s) to collect the reverse transcription master mix at the bottom of the tube and place on ice.
5. Transfer 9 μl of reverse transcription master mix to each RNA sample. Mix thoroughly by pipetting up and down two to three times, then flicking the tube three to four times, and centrifuge briefly to collect the reaction in the bottom of the tube.
6. Place the samples in a 42°C incubator. Incubate reactions for 2 h at 42°C.
7. After the incubation, centrifuge briefly (~5 s) to collect the reaction mixture at the bottom of the tube. Place the tubes on ice and immediately proceed second strand cDNA Synthesis.

3.4.2. Second Strand cDNA Synthesis

1. On ice, prepare a second strand master mix in a nuclease-free tube: for a single reaction mix 63 μl RNase-free water, 10 μl 10× second strand buffer, 4 μl dNTP mix, 2 μl DNA polymerase, and 1 μl RNase H.
2. Mix well by gently vortexing. Centrifuge briefly (~5 s) to collect the mixture at the bottom of the tube and place on ice.
3. Transfer 80 μl of second strand master mix to each sample. Mix thoroughly by pipetting up and down two to three times, then flicking the tube three to four times, and centrifuge briefly to collect the reaction in the bottom of the tube.
4. Place the tubes in a 16°C thermal cycler. It is important to cool the thermal cycler block to 16°C before adding the reaction tubes because subjecting the reactions to temperatures > 16°C will compromise cRNA yield. Incubate 2 h in a 16°C thermal cycler.
5. After the 2 h incubation at 16°C, place the reactions on ice and proceed to cDNA purification (Section 3.4.3), or immediately freeze reactions at –20°C. Do not leave the reactions on ice for more than 1 h.
6. This is a potential overnight stopping point (at –20°C), but it is better to complete the cDNA purification before stopping.

3.4.3. cDNA Purification

1. Preheat RNase-free water to 50–55°C.
2. Check the cDNA binding buffer for precipitation before using it. If a precipitate is visible, redissolve it by warming the solution to 37°C for up to 10 min and vortexing vigorously. Cool to room temperature before use.
3. Add 250 μl of cDNA binding buffer to each sample, and mix thoroughly by pipetting up and down two to three times, then flicking the tube three to four times. Follow up with a quick spin to collect the reaction mixture in the bottom of the tube. Proceed quickly to the next step.
4. Check that the cDNA filter cartridge is firmly seated in its wash tube (supplied).
5. Pipet the cDNA sample/cDNA binding buffer onto the center of the cDNA filter cartridge.
6. Centrifuge for ~1 min at 10,000 × g, or until the mixture is through the filter. Discard the flow-through and replace the cDNA filter cartridge in the wash tube.
7. Apply 500 μl wash buffer to each cDNA filter cartridge.
8. Centrifuge for ~1 min at 10,000 × g, or until all the wash buffer is through the filter.
9. Discard the flow-through and spin the cDNA filter cartridge for an additional minute to remove trace amounts of wash buffer.
10. Transfer cDNA filter cartridge to a cDNA elution tube.
11. Elute cDNA with a total of 19 μl 50–55°C RNase-free water. It is important to use RNase-free water that is at 50–55°C for

the cDNA elution. Colder water will be less efficient at eluting the cDNA, and using hotter water (≥ 58°C) may result in reduced cRNA yield.

12. Apply 10 μl of RNase-free water (preheated to 50–55°C) to the center of the filter in the cDNA filter cartridge.
13. Leave at room temperature for 2 min and then centrifuge for ~1.5 min at 10,000 × *g*, or until all the RNase-free water is through the filter.
14. Apply a second aliquot of 9 μl preheated RNase-free water and centrifuge for 2 min. The double-stranded cDNA will now be in the eluate (~17.5 μl).
15. Proceed directly to next section to synthesize cRNA, or store the cDNA at –20°C.

3.4.4. *In Vitro* Transcription to Synthesize cRNA

1. At room temperature, prepare an IVT master mix by adding the following reagents to a nuclease-free tube: for a single 25 μl reaction mix 2.5 μl T7 10× reaction buffer, 2.5 μl T7 enzyme mix, and 2.5 μl biotin–NTP mix.
2. Mix well by gently vortexing. Centrifuge briefly (~5 s) to collect the IVT master mix at the bottom of the tube and place on ice.
3. Transfer 7.5 μl of IVT master mix to each cDNA sample (volume ~17.5 μl). If less than 17.5 μl are eluted fill up with RNAse-free water to reach volume of 17.5 μl. Mix thoroughly by pipetting up and down two to three times, then flicking the tube three to four times, and centrifuge briefly to collect the reaction mixture in the bottom of the tube. Once assembled, place the tubes at 37°C.
4. Incubate reactions for 16 h. It is important to maintain a constant 37°C incubation temperature. It is recommended to incubate in a hybridization oven because it is extremely important that condensation does not form inside the tubes; his would change the reagent concentrations and reduce yield. Alternatively you can use a thermal cycler with heated lid temperature set to 40°C.
5. Stop the reaction by adding 75 μl RNase-free water to each cRNA sample to bring the final volume to 100 μl. Mix thoroughly by gentle vortexing. Proceed to the cRNA purification step (Section 3.4.5) or store at –20°C.

3.4.5. cRNA Purification

1. Preheat RNase-free water to 50–60°C.
2. Assemble cRNA filter cartridges and tubes and add 350 μl of cRNA binding buffer to each cRNA sample. Proceed to the next step immediately.
3. Add 250 μl of 100% ethanol to each cRNA sample, and mix by pipetting the mixture up and down three times. Do not vortex to mix and do not centrifuge.

4. Pipet each sample mixture from step 3 onto the center of the filter in the cRNA filter cartridge.
5. Centrifuge for ~1 min at 10,000 × *g*. Continue until the mixture has passed through the filter.
6. Discard the flow-through and replace the cRNA filter cartridge back into the cRNA collection tube.
7. Apply 650 μl wash buffer to each cRNA filter cartridge.
8. Centrifuge for ~1 min at 10,000 × *g*, or until all the wash buffer is through the filter.
9. Discard the flow-through and spin the cRNA filter cartridge for an additional ~1 min to remove trace amounts of wash buffer.
10. Transfer filter cartridge(s) to a fresh cRNA collection tube.
11. To the center of the filter, add 100 μl RNase-free water (preheated to 50–60°C). Leave at room temperature for 2 min and then centrifuge for ~1.5 min at 10,000 × *g*, or until the RNase-free water is through the filter.
12. The cRNA will now be in the cRNA collection tube in ~100 μl of RNase-free water.
13. The concentration of cRNA is determined as described in Section 3.3.
14. The size distribution of cRNA can be checked by denaturing agarose gel electrophoresis (1% w/v) and ethidium bromide staining.

3.5. Hybridization of Bead Chips

The BeadStation 500X system uses a "direct hybridization" assay, whereby gene-specific probes are used to detect labeled RNAs. Each bead in the array contains a 50-mer, sequence-specific oligo probe synthesized in-house using Illumina's Oligator™ technology. The most consistent results are achieved by hybridizing equivalent amounts of cRNA on each array. An appropriate volume of cRNA from each sample is aliquoted into the hybridization tube(s). The here described protocol is according to the manufacturer´s instructions and is applicable only to six-sample Beadchips with IntelliHyb Seal.

1. Preheat the oven (with rocking platform) to 58°C.
2. Prepare cRNA samples (dried down in a vacuum centrifuge, if necessary to achieve required concentration). To 1.5 μg cRNA, add RNase-free water up to 10 μL and mix. Leave at room temperature for 10 min to resuspend cRNA.
3. Place the GEX-HYB and GEX-HCB tubes in the 58°C oven for 10 min to dissolve any salts that may have precipitated in storage. Inspect the solution; if any salts remain undissolved, incubate at 58°C for another 10 min. After allowing to cool to room temperature, mix thoroughly before using.
4. Add 20 μL GEX-HYB to each cRNA sample.
5. Place Illumina Hyb chamber gaskets into BeadChip Hyb chamber.

6. Dispense 200 μL GEX-HCB into each of the two humidifying buffer reservoirs in each Hyb chamber. Only add buffer to chambers that will be used.
7. Seal Hyb chamber with lid and keep on bench at room temperature (~22°C) until ready to load BeadChips into Hyb chamber.
8. Remove all BeadChips from their packages.
9. Holding BeadChip by cover seal tab with tweezers using powder-free gloved hands, slide BeadChip into Hyb chamber insert such that the barcode lines up with barcode symbol on the insert.
10. Preheat the assay sample at 65°C for 5 min.
11. Briefly vortex, then briefly centrifuge to collect the liquid in the bottom of the tube. Allow sample to cool to room temperature before using. Pipet sample immediately after cooling to room temperature.
12. Load Hyb chamber inserts containing BeadChips into the Hyb chamber with barcode on BeadChip aligned to barcode symbol on the Hyb chamber.
13. Dispense 30 μL assay sample onto the large sample port of each array.
14. Seal lid onto the Hyb chamber carefully to avoid dislodging the Hyb chamber insert(s).
15. Incubate for 16–20 h at 58°C with rocker speed at 5 (manufacturer's unit).
16. Prepare 1× high-temp wash buffer (add 50 ml 10× stock to 450 ml RNase-free water).
17. Place waterbath insert into heat block, and add 500 ml prepared 1× high-temp wash buffer.
18. Set heat block temperature to 55°C and pre-warm high-temp wash buffer to that temperature.
19. Close heat block lid and leave overnight.

3.6. Washing and Scanning

1. Make wash E1BC solution: add 7 ml E1BC buffer to 2.5 l RNase-free water.
2. Pre-warm block E1 buffer (4 ml/chip) to room temperature.
3. Prepare block E1 buffer (2 ml/chip) with streptavidin-Cy3 (2 μL of 1 mg/ml stock per chip). Use a single conical tube for all BeadChips. Store in dark until detection step.
4. Remove Hyb chamber from oven and disassemble.
5. Remove cover seal from the BeadChip with Beadchip submerged in ~1 l E1BC-buffer (use glass dish for this step). Using tweezers or powder-free gloved hands, place the BeadChip into the slide rack submerged in the staining dish containing 250 ml wash E1BC solution.
6. Repeat disassembly and placement in E1BC solution for all BeadChips.
7. Using the slide rack handle, transfer the rack into the Hybex waterbath insert containing high-temp wash buffer.

8. Incubate static for 10 min with the Hybex lid closed.
9. During the 10-min high-temp wash buffer incubation, add fresh 250 ml wash E1BC solution to a clean staining dish.
10. After the 10-min high-temp wash buffer incubation is complete, immediately transfer the slide rack into the staining dish containing E1BC.
11. Briefly agitate using rack, then shake on orbital shaker for 5 min at the highest speed possible without allowing solution to splash out of dish.
12. Transfer rack to a clean staining dish containing 250 ml 100% ethanol (use fresh from ethanol source bottle).
13. Briefly agitate using rack handle, then shake on orbital shaker for 10 min.
14. Transfer rack to a clean staining dish containing fresh 250 ml wash E1BC solution.
15. Briefly agitate using rack handle, then shake on orbital shaker for 2 min
16. Pipet 4 ml block E1 buffer into the wash tray(s).
17. Transfer the BeadChip, face up into BeadChip wash tray(s) on rocker.
18. Rock at medium speed for 10 min.
19. Pipet 2 ml block E1 buffer + streptavidin-Cy3 into fresh wash tray(s).
20. Transfer the BeadChip, face up into wash tray(s) on rocker.
21. Place cover on tray and rock at medium speed for 10 min.
22. Add 250 ml of wash E1BC solution to a clean staining dish.
23. Transfer the BeadChip to the slide rack submerged in the staining dish.
24. Briefly agitate using rack, and then shake at room temperature on orbital shaker for 5 min with Beadchips protected from light.
25. Prepare centrifuge with plateholders, paper towels, and balance rack. Set speed to 275 rcf.
26. Transfer rack of BeadChips from staining dish to centrifuge and spin at room temperature for 4 min.
27. Store dry chips in the dark until scanned.
28. BeadChips are imaged using the Illumina BeadArray reader, a two-channel, 0.8 μm resolution confocal laser scanner. The BeadArray reader, using SentrixScan menu-driven software, automatically scans up to three different BeadChips or a SAM, hands-free, at one or at two wavelengths and creates an image file for each channel.

3.7. Data Extraction Using Illumina BeadStudio Software

Image analysis and data extraction is performed using Illumina BeadStudio software.

1. Using the gene expression module of BeadStudio, load data into BeadStudio.

2. Make sure to choose "no normalization" when asked for the normalization method to use.
3. For further analysis using IlluminaGUI, a data file has to be exported. Mark the "Group gene profile" or the "Sample gene profile" on the right side and use the column chooser method to create a file with the following columns, ProbeID, AvgSignal Array1, Detection Pval Array1, AvgSignal Array2, Detection Pval Array2, and so on.
4. Export a tab-delimited file using the "export data to file" method.

3.8. Data Analysis Using IlluminaGUI

We have implemented a software package called IlluminaGUI which includes a graphical user interface and can be used for analyzing microarray data from the Illumina BeadChip platform *(36)*. All key components of a microarray experiment, including quality control, normalization, inference and classification methods are provided in a 'point and click' approach. After downloading and installing the package the first step is to load data into IlluminaGUI. Detailed descriptions of both the installation process and the loading process can be found on the corresponding webpage (www. IlluminaGUI.dnsalias.org).

3.8.1. Quality Control

Quality control of microarray data is the first and probably one of the most important steps in a microarray analysis. Using IlluminaGUI's quality control button, quality of the data can be examined in three different ways:

1. Displaying a control summary plot based on built-in controls: The control summary plot displays different built-in controls, namely hybridization controls, negative versus housekeeping controls, controls for high and low stringency as well as for labeling and background.
2. Displaying basic diagnostic plots: Diagnostic plots include box plots, MA plots *(37)* and pairwise scatter plots. Box plots can be used to determine the distribution of intensity signals across an array, thereby verifying the comparability of all arrays within an experiment. MA and pairwise scatter plots are especially useful for the detection of the intensity-dependent differences of signals between any two arrays in the experiment.
3. Determination of absent resp. present status of probes: Determination of absent resp. present status of probes can be used to investigate sensitivity differences of arrays in an experiment. IlluminaGUI reports the percentage of present probes on each array, thereby indicating a probe as present, if the detection *p*-value calculated by the Illumina software <0.05. Within an experiment, similar percentages for each array should be achieved.
4. If any of the quality control plots indicates an outlier in the data set, the array should be removed from further analysis.

3.8.2. Normalization

During a microarray experiment, different sources of systematic variation can affect the measured gene expression levels, including unequal quantities of starting material, differences in labeling or detection efficiencies between one experiment and the other. A normalization process is therefore used to remove such variation to be able to detect biological differences between samples. In IlluminaGUI one can choose from three different normalization techniques which are all based on different assumptions concerning the nature of the raw data. These are the quantiles-method *(38)*, the vsn-method *(39)* and the qspline-method *(40)*. Depending on the data set, a different normalization technique gives better results. As a default method, we are using the quantiles-method since it has shown to perform best over a wide variety of data sets. However, all mentioned normalization techniques can be tested within an analysis and the best performing technique can then be used for further analysis.

3.8.3. Inference

IlluminaGUI provides several statistical methods for identifying differentially expressed genes. Generally we are using the combined *t*-test/fold change analysis. During this analysis the user can choose different criteria a gene must meet to be called differentially expressed. These criteria include the fold change, i.e., the fold difference of means between the investigated groups, the *p*-value which gives a statistical measure of the significant difference of the groups, and the difference in means, i.e., the actual difference of means. Our default criteria are a fold change of 2, a *p*-value of <0.05 and a difference of means of >100.

3.8.4. Classification

Classification is an important question in microarray experiments, for purposes of classifying biological samples and predicting clinical or other outcomes using gene expression data. One discriminates between unsupervised and supervised methods of classification:

1. Unsupervised classification, also known as cluster analysis or clustering: Clustering is the classification of objects into different groups, or more precisely, the partitioning of a data set into subsets (clusters), so that the data in each subset (ideally) share some common trait – often proximity according to some defined distance measure class discovery. In our experiments, we are typically using the hierarchical clustering method in IlluminaGUI and are visualizing the results using a heatmap.
2. Supervised classification, also known as class prediction: In supervised classification the class of each sample in the data set is predefined. The task is to understand the basis for the classification from this data set (which is called training or learning set) which is achieved by using various classification algorithms. This information is then used to classify future

samples into one of the predefined classes. In IlluminaGUI, we are generally using two different methods, nearest shrunken centroids (PAM) *(41)* and support vector machines (SVMs). With both methods, the user obtains written as well as graphical outputs of the results. For example, when using PAM, the accuracy of each prediction, the overall accuracy and the certainty of a prediction can be exported to an Excel file. In addition, a probability plot displaying the result of all predictions is created.

3.9. RNA Preparation Alternatives

Blood is a heterogeneous tissue consisting of different cell types: erythrocytes, granulocytes, and other PBMC. The complex cellular nature of blood samples makes it difficult to detect differential gene expression that occurs in only a subset of these cell types. Also, whole blood mRNA consists of a relatively large proportion of globin mRNA transcripts which might represent as much as 70% of the total mRNA population. These "unwanted" globin transcripts effectively dilute the mRNA population and decrease the sensitivity of detecting less abundant mRNAs using microarray technology. Evidence of the dilution of mRNA by highly abundant globin mRNA has been observed on Affymetrix® GeneChip® arrays, where expression profiles from whole blood RNA show decreased percent present calls and increased variability in comparison to RNA from the PBMC fraction of whole blood *(29)*. Therefore, several approaches have been developed to remove the unwanted globin mRNA transcripts from whole blood RNA samples, prepared with the PAXgene system.

3.9.1. Globin Reduction by RNase H Digestion

Globin reduction can be performed according to the Affymetrix technical note *(42)*.

1. Prepare the 10× globin reduction oligo mix with the following final concentration:

   α1: 5′-TGCAGGAAGGGGAGGAGGGGCTG-3′ (7.5 μM);
   α2: 5′-TGCAAGGAGGGGAGGAGGGCCCG-3′ (7.5 μM);
   ß: 5′-CCCCAGTTTAGTAGTTGGACTTAGGG-3′ (20 μM).

2. Prepare the hybridization mix: 3–10 μg total RNA from whole blood, 2 μl 10× globin reduction oligo mix, 1 μl 10× Oligo Hyb buffer, with RNase-free water to a final volume of 10 μL.
3. Incubate in a thermal cycler at 70°C for 5 min, and then cool to 4°C.
4. Immediately proceed to RNaseH digestion.
5. Before starting, dilute an appropriate amount of RNase H (10 U/μL) tenfold to 1 U/μL with 1× RNase H buffer.

6. Prepare the RNase H reaction mix: 2 μl 10× RNase H buffer, 1 μl SUPERase•In™, 2 μl diluted RNase H (1 U/μL), 5 μl RNase-free water.
7. Add 10 μL of the RNase H reaction mix to each RNA:globin reduction oligo hybridization sample and mix thoroughly.
8. Incubate at 37°C for 10 min and cool to 4°C.
9. As soon as the RNase H digestion is complete, add 1 μL of 0.5 M EDTA to each sample to stop the reaction and proceed immediately to the cleanup step. Important: Do not leave the samples in the RNase H digestion step at 4°C for an extended period of time. Prolonged incubation may result in undesired non-specific digestion of the sample.
10. Purify RNAse H treated RNA samples with IVT cRNA cleanup spin columns from the Affymetrix GeneChip sample cleanup module.
11. Add 80 μl of RNase-free water to the globin-depleted RNA sample and mix by vortexing for 3 s.
12. Add 350 μl IVT cRNA binding buffer to the sample and mix by vortexing for 3 s.
13. Add 250 μl ethanol (96–100%) to the lysate and mix well by pipetting. Do not centrifuge.
14. Apply sample (700 μl) to the IVT cRNA cleanup spin column sitting in a 2 ml collection tube. Centrifuge for 15 s at $\geq 8,000 \times g$ ($\geq 10,000$ rpm). Discard flow-through and collection tube.
15. Transfer the spin column into a new 2 ml collection tube (supplied). Pipet 500 μl IVT cRNA wash buffer onto the spin column. Centrifuge for 15 s at $\geq 8,000 \times g$ ($\geq 10,000$ rpm) to wash. Discard flow-through.
16. Pipet 500 μl 80% (v/v) ethanol onto the spin column and centrifuge for 15 s at $\geq 8,000 \times g$ ($\geq 10,000$ rpm). Discard flow-through.
17. Open the cap of the spin column and centrifuge for 5 min at maximum speed ($< 25,000 \times g$). Discard flow-through and collection tube. Place columns into the centrifuge using every second bucket. Position caps over the adjoining bucket so that they are oriented in the opposite direction to the rotation, i.e., if the microcentrifuge rotates in a clockwise direction, orient the caps in a counterclockwise direction. This avoids damage of the caps. Centrifugation with open caps allows complete drying of the membrane. Transfer spin column into a new 1.5 ml collection tube (supplied), and pipet 14 μl of RNase-free water directly onto the spin column membrane. Ensure that the water is dispensed directly onto the membrane. Centrifuge 1 min at maximum speed ($\leq 25,000 \times g$) to elute.
18. Collect the eluate, apply again to the center of column, and spin at $\geq 8,000 \times g$ ($\geq 10,000$ rpm) for another

minute. The final recovery volume after the second spin is approximately 13 μL and the recovery of the cleanup step is ~75%.

19. The quality and integrity of the RNAse H treated RNA samples can be assessed as described under Section 3.3 or alternatively can be assessed by Agilent 2100 Bioanalyzer analysis prior to cDNA synthesis and in vitro transcription.

3.9.2. Globin Reduction by Magnetic Beads (Ambion GlobinClear Kit)

1. Thaw globin capture oligos and RNA samples.
2. Preheat hybridization oven to 50°C.
3. Preincubate 2 × hybridization buffer and streptavidin bead buffer at least 15 min at 50°C before use. Vortex well before use. Within this time start to prepare streptavidin magnetic beads.
4. In a 1.5 ml tube, combine per reaction 4 μl RNA bead buffer with 10 μl RNA binding beads and mix briefly. Then add 6 μl 100% isopropanol and mix thoroughly by vortexing. Store at room temperature. This mixture is the bead resuspension mix.
5. For preparation of streptavidin magnetic beads, you need 30 μl of streptavidin magnetic beads for each sample; calculate the volume of beads needed for the samples being processed that day.
6. Vortex the tube of the streptavidin magnetic beads to suspend the settled beads, and transfer the volume needed into a 1.5 ml non-stick tube.
7. Briefly centrifuge (<2 s) at low speed (< 1,000 × *g*) to collect the mixture at the bottom of the tube.
8. Place the tube on a magnetic stand to capture the streptavidin magnetic beads. Leave the tube on the magnetic stand until the mixture becomes transparent (~3–5 min), indicating that capture is complete.
9. Carefully aspirate the supernatant using a pipet without disturbing the streptavidin magnetic beads. Discard the supernatant, and remove the tube from the magnetic stand.
10. Add streptavidin bead buffer to the streptavidin magnetic beads; use a volume equal to the original volume of streptavidin magnetic beads. Vortex vigorously until the beads are resuspended.
11. Place the prepared streptavidin magnetic beads beads at 50°C (hybridization oven recommended) and immediately proceed to the next step. The beads should remain at 50°C for at least 15 min before they are used in globin removal step.
12. For hybridization of globin mRNA and globin capture oligonucleotides combine the following in a 1.5 ml non-stick tube: 1–10 μg RNA in a maximum volume of 14 μl, 1 μl of

capture oligo mix, If necessary, add RNase-free water to the sample mixture to a final volume of 15 μl.

13. Add 15 μl of 50°C 2× hybridization buffer to the sample.
14. Vortex briefly to mix and centrifuge briefly at low speed to collect the contents in the bottom of the tube.
15. Place the sample in a prewarmed 50°C incubator and allow the globin capture oligo mix to hybridize to the globin mRNA for 15 min. Both the time and temperature of this incubation are important for efficient globin mRNA depletion. Proceed to Removal of Globin mRNA:
16. Remove the prepared streptavidin magnetic beads from the 50°C incubator, and resuspend them by gentle vortexing. Briefly centrifuge (<2 s) at low speed (< 1,000 × *g*) to collect the mixture at the bottom of the tube.
17. Add 30 μl of prepared streptavidin magnetic beads to each RNA sample.
18. Vortex to mix well and centrifuge briefly at low speed as in the previous steps to collect the contents in the bottom of the tube.
19. Flick the tube very gently to resuspend the beads, being careful to keep the contents at the bottom of the tube.
20. Place the RNA bead mixture at 50°C (hybridization oven recommended) and incubate for 30 min.
21. Remove sample from the incubator, and vortex briefly to mix. Then centrifuge briefly at low speed as in the previous steps to collect the contents in the bottom of the tube.
22. Capture the streptavidin magnetic beads on a magnetic stand. Leave the tube on the magnetic stand until the mixture becomes transparent (~3–5 min), indicating that capture is complete.
23. Carefully draw up the supernatant, which contains the globin mRNA depleted RNA, using a pipet without disturbing the streptavidin magnetic beads. Transfer the RNA to a new 1.5 ml non-stick tube supplied with the kit, and place on ice. The supernatant contains the GLOBINclear RNA; do not discard the supernatant.

    Proceed to Purification of the GLOBINClear RNA:
24. Prewarm the elution buffer to 58°C.
25. Vortex the bead resuspension mix and then immediately dispense 20 μl to each sample. It is important to resuspend the beads thoroughly before adding them to the samples.
26. Vigorously vortex the sample for 10 s to fully mix the reagents, and to allow the RNA binding beads to bind the RNA.
27. Briefly centrifuge (<2 s) at low speed (< 1,000 × *g*) to collect the mixture at the bottom of the tube.
28. Capture the RNA binding beads by placing the tube on a magnetic stand. Leave the tube on the magnetic stand until

the mixture becomes transparent (~3–5 min), indicating that capture is complete.

29. Carefully aspirate the supernatant using a pipet without disturbing the RNA binding beads. Discard the supernatant. It is important to remove as much of the supernatant as possible at this step.
30. Remove the tube from the magnetic stand. It is critical for effective washing to remove the tube from the magnetic stand before adding the RNA wash solution.
31. Add 200 μl RNA wash solution to each sample and vortex for 10 s. The RNA binding beads may not fully disperse during this step; this is expected, and it will not affect RNA purity or yield.
32. Briefly centrifuge (<2 s) at low speed (< 1,000 × *g*) to collect the mixture at the bottom of the tube.
33. Capture the RNA binding beads on a magnetic stand as in the previous magnetic bead capture steps.
34. Carefully aspirate and discard the supernatant, and remove the tube from the magnetic stand.
35. Briefly centrifuge the tube as in previous steps and place it back on the magnetic stand.
36. Remove any liquid in the tube with a small-bore pipet tip.
37. Remove the tube from the magnetic stand and allow the beads to air-dry for 5 min with the caps left open. Do not air-dry the beads for >5 min.
38. Add 30 μl warm (58°C) Elution buffer to each sample, and vortex vigorously for ~10 s to thoroughly resuspend the RNA binding beads.
39. Incubate the mixture at 58°C for 5 min. This incubation is critical for optimal RNA recovery.
40. Vortex the sample vigorously for ~10 s to thoroughly resuspend the RNA binding beads and centrifuge briefly at low speed as in previous steps to collect the mixture at the bottom of the tube.
41. Capture the RNA binding beads on a magnetic stand as in the previous magnetic bead capture steps. Be especially careful at this step to avoid disturbing the RNA binding beads when collecting the supernatant. The purified RNA will be in the supernatant.
42. Transfer the supernatant containing the RNA to a 1.5 ml non-stick tube. Store the purified RNA at –20 or –80°C. Frequently some of the RNA binding beads are carried over to the eluate, tinting it brownish. The RNA binding beads have been shown to have no effect on absorbance readings or downstream enzymatic applications. If desired, however, magnetic bead contamination can usually be reduced by recapturing the beads on a magnetic stand for 3–5 min, and transferring the RNA solution to a new tube.

3.10. Microarray Procedure Alternatives Using Affymetrix Arrays

3.10.1. Preparation of Poly-A RNA Controls for One-Cycle cDNA Synthesis

1. Each eukaryotic GeneChip probe array contains probe sets for several *Bacillus subtilis* genes that are absent in eukaryotic samples (*lys*, *phe*, *thr*, and *dap*). These poly-A RNA controls are in vitro synthesized, and the polyadenylated transcripts for the *B. subtilis* genes are premixed at staggered concentrations. The concentrated poly-A control stock can be diluted with the poly-A control dilution buffer and spiked directly into RNA samples (spike-in controls) to achieve the final concentrations (referred to as a ratio of copy number) as follows: *lys*, 1:100,000; *phe*, 1:50,000; *thr*, 1:25,000; *dap*, 1:7,500.
2. The controls are then amplified and labeled together with the samples. Examining the hybridization intensities of these controls on GeneChip arrays helps to monitor the labeling process independently from the quality of the starting RNA samples.
3. The poly-A RNA control stock and poly-A control dilution buffer are provided with the kit to prepare the appropriate serial dilutions based on **Table 13.1**. This is a guideline when 1, 5, or 10 μg of total RNA is used as starting material (*see* **Note 4**).
4. For example, to prepare the poly-A RNA dilutions for 5 μg of total RNA: add 2 μl of the poly-A control stock to 38 μl of poly-A control dilution buffer for the first dilution (1:20).
5. Mix thoroughly and spin down to collect the liquid at the bottom of the tube. This dilution can be stored up to six weeks at –20°C and frozen–thawed up to eight times.
6. Add 2 μl of the first dilution to 98 μl of poly-A control dilution buffer to prepare the second dilution (1:50).
7. Mix thoroughly and spin down to collect the liquid at the bottom of the tube.
8. Add 2 μl of the second dilution to 18 μl of poly-A control dilution buffer to prepare the third dilution (1:10).
9. Mix thoroughly and spin down to collect the liquid at the bottom of the tube.
10. Add 2 μl of the third dilution to 5 μg of sample total RNA.

Table 13.1
Serial dilutions of the poly-A RNA control stock for different amounts of total RNA used

Starting amount of total RNA (μg)	First dilution	Second dilution	Third dilution	Spike-in volume (μl)
1	1:20	1:50	1:50	2
5	1:20	1:50	1:10	2
10	1:20	1:50	1:5	2

3.10.2. First Strand cDNA Synthesis

1. Place 1–15 μg total RNA in a maximum total volume of 7 μl (for 8.1–15 μg total RNA) or 8 μl (1–8 μg total RNA) in a 1.5 ml PCR tube.
2. Add 2 μl of the appropriately diluted poly-A RNA controls.
3. Add 2 μl of 50 μM T7-oligo(dT) primer.
4. Add RNase-free water to a final volume of 11 μl (for 8.1–15 μg total RNA) or 12 μl (for 1–8 μg total RNA).
5. Gently flick the tube a few times to mix, and then centrifuge briefly (~5 s) to collect the reaction at the bottom of the tube.
6. Incubate the reaction for 10 min at 70°C.
7. Cool the sample at 4°C for at least 2 min.
8. Centrifuge the tube briefly (~5 s) to collect the sample at the bottom of the tube.
9. In a separate tube, assemble the first strand master mix. Prepare sufficient first strand master mix for all of the RNA samples. When there are more than two samples, it is prudent to include additional material to compensate for potential pipetting inaccuracy or solution lost during the process. The following recipe is for a single reaction: 4 μl 5× first strand reaction mix, 2 μl 0.1 M DTT, 1 μl 10 mM dNTP.
10. Mix well by flicking the tube a few times. Centrifuge briefly (~5 s) to collect the master mix at the bottom of the tube.
11. Transfer 7 μl of first strand master mix to each RNA/T7-oligo(dT) primer mix for a final volume of 18 or 19 μl. Mix thoroughly by flicking the tube a few times. Centrifuge briefly (~5 s) to collect the reaction at the bottom of the tube, and immediately place the tubes at 42°C.
12. Incubate for 2 min at 42°C.
13. Add the appropriate amount of SuperScript II to each RNA sample for a final volume of 20 μl. For 1–8 μg total RNA – 1 μl SuperScript II. For 8.1–15 μg of total RNA – 2 μl SuperScript II.
14. Mix thoroughly by flicking the tube a few times. Centrifuge briefly (~5 s) to collect the reaction at the bottom of the tube, and immediately place the tubes at 42°C.
15. Incubate for 1 h at 42°C; then cool the sample for at least 2 min at 4°C. After incubation at 4°C, centrifuge the tube briefly (~5 s) to collect the reaction at the bottom of the tube and immediately proceed to second strand cDNA synthesis.

3.10.3. Second Strand cDNA Synthesis

1. In a separate tube, assemble second strand master mix. Prepare sufficient second strand master mix for all of the samples. When there are more than two samples, it is prudent to include additional material to compensate for potential pipetting inaccuracy or solution lost during the process. The following recipe is for a single reaction: 91 μl RNAse-free water, 30 μl 5× second strand reaction mix, 3 μl dNTP (10 mM

each), 1 μl *E. coli* DNA ligase, 4 μl *E. coli* DNA-polymerase I, 1 μl RNase H.

2. Mix well by gently flicking the tube a few times. Centrifuge briefly (~5 s) to collect the solution at the bottom of the tube.
3. Add 130 μl of second strand master mix to each first strand synthesis sample from first strand cDNA synthesis for a total volume of 150 μl.
4. Gently flick the tube a few times to mix, and then centrifuge briefly (~5 s) to collect the reaction at the bottom of the tube.
5. Incubate for 2 h at 16°C.
6. Add 2 μl of T4 DNA polymerase to each sample and incubate for 5 min at 16°C.
7. After incubation with T4 DNA polymerase add 10 μl of EDTA, 0.5 M and proceed to cleanup of double-stranded cDNA or store at –20°C.
8. Do not leave the reactions at 4°C for long periods of time.

3.10.4. Cleanup of Double-Stranded cDNA with Sample Cleanup Module

cDNA wash buffer is supplied as a concentrate. Before using for the first time, add 24 ml of ethanol (96–100%), as indicated on the bottle, to obtain a working solution. All steps of the protocol should be performed at room temperature. During the procedure, work without interruption.

1. Add 600 μl cDNA binding buffer to the 162 μl final double-stranded cDNA synthesis preparation. Mix by vortexing for 3 s.
2. Check that the color of the mixture is yellow (similar to cDNA binding buffer without the cDNA synthesis reaction). If the color of the mixture is orange or violet, add 10 μl of 3 M sodium acetate, pH 5.0, and mix. The color of the mixture will turn to yellow.
3. Apply 500 μl of the sample to the cDNA cleanup spin column sitting in a 2 ml collection tube, and centrifuge for 1 min at 8,000 × *g* (≥ 10,000 rpm). Discard flow-through.
4. Reload the spin column with the remaining mixture (262 μl) and centrifuge as step 3. Discard flow-through and collection tube.
5. Transfer spin column into a new 2 ml collection tube (supplied). Pipet 750 μl cDNA wash buffer onto the spin column. Centrifuge for 1 min at 8,000 × *g* (≥ 10,000 rpm). Discard flow-through.
6. Open the cap of the spin column and centrifuge for 5 min at maximum speed (≤ 25,000 ×*g*). Discard flow-through and collection tube.
7. Place columns into the centrifuge using every second bucket. Position caps over the adjoining bucket so that they are oriented in the opposite direction to the rotation, i.e., if the microcentrifuge rotates in a clockwise direction, orient the caps in a counterclockwise direction. This avoids damage of

the caps. Centrifugation with open caps allows complete drying of the membrane.

8. Transfer spin column into a 1.5 ml collection tube, and pipet 14 μl of cDNA Elution buffer directly onto the spin column membrane. Incubate for 1 min at room temperature and centrifuge 1 min at maximum speed ($\leq 25,000 \times g$) to elute. Ensure that the cDNA elution buffer is dispensed directly onto the membrane. The average volume of eluate is 12 μl from 14 μl elution buffer.

3.10.5. Synthesis of Biotin-Labeled cRNA with GeneChip IVT Labeling Kit

1. Determine the amount of cDNA used for each IVT reaction as follows: for initial total RNA input of 10 ng to 8 μg use complete cDNA (~12 μl); for initial total RNA input of 8.1–15 μg use 6 μl cDNA.
2. Transfer the needed amount of template cDNA to RNase-free microfuge tubes and add the following reaction components: 4 μl 10× IVT labeling buffer, 12 μl IVT labeling NTP-mix, 4 μl IVT labeling enzyme mix, with RNase-free water adjust to final volume of 40 μl (*see* **Note 5**).
3. Carefully mix the reagents and collect the mixture at the bottom of the tube by brief (~5 s) microcentrifugation.
4. Incubate at 37°C for 16 h. To prevent condensation that may result from water bath-style incubators, incubations are best performed in oven incubators for even temperature distribution, or in a thermal cycler.
5. Store labeled cRNA at –20 or –70°C if not purifying immediately. Alternatively, proceed to cleanup and quantification of biotin-labeled cRNA.

3.10.6. Cleanup of Biotin-Labeled cRNA with Sample Cleanup Module

Save an aliquot (1 μl) of the unpurified IVT product for analysis by gel electrophoresis. IVT cRNA wash buffer is supplied as a concentrate. Before using for the first time, add 20 ml of ethanol (96–100%), as indicated on the bottle, to obtain a working solution, and checkmark the box on the left-hand side of the bottle label to avoid confusion(*see* **Note 6**). All steps of the protocol should be performed at room temperature. During the procedure, work without interruption.

1. Add 60 μl of RNase-free water to the in vitro transcription reaction and mix by vortexing for 3 s.
2. Add 350 μl IVT cRNA binding buffer to the sample and mix by vortexing for 3 s.
3. Add 250 μl ethanol (96–100%) to the lysate, and mix well by pipetting. Do not centrifuge.
4. Apply sample (700 μl) to the IVT cRNA cleanup spin column sitting in a 2 ml collection tube. Centrifuge for 15 s at $\geq 8,000 \times g$ ($\geq 10,000$ rpm). Discard flow-through and collection tube.

5. Transfer the spin column into a new 2 ml collection tube (supplied). Pipet 500 μl IVT cRNA wash buffer onto the spin column. Centrifuge for 15 s at $\geq 8,000 \times g$ ($\geq$10,000 rpm) to wash. Discard flow-through.
6. Pipet 500 μl 80% (v/v) ethanol onto the spin column and centrifuge for 15 s at $\geq 8,000 \times g$ ($\geq$ 10,000 rpm). Discard flow-through.
7. Open the cap of the spin column and centrifuge for 5 min at maximum speed ($\leq 25,000 \times g$). Discard flow-through and collection tube. Place columns into the centrifuge using every second bucket. Position caps over the adjoining bucket so that they are oriented in the opposite direction to the rotation, i.e., if the microcentrifuge rotates in a clockwise direction, orient the caps in a counterclockwise direction. This avoids damage of the caps. Centrifugation with open caps allows complete drying of the membrane.
8. Transfer spin column into a new 1.5 ml collection tube (supplied), and pipet 11 μl of RNase-free water directly onto the spin column membrane. Ensure that the water is dispensed directly onto the membrane. Centrifuge 1 min at maximum speed ($\leq 25,000 \times g$) to elute.
9. Pipet 10 μl of RNase-free water directly onto the spin column membrane. Ensure that the water is dispensed directly onto the membrane. Centrifuge 1 min at maximum speed ($\leq 25,000 \times g$) to elute.
10. The minimum concentration for purified cRNA is 0.6 μg/μl for the following fragmentation reaction in fragmenting the cRNA for target preparation.

3.10.7. cRNA Quantification

1. Use spectrophotometric analysis to determine the cRNA yield. For photometric quantification of the purified cRNA, we recommend dilution of the eluate between 1:100- and 1:200-fold. Apply the convention that one absorbance unit at 260 nm equals 40 μg/ml cRNA.
2. Check the absorbance at 260 and 280 nm to determine sample concentration and purity. Maintain the A_{260}/A_{280} ratio close to 2.0 for pure RNA (ratios between 1.9 and 2.3 are acceptable).
3. For quantification of cRNA when using total RNA as starting material, an adjusted cRNA yield must be calculated to reflect carryover of unlabeled total RNA. Using an estimate of 100% carryover, determine the adjusted cRNA yield as follows:

$$\text{adjusted cRNA yield} = \text{RNA}_\text{m} - (\text{total RNA}i * y)$$

RNA_m, amount of cRNA measured after IVT (μg)
total RNAi, starting amount of total RNA (μg)
y, fraction of cDNA reaction used in IVT.

Assuming 12 μl resuspension/elution volume for cDNA use the following y-values: $y = 1$ for 12 μl, $y = 0.83$ for 10 μl, $y = 0.42$ for 5 μl, $y = 0.33$ for 3.3 μl, $y = 0.21$ for 2.5 μl, $y = 0.17$ for 2 μl (*see* **Note 7**).

3.10.8. Checking Unfragmented Samples by Gel Electrophoresis

1. Save an aliquot (1 μg) of the purified IVT product for analysis by gel electrophoresis. Gel electrophoresis of the IVT product is done to estimate the yield and size distribution of labeled transcripts. Parallel gel runs of unpurified and purified IVT product can help determine the extent of a loss of sample during the cleanup process.
2. Mix RNA (samples or markers) with loading dye and TE-buffer and heat to 65°C for 5 min before loading on the gel.
3. Ethidium bromide can be used to visualize the RNA in the gel.

3.10.9. Fragmentation of cRNA for Target Preparation

1. The cRNA used in the fragmentation procedure should be sufficiently concentrated to maintain a small volume during the procedure. This will minimize the amount of magnesium in the final hybridization cocktail. The cRNA must be at a minimum concentration of 0.6 μg/μl. Fragment an appropriate amount of cRNA for hybridization cocktail and gel analysis. Use 1 μg of the fragmeted cRNA for gel-analysis.
2. The cRNA must be at a minimum concentration of 0.6 μg/μl. If it is not, it can be concentrated by vacuum centrifugation.
3. Add 2 μl of 5× fragmentation buffer for every 8 μl of RNA plus H_2O. The fragmentation buffer has been optimized to break down full-length cRNA to 35–200 bases fragments by metal-induced hydrolysis. The final concentration of RNA in the fragmentation mix can range from 0.5 to 2 μg/μl. Example for fragmentation of 20 μg cRNA mix components as follows: 20 μg cRNA in a volume of 1–32 μl, 8 μl 5× fragmentation buffer, with RNase-free water adjust to 40 μl total volume.
4. Incubate at 94°C for 35 min. Put on ice following the incubation.
5. Save an aliquot (1 μg) for gel analysis. The standard fragmentation procedure should produce a distribution of RNA fragment sizes from approximately 35–200 bases.
6. Store undiluted, fragmented sample RNA at –20°C until ready to perform the hybridization.

3.10.10. Array Hybridization

1. Equilibrate probe array to room temperature immediately before use. It is important to allow the array to normalize to room temperature completely. Meanwhile, mix the components as given in **Table 13.2** (*see* **Note 8**). Scale-up volumes for hybridization to multiple probe arrays.
2. Heat the hybridization cocktail to 99°C for 5 min in a heat block.

Table 13.2
Components and volumes for target hybridization to standard arrays

Component	Volume (standard array)	Final concentration
Fragmented cRNA	15 μg	0.05 μg/μl
Oligonucleotide B2 (3 nM)	5 μl	50 pM
Eukaryotic hybridization controls (20×)	15 μl	1.5, 5, 25, and 100 pM
Herring sperm DNA 10 mg/ml	3 μl	0.1 mg/ml
Acetylated BSA 50 mg/ml	3 μl	0.5 mg/ml
Hybridization buffer (2×)	150 μl	1×
H_2O	To final of 300 μl	

3. Meanwhile, wet the array by filling it through one of the septa with appropriate volume 1× hybridization buffer (**Table 13.3**).
4. Incubate the probe array filled with 1× hybridization buffer at 45°C for 10 min with rotation (60 rpm).
5. Transfer the hybridization cocktail that has been heated at 99°C to a 45°C heat block or hybridization oven for 5 min.
6. Spin hybridization cocktail(s) at maximum speed in a microcentrifuge for 5 min to remove any insoluble material from the hybridization mixture.
7. Remove the buffer solution from the probe array cartridge and fill with appropriate volume (*see* **Table 13.3**) of the clarified hybridization cocktail avoiding any insoluble matter in the volume at the bottom of the tube.

Table 13.3
Appropriate volume of 1× hybridization buffer for different array formats

Array format	Hybridization volume (μl)	Total fill volume (μl)
Standard	200	250
Midi	130	160
Mini	80	100
Micro	80	100

8. Place probe array in rotisserie box in 45°C oven. Rotate at 60 rpm.
9. Hybridize for 16 h.

3.10.11. Washing, Staining, and Scanning of the Arrays

1. After 16 h of hybridization, remove the hybridization cocktail from the probe array and set it aside in a microcentrifuge tube. Store on ice during the procedure or at –80°C for long-term storage.
2. Fill the probe array completely with the appropriate volume of non-stringent wash buffer (**Table 13.3**).
3. Defining file locations: Launch Microarray Suite from the workstation and select Tools ⇒ Defaults ⇒ Database ⇒ LIMS-Mode
4. Entering experiment name and information. Select Run ⇒ Experiment Info from the menu bar, fill in all necessary information. Save the experiment by selecting Save.
5. Prime the fluidic station: Change the intake buffer reservoir A to non-stringent wash buffer and intake buffer reservoir B to stringent wash buffer. For the water reservoir use autoclaved, distilled water (Ampuwa).
6. Select Run ⇒ Fluidics from the menu bar. Choose the Prime protocol from the protocol drop-down list for the respective modules. Click run for each module to begin priming. Repeat the Prime protocol for a second time.
7. Prepare the SAPE stain solution (volumes given are sufficient for one probe array): 600 μl 2 × MES stain buffer, 48 μl acetylated BSA, 12 μl SAPE (1 mg/ml), 540 μl RNase-free water (*see* **Note 9**).
8. Mix well and centrifuge for 5 min at room temperature at maximum speed. Remove supernatant without carryover of insoluble material at the bottom of the tube and divide into two aliquots of 600 μl each to be used for stains 1 and 3, respectively.
9. Prepare the antibody solution mix: 300 μl 2 × MES stain buffer, 24 μl acetylated BSA, 6 μl 10 mg/ml normal goat IgG, 3.6 μl 0.5 mg/ml biotinylated antibody, 266.4 μl RNase-free water.
10. For washing and staining of microarrays, select the correct experiment name in the drop-down Experiment list in the Fluidic station dialog box on the workstation. Chose the appropriate Washing and Staining program (Test3: Micro1v1; U133A: EukGeWS2v4) and run. Follow the instructions on the LCD display of the fluidic station/workstation.
11. When washing/staining program has finished check the probe array window for large bubbles or air pockets. If bubbles are present, the array should be filled with non-stringent wash buffer A manually.

12. Shutting down the fluidics station. If the fluidic station will not be used at this day, place wash lines into a bottle filled with deionized water. Choose the Shutdown protocol from the protocol drop-down list and click Run. Repeat the Shutdown protocol a second time with water and a third time without fluids to remove residual water.
13. To scan the microarrays switch on laser before starting the workstation. Make sure that the scanner is warmed up prior to scanning by turning the laser on at least 15 min before use.
14. If probe array was stored at 4°C, equilibrate to room temperature before scanning.
15. If necessary, clean the glass surface of probe array with a non-abrasive towel or tissue before scanning. Do not use alcohol to clean glass.
16. Choose the experiment from the drop-down list to be scanned, click Run and follow the instructions.
17. If the scanner will not be used for longer time periods or after finishing at this day, choose "put laser off". When the cooling fan has turned down (after a few minutes) it is sure to switch off the scanner.

3.10.12. Data Extraction and Statistics

For Affymetrix data collection and assessment, Affymetrix Microarray Analysis Suite Version 5.0 program (MAS5.0) and dCHIP 1.3® can be used. According to standard microarray analysis methods, for data analysis using MAS5.0 the fluorescence intensity of each array is scaled to an overall intensity of 100 to enable comparison of all hybridization data. In dCHIP Affymetrix CEL files are imported and normalized to the median intensity using the perfect match (PM) model.

4. Notes

1. PAXgene™ blood RNA spin columns and buffers can be stored dry at room temperature (15–25°C) for up to 1 year. The proteinase K solution is stable for at least 1 year after delivery when stored at room temperature. To store for more than 1 year, it is suggested keeping the proteinase K at 2–8°C.
2. The cDNA binding buffer may form a precipitate if stored below room temperature. If a precipitate is visible, redissolve it by warming the solution to 37°C for up to 10 min and vortexing vigorously. Cool to room temperature before use.
3. Once the SA-Cy3 has been reconstituted, it is recommended to make 20 μL aliquots. Store these aliquots at –20°C. Once the aliquots have been thawed for use, store them in the dark at 4°C for up to one month. Do not freeze and thaw repeatedly. Protect Cy3 from light.

4. For starting sample amounts other than those listed here, calculations are needed in order to perform the appropriate dilutions to arrive at the same proportionate final concentration of the spike-in controls in the samples. Avoid pipetting solutions less than 2 μl in volume to maintain precision and consistency when preparing the dilutions.
5. If more than one IVT reaction is to be performed, a master mix can be prepared by multiplying the reagent volumes by the number of reactions. Do not assemble the reaction on ice, since spermidine in the 10× IVT labeling buffer can lead to precipitation of the template cDNA.
6. IVT cRNA binding buffer may form a precipitate upon storage. If necessary, redissolve by warming in a water bath at 30°C, and then place the buffer at room temperature.
7. Example: Starting with 10 μg total RNA, 50% of the cDNA reaction is added to the IVT, giving a yield of 50 μg cRNA. Therefore, adjusted cRNA yield = 50 μg cRNA – ((10 μg total RNA) × (0.5 cDNA reaction)) = 45.0 μg. Use adjusted yield for fragmentation procedure.
8. It is mandatory that frozen stocks of 20× GeneChip® eukaryotic hybridization control cocktail be heated to 65°C for 5 min before first use to completely resuspend the cRNA and then aliquot in small volumes (5 μl aliquots and 15 μl aliquots that can be stored at −20°C). The GeneChip® eukaryotic hybridization control cocktail can maximal freeze/thawed three times.
9. Streptavidin phycoerythrin (SAPE) should be stored in the dark at 4°C, either foil-wrapped or kept in an amber tube. Remove SAPE from refrigerator and tap the tube to mix well before preparing stain solution. Do not freeze concentrated SAPE or diluted SAPE stain solution. Always prepare the SAPE stain solution immediately before use.

Acknowledgments

This work was mainly supported by the Alexander von Humboldt Foundation via a Sofja-Kovalevskaja Award to JLS, JLS is a member of the National Genome Research Network (NGFN) in Germany.

References

1. Branca, M. (2003) Genetics and medicine. Putting gene arrays to the test. *Science* 300, 238.
2. Schubert, C. M. (2003) Microarray to be used as routine clinical screen. *Nat Med* 9, 9.
3. Alizadeh, A. A., Eisen, M. B., Davis, R. E., Ma, C., Lossos, I. S., Rosenwald, A., Boldrick, J. C., Sabet, H., Tran, T., Yu, X., Powell, J. I., Yang, L., Marti, G. E., Moore, T., Hudson, J. Jr., Lu, L,, Lewis, D. B., Tibshirani, R.,

Sherlock, G., Chan, W. C,, Greiner, T. C., Weisenburger, D. D., Armitage, J. O., Warnke, R., Levy, R., Wilson, W., Grever, M. R., Byrd, J. C., Botstein, D., Brown, P. O., Staudt, L. M. (2000) Distinct types of diffuse large B-cell lymphoma identified by gene expression profiling. *Nature* 403, 503–511.
4. Bittner, M., Meltzer, P., Chen, Y., Jiang, Y., Seftor, E., Hendrix, M., Radmacher, M., Simon, R., Yakhini, Z., Ben-Dor, A., Sampas, N., Dougherty, E., Wang, E., Marincola, F., Gooden, C., Lueders, J., Glatfelter, A., Pollock, P., Carpten, J., Gillanders, E., Leja, D., Dietrich, K., Beaudry, C., Berens, M., Alberts, D., Sondak, V. (2000) Molecular classification of cutaneous malignant melanoma by gene expression profiling. *Nature* 406, 536–540.
5. Golub, T. R., Slonim, D. K., Tamayo, P., Huard, C., Gaasenbeek, M., Mesirov, J. P., Coller, H., Loh, M. L., Downing, J. R., Caligiuri, M. A., Bloomfield, C. D., Lander, E. S. (1999) Molecular classification of cancer: class discovery and class prediction by gene expression monitoring. *Science* 286, 531–537.
6. Bullinger, L., Döhner, K., Bair, E., Fröhling, S., Schlenk, R. F., Tibshirani, R., Döhner, H., Pollack, J. R. (2004) Use of gene-expression profiling to identify prognostic subclasses in adult acute myeloid leukemia. *N Engl J Med* 350, 1605–1616.
7. Rosenwald, A., et al., (2002) The use of molecular profiling to predict survival after chemotherapy for diffuse large-B-cell lymphoma. *N Engl J Med* 346, 1937–1947.
8. Shipp, M. A., Ross, K. N., Tamayo, P., Weng, A. P., Kutok, J. L., Aguiar, R. C., Gaasenbeek, M., Angelo, M., Reich, M., Pinkus, G. S., Ray, T. S., Koval, M. A., Last, K. W., Norton, A., Lister, T. A., Mesirov, J., Neuberg, D. S., Lander, E. S., Aster, J. C., Golub, T. R. (2002) Diffuse large B-cell lymphoma outcome prediction by gene-expression profiling and supervised machine learning. *Nat Med* 8, 68–74.
9. Valk, P. J., Verhaak, R. G., Beijen, M. A., Erpelinck, C. A., Barjesteh van Waalwijk van Doorn-Khosrovani, S., Boer, J. M., Beverloo, H. B., Moorhouse, M. J., van der Spek, P. J., Löwenberg, B., Delwel, R. (2004) Prognostically useful gene-expression profiles in acute myeloid leukemia. *N Engl J Med* 350, 1617–1628.
10. van de Vijver, M. J., He, Y. D., van't Veer, L. J., Dai, H., Hart, A. A., Voskuil, D. W., Schreiber, G. J., Peterse, J. L., Roberts, C., Marton, M. J., Parrish, M., Atsma, D., Witteveen, A., Glas, A., Delahaye, L., van der Velde, T., Bartelink, H., Rodenhuis, S., Rutgers, E. T., Friend, S. H., Bernards, R. (2002) A gene-expression signature as a predictor of survival in breast cancer. *N Engl J Med* 347, 1999–2009.
11. Gerhold, D. L., Jensen, R. V., Gullans, S. R. (2002) Better therapeutics through microarrays. *Nat Genet* 32 suppl, 547–551.
12. Gunther, E. C., Stone, D. J., Gerwien, R. W., Bento, P., Heyes, M. P. (2003) Prediction of clinical drug efficacy by classification of drug-induced genomic expression profiles in vitro. *Proc Natl Acad Sci USA* 100, 9608–9613.
13. Scherf, U., Ross, D. T., Waltham, M., Smith, L. H., Lee, J. K., Tanabe, L., Kohn, K. W., Reinhold, W. C., Myers, T. G., Andrews, D. T., Scudiero, D. A., Eisen, M. B., Sausville, E. A., Pommier, Y., Botstein, D., Brown, P. O., Weinstein, J. N. (2000) A gene expression database for the molecular pharmacology of cancer. *Nat Genet* 24, 236–244.
14. Bennett, L., Palucka, A. K., Arce, E., Cantrell, V., Borvak, J., Banchereau, J., Pascual, V. (2003) Interferon and granulopoiesis signatures in systemic lupus erythematosus blood. *J Exp Med* 197, 711–723.
15. Burczynski, M. E., Twine, N. C., Dukart, G., Marshall, B., Hidalgo, M., Stadler, W. M., Logan, T., Dutcher, J., Hudes, G., Trepicchio, W. L., Strahs, A., Immermann, F., Slonim, D. K., Dorner, A. J. (2005) Transcriptional profiles in peripheral blood mononuclear cells prognostic of clinical outcomes in patients with advanced renal cell carcinoma. *Clin Cancer Res* 11, 1181–1189.
16. Palucka, A. K., Blanck, J. P., Bennett, L., Pascual, V., Banchereau, J. (2005) Cross-regulation of TNF and IFN-alpha in autoimmune diseases. *Proc Natl Acad Sci USA* 102, 3372–3377.
17. Pascual, V., Allantaz, F., Arce, E., Punaro, M., Banchereau, J. (2005) Role of interleukin-1 (IL-1) in the pathogenesis of systemic onset juvenile idiopathic arthritis and clinical response to IL 1 blockade. *J Exp Med* 201, 1479–1486.
18. Osman, I., Bajorin, D. F., Sun, T. T., Zhong, H., Douglas, D., Scattergood, J., Zheng, R., Han, M., Marshall, K. W., Liew, C. C. (2006) Novel blood biomarkers of human urinary bladder cancer. *Clin Cancer Res* 12, 3374–3380.
19. Ramilo, O., Allman, W., Chung, W., Mejias, A., Ardura, M., Glaser, C., Wittkowski, K. M., Piqueras, B., Banchereau, J., Palucka, A. K., Chaussabel, D. (2007) Gene expression patterns in blood leukocytes discriminate patients with acute infections. *Blood* 109, 2066–2077.

20. Brazma, A., Hingamp, P., Quackenbush, J., Sherlock, G., Spellman, P., Stoeckert, C., Aach, J., Ansorge, W., Ball, C. A., Causton, H. C., Gaasterland, T., Glenisson, P., Holstege, F. C., Kim, I. F., Markowitz, V., Matese, J. C., Parkinson, H., Robinson, A., Sarkans, U., Schulze-Kremer, S., Stewart, J., Taylor, R., Vilo, J., Vingron, M. (2001) Minimum information about a microarray experiment (MIAME)-toward standards for microarray data. *Nat Genet* 29, 365–371.
21. Bammler, T., et al., (2005) Standardizing global gene expression analysis between laboratories and across platforms. *Nat Methods* 2, 351–356.
22. Canales, R. D., Luo, Y., Willey, J. C., Austermiller, B., Barbacioru, C. C., Boysen, C., Hunkapiller, K., Jensen, R. V., Knight, C. R., Lee, K. Y., Ma, Y., Maqsodi, B., Papallo, A., Peters, E. H., Poulter, K., Ruppel, P. L., Samaha, R. R., Shi, L., Yang, W., Zhang, L., Goodsaid, F. M. (2006) Evaluation of DNA microarray results with quantitative gene expression platforms. *Nat Biotechnol* 24, 1115–1122.
23. Guo, L., Lobenhofer, E. K., Wang, C., Shippy, R., Harris, S. C., Zhang, L., Mei, N., Chen, T., Herman, D., Goodsaid, F. M., Hurban, P., Phillips, K. L., Xu, J., Deng, X., Sun, Y. A., Tong, W., Dragan, Y.P., Shi, L. (2006) Rat toxicogenomic study reveals analytical consistency across microarray platforms. *Nat Biotechnol* 24, 1162–1169.
24. Irizarry, R. A., Warren, D., Spencer, F., Kim, I. F., Biswal, S., Frank, B. C., Gabrielson, E., Garcia, J. G., Geoghegan, J., Germino, G., Griffin, C., Hilmer, S. C., Hoffman, E., Jedlicka, A. E., Kawasaki, E., Martínez-Murillo, F., Morsberger, L., Lee, H., Petersen, D., Quackenbush, J., Scott, A., Wilson, M., Yang, Y., Ye, S.Q., Yu, W. (2005) Multiple-laboratory comparison of microarray platforms. *Nat Methods* 2, 345–350.
25. Larkin, J. E., Frank, B. C., Gavras, H., Sultana, R., Quackenbush, J. (2005) Independence and reproducibility across microarray platforms. *Nat Methods* 2, 337–344.
26. Patterson, T. A., Lobenhofer, E. K., Fulmer-Smentek, S. B., Collins, P. J., Chu, T. M., Bao, W., Fang, H., Kawasaki, E. S., Hager, J., Tikhonova, I. R., Walker, S. J., Zhang, L., Hurban, P., de Longueville, F., Fuscoe, J. C., Tong, W., Shi, L., Wolfinger, R.D. (2006) Performance comparison of one-color and two-color platforms within the MicroArray Quality Control (MAQC) project. *Nat Biotechnol* 24, 1140–1150.
27. Shi, L., et al., (2006) The MicroArray Quality Control (MAQC) project shows inter- and intraplatform reproducibility of gene expression measurements. *Nat Biotechnol* 24, 1151–1161.
28. Tong, W., Lucas, A. B., Shippy, R., Fan, X., Fang, H., Hong, H., Orr, M. S., Chu, T. M., Guo, X., Collins, P. J., Sun, Y. A., Wang, S. J., Bao, W., Wolfinger, R. D., Shchegrova, S., Guo, L., Warrington, J. A., Shi, L. (2006) Evaluation of external RNA controls for the assessment of microarray performance. *Nat Biotechnol* 24, 1132–1139.
29. Debey, S., Schoenbeck, U., Hellmich, M., Gathof, B. S., Pillai, R., Zander, T., Schultze, J. L. (2004) Comparison of different isolation techniques prior gene expression profiling of blood derived cells: impact on physiological responses, on overall expression and the role of different cell types. *Pharmacogenomics J* 4, 193–207.
30. Debey, S., Zander, T., Brors, B., Popov, A., Eils, R., Schultze, J. L. (2006) A highly standardized, robust, and cost-effective method for genome-wide transcriptome analysis of peripheral blood applicable to large-scale clinical trials. *Genomics* 87, 653–664.
31. Rainen, L., Oelmueller, U., Jurgensen, S., Wyrich, R., Ballas, C., Schram, J., Herdman, C., Bankaitis-Davis, D., Nicholls, N., Trollinger, D., Tryon, V. (2002) Stabilization of mRNA expression in whole blood samples. *Clin Chem* 48, 1883–1890.
32. Cobb, J. P., et al., (2005) Application of genome-wide expression analysis to human health and disease. *Proc Natl Acad Sci USA* 102, 4801–4806.
33. Kacharmina, J. E., Crino, P. B., Eberwine, J. (1999) Preparation of cDNA from single cells and subcellular regions. *Methods Enzymol* 303, 3–18.
34. Pabon, C., Modrusan, Z., Ruvolo, M. V., Coleman, I. M., Daniel, S., Yue, H., Arnold, L. J. Jr. (2001) Optimized T7 amplification system for microarray analysis. *Biotechniques* 31, 874–879.
35. Van Gelder, R. N., von Zastrow, M. E., Yool, A., Dement, W. C., Barchas, J. D., Eberwine, J. H. (1990) Amplified RNA synthesized from limited quantities of heterogeneous cDNA. *Proc Natl Acad Sci USA* 87, 1663–1667.
36. Schultze, J. L., Eggle, D. (2007) IlluminaGUI: graphical user interface for analyzing gene expression data generated on the Illumina platform. *Bioinformatics* 23, 1431–1433.
37. Dudoit, S., et al., (2002) Statistical methods for identifying differentially expressed genes in replicated cDNA microarray experiments. *Statistica Sinica* 12, 111–139.

38. Bolstad, B. M., Irizarry, R. A., Astrand, M., Speed, T. P. (2003) A comparison of normalization methods for high density oligonucleotide array data based on variance and bias. *Bioinformatics* 19, 185–193.
39. Huber, W., von Heydebreck, A., Sültmann, H., Poustka, A., Vingron, M. (2002) Variance stabilization applied to microarray data calibration and to the quantification of differential expression. *Bioinformatics* 18 Suppl (1), S96–104.
40. Workman, C., Jensen, L. J., Jarmer, H., Berka, R., Gautier, L., Nielser, H. B., Saxild, H. H., Nielsen, C., Brunak, S., Knudsen, S. (2002) A new non-linear normalization method for reducing variability in DNA microarray experiments. *Genome Biol* 3, research0048.
41. Tibshirani, R., Hastie, T., Narasimhan, B., Chu, G. (2002) Diagnosis of multiple cancer types by shrunken centroids of gene expression. *Proc Natl Acad Sci USA* 99, 6567–6572.
42. Affymetrix Technical Note: Globin Reduction Protocol: A Method for Processing Whole Blood RNA Samples for Improved Array Results. (2003).

Chapter 14

RNA Profiling in Peripheral Blood Cells by Fluorescent Differential Display PCR

Martin Steinau and Mangalathu S. Rajeevan

Abstract

The differential display-polymerase chain reaction (DD-PCR) technique is a unique, sequence independent tool for mRNA profiling and relative quantification. It is particularly suited for clinical samples yielding limited amounts of RNA. Unlike closed systems like microarray-based platforms, DD-PCR can be used to detect expression changes in known and novel transcripts, alternate splice products and to identify non-human transcripts. This chapter details fluorescent DD-PCR protocols that were optimized for peripheral blood mononuclear cells (PBMC). Subpopulations of mRNAs are reverse transcribed with two-base anchored oligo dT primers, amplified in combination with arbitrary primers and after gel separation visualized by fluorescent tags on the primers. Besides the DD-PCR itself, methods are described for subsequent extraction, amplification, and sequencing of DNA from bands of interest to identify the corresponding genes.

Key words: Differential display PCR, fluorescent detection, blood RNA profiling, PBMC.

1. Introduction

Differential display-polymerase chain reaction (DD-PCR), introduced by Liang and Pardee in the early 1990s, is an effective platform applicable to a variety of samples from different species for genome-wide profiling of known and novel gene transcripts *(1, 2)*. The original DD-PCR protocol has been greatly improved in terms of its' reliability and ability to detect small differences in gene expression. Since DD-PCR can be performed with small amounts of total RNA (<100 ng), it is suitable for transcription profiling of clinical samples like peripheral blood mononuclear cells (PBMCs) *(3)* known to yield limited amounts of RNA. Unlike closed chip-based systems DD-PCR also offers the

Peter Bugert (ed.), *DNA and RNA Profiling in Human Blood: Methods and Protocols, vol. 496*

DOI 10.1007/978-1-59745-553-4_14 Springerprotocols.com

possibility to discover new transcripts, variants or foreign RNAs from pathogens.

In order to reduce the complexity of amplified cDNA products, subpopulations of mRNA are synthesized into cDNA by the use of 3′ two-base anchored oligo (dT) primers (AP primer) that target the two nucleotides immediately upstream of the poly-A tail. The resulting portion of cDNA is then amplified with the same AP primer in combination with a set of randomly chosen arbitrary primer (ARP), so that each DD-PCR generates products from 50 to 100 mRNA transcripts in a single reaction. It is estimated that 200 PCR reactions employing ten two-base AP and 20 ARP combinations can provide nearly complete coverage of the human RNA profile. Besides DD-PCR itself, procedures are described for subsequent sequence recovery and corresponding gene identification of differentially displayed messages. The methods described were optimized specifically for RNA from human PBMCs *(4)*. It is understood that all DD-PCR results should be validated by other methods such as real-time PCR *(5)* which is not described here.

2. Materials

1. RNAaqueous Kit (Ambion/Applied Biosystems, Austin TX)
2. Formaldehyde, 37% aqueous solution
3. Agarose (electrophoresis grade)
4. Small electrophoresis equipment (3′ × 4′)
5. 3-(*N*-morpholino)propanesulfonic acid (MOPS) 10×
6. Tris/borate/EDTA (TBE) buffer 10×
7. RNA sample loading buffer with ethidium bromide (Sigma-Aldrich)
8. RNA Ladder 0.1–2 kb (Invitrogen, Carlsbad, CA)
9. UV/vis spectrophotometer
10. Message clean DNase I (GenHunter, Knoxville, TN)
11. RNasin (Promega, Madison WI)
12. Superscript II reverse transcriptase (Invitrogen)
13. Platinum *Taq* DNA polymerase, 5 U/μl (Invitrogen)
14. dNTP mixture 2.5 mM each: dATP, dCTP, dGTP, dTTP (Invitrogen)
15. Hieroglyph mRNA profile kit for differential display analysis (Genomyx Corp., Foster City, CA)
16. FluoroDD TMR-fluorescent anchored primer adapter kit for the Hieroglyph mRNA profile kit system for differential display analysis (Genomyx Corp.)
17. Hr-1000 5.6% denaturing clear gel (Beckman Coulter, Fullerton, CA)
18. Sodium hydroxide (NaOH), 10 N in H_2O

19. Glass shield (Beckman Coulter)
20. Ammonium persulfate (APS), 10%
21. Tetramethylethylenediamine (TEMED)
22. Programmable sequencing gel station with 60 cm SC-Glass plate set and SC-optimized comb/spacer kit, 60 cm, 250 μm, 48 well (Beckman Coulter)
23. Fluorodd Grid (Beckman Coulter)
24. Programmable thermocycler: GeneAmp PCR System 9700 (Applied Biosystems, Foster City, CA)
25. PCR cooler (Eppendorf, Westbury, NY)
26. Fluorescent Image Scanning Unit FMBIO III (MiraiBio Inc., Alameda, CA)
27. Graphics editor software – Adobe Photoshop (Adobe Systems Inc., San Jose, CA)
28. Centricon YM-100 (Millipore, Billerica, MA)
29. Expand High Fidelity PCR System (Roche Applied Science, Indianapolis, IN)

3. Methods

The methods described here include *(1)* RNA isolation from PBMCs, *(2)* DNase treatment, *(3)* reverse transcription (RT), *(4)* DD-PCR, *(5)* product separation in a sequencing gel *(6)* band recovery, *(7)* re-amplification, and *(8)* sequencing and identification.

3.1. Sample Collection and RNA Isolation

Peripheral blood samples were collected by venepuncture into tubes containing heparin. PBMCs were isolated immediately using Ficoll gradients and stored in RNAaqueous lysis buffer at –80°C. Total RNA was extracted with the RNAaqueous kit (Ambion, Austin, TX) following the manufacturers directions. RNA was stored at –80°C until further use.

For a successful DD-PCR study, it is essential to ensure that the RNA from all samples is undegraded, and equal quantities are subjected to the procedure. Therefore, RNA quality was verified visually by using denaturing formaldehyde gel (1%) electrophoresis:

1. For a 3" × 4" gel, mix 0.5 g agarose and 36 ml diethylpyrocarbonate (DEPC)-treated water and heat until agarose has completely dissolved. Add 9 ml of formaldehyde and 6 ml 10× MOPS, mix thoroughly and cast the gel in a chemical fume hood.
2. To 1 μl RNA sample, add 8 μl DEPC H_2O and 1 μl of 10× RNA loading dye with ethidium bromide. Denature for 10 min at 70°C and set tubes on ice for 2 min.

3. Briefly spin the tubes to remove condensation. Load the samples and an RNA molecular weight marker into the wells of the gel and run for 90 min at 90V.
4. RNA quality can be estimated by the intactness of the 28S and 18S ribosomal bands. Intensity of these ribosomal bands should appear in 2:1 ratio for undegraded total RNA.

RNA concentration was determined with the ND-1000 (NanoDrop Technologies, Wilmington, DE) UV/Vis spectrophotometer. Equal dilution to a concentration of 50 to 100 ng/μl in 1× TE buffer is most suitable for the following downstream application. RNA yield from PBMCs is relatively low and can degrade quickly if improperly handled. To avoid repeated freeze–thaw cycles, these dilutions should be prepared immediately prior to processing.

3.2. DNase Treatment

Since DD-PCR is a very sensitive method, and employs random priming, great care should be taken to remove all traces of DNA in the samples to avoid false-positive results. Treatment of total RNA with DNase I and the subsequent reverse transcription (Section 3.3) were carried out as a single-tube reaction. All incubation steps were performed in a thermocycler programmed for the relevant times and temperatures as outlined in step 2. To keep the sample at 4°C on the bench a PCR cooler was used.

As for all work with RNA, a clean RNase-free work environment and gloves are essential. RNA samples, DNase I and 10× buffer should be thawed on ice and gently resuspended prior to use.

1. Dilute sufficient amount of DNase I 1:10 in 1× DNase I buffer to 1 U/μl. In a 0.2 ml thin-wall PCR tube, mix: 0.5 μl 10× DNase I buffer, 1 μl DNase I (1 U) with 100 ng total RNA and add DEPC H_2O to bring the total reaction volume to 5 μl.
2. Incubate for 30 min at 37°C, followed by 2 min at 70°C in a thermocycler. Then touch-spin to remove condensation and place into the PCR cooler. Remove 1 μl of DNase I-treated sample at this step, and dilute with water to 10 μl. This aliquot will serve as "no-RT control" from each sample and should be kept at 4°C. Proceed directly to reverse transcription (Section 3.3).

Effectiveness of DNase I treatment should be verified by processing the no-RT control through the DD-PCR reaction (Section 3.4) and should not produce any visible products on the sequencing gel (Section 3.5).

3.3. Reverse Transcription

1. Dilute the dNTP mix to 250 μM prior to use. Do not vortex primer or dNTPs since this may result in degradation (*see* **Note 1**).

Table 14.1
Anchored primers (AP) ***(6)***

AP primer	Sequence (5′ to 3′)
T7(dT$_{12}$)AP1	5′ ACGACTCACTATAGGGCTTTTTTTTTTTT**GA** 3′
T7(dT$_{12}$)AP2	5′ ACGACTCACTATAGGGCTTTTTTTTTTTT**GC** 3′
T7(dT$_{12}$)AP3	5′ ACGACTCACTATAGGGCTTTTTTTTTTTT**GG** 3′
T7(dT$_{12}$)AP5	5′ ACGACTCACTATAGGGCTTTTTTTTTTTT**GT** 3′
T7(dT$_{12}$)AP6	5′ ACGACTCACTATAGGGCTTTTTTTTTTTT**CA** 3′
T7(dT$_{12}$)AP7	5′ ACGACTCACTATAGGGCTTTTTTTTTTTT**CC** 3′
T7(dT$_{12}$)AP8	5′ ACGACTCACTATAGGGCTTTTTTTTTTTT**CG** 3′
T7(dT$_{12}$)AP9	5′ ACGACTCACTATAGGGCTTTTTTTTTTTT**AA** 3′
T7(dT$_{12}$)AP10	5′ ACGACTCACTATAGGGCTTTTTTTTTTTT**AC** 3′
T7(dT$_{12}$)AP11	5′ ACGACTCACTATAGGGCTTTTTTTTTTTT**AG** 3′
T7(dT$_{12}$)AP12	5′ ACGACTCACTATAGGGCTTTTTTTTTTTT**AT** 3′
T7(dT$_{12}$)AP12	5′ ACGACTCACTATAGGGCTTTTTTTTTTTT**CT** 3′

In addition to the regular primer oligos an additional set of APs is needed with 5′ TMR label; bold bases indicate the two anchoring sequences that target the nucleotides immediately upstream of the poly-A tail. The sequence of the T7 primer sequence (for re-amplification (**Section 3.7**)) is underlined.

2. To the tube containing the remaining 4 μl DNase-treated sample add 1 μl of an anchor primer (2 mM) (**Table 14.1**), incubate at 70°C for 5 min, then set back into the PCR cooler.
3. During the incubation period, the reagents for the reverse transcription reaction should be prepared on ice, as a master mix for all samples per cDNA type (with a particular AP). For each RT reaction mix 1.55 μl H_2O, 1 μl 5x RT buffer, 1 μl dNTPs (250 μM), 1 μl DTT (0.1 M), 0.25 μl RNasin, 0.2 μl reverse transcriptase.
4. Add 5 μl of RT master mix and mix well. Incubate in a thermocycler under the following conditions: 5 min at 40°C, 50 min at 50°C, and 15 min at 70°C. After the synthesis step is completed, remove sample, touch-spin, and dilute fourfold with 30 μl DEPC water. Keep cDNAs at –20°C until DD-PCR is performed. The type of cDNA, i.e., which AP was used should be carefully recorded.

3.4. Fluorescent DD-PCR

Amplification of cDNAs was achieved with the same AP (version with 5′-end tetramethylrhodamine label) that had been used for its synthesis and any one of the ARPs. Since detection of differential expression involves comparison of RNA profiles of different samples generated under identical condition, a particular type of cDNA from all samples should be amplified with the same ARP at the same time. However, a sample set can be processed with several ARPs in parallel. Excessive light exposure should be avoided to prevent photo-bleaching of the TMR label (*see* **Note 1**).

1. Thaw cDNA samples (**Section 3.3**) on PCR cooler and pre-chill 0.2 ml thin-wall PCR tubes for the DD-PCR. Each sample should be amplified in duplicates (*see* **Note 2**).
2. Prepare all biochemical components as a master mix. Each DD-PCR reaction contains 2.7 μl dH_2O, 1 μl 10× PCR buffer (without $MgCl_2$), 0.75 μl $MgCl_2$ (50 mM), 2 μl dNTPs (250 μM), 1.75 μl ARP, 0.7 μl TMR-AP (2 μM), and 0.1 μl platinum *Taq* (5 U/μl). Mix by pipetting up and down and aliquot. Add 1 μl of cDNA to 9 μl PCR master mix in the prepared PCR tubes. Touch-spin before inserting the tubes into the thermocycler.
3. Amplification is performed under the following conditions: 95°C for 2 min; four cycles of 92°C for 15 s, 50 °C for 30 s, 72°C for 2 min; 25 cycles of 92°C for 15 s, 60°C for 30 s, 72°C for 2 min; 72°C for 7 min; 4°C until processing.

3.5. High-Resolution (HR) Sequencing Gel

Separation of the fluorescent DD-PCR products was accomplished by a 24" vertical sequencer in a 5.6% clear denaturing polyacrylamide gel matrix. Some preparations need to be made to allow the location and recovery of DD-PCR bands that might be of interest later (**Section 3.6**): *(1)* Visual marks for the alignment of the gel image with the grid (*see* **Note 3**). *(2)* To allow access to the gel, the glass plates were individually treated to facilitate gel adhesion to the unnotched plate only and allowing detachment of the other.

3.5.1. Unnotched Plate

This bottom plate will support the gel when bands of interest are excised (Section 3.6.) and bands are located with the help of marks made on the plate by aligning it with locations on a "physical grid".

1. Position glass exactly over the transparent physical grid with the outside facing up. Above designated spots on the grid (in upper left and lower right corners) place markings onto the glass plate that line up with marking on the grid (clear whole punch reinforcement labels over the A1 and H19 markers work well).
2. Clean plate under running water, rubbing of all remaining dirt or gel fragments and dry both sides of the glass with lint-free wipes.
3. Prepare 250 ml 4 N NaOH. Treat the inner side of the glass with 200 ml 4M NaOH rubbing in circular motion with a paper towel. Overlay the entire plate with additional NaOH and let soak for 5 min.

4. Thoroughly wash of the NaOH with running water for 2–5 min. Dry with wipes and clean with 95% EtOH. Ensure glass plate is absolutely clean and dry.

3.5.2. Notched Upper Plate

1. Clean plate with running water, 95% EtOH and dry.
2. Apply a thin film of glass shield wiping in circular motion over the entire surface plate using lint-free wipes. Ensure that the edges are treated too. Wash of excessive glass shield and clean glass plate with 95% EtOH.
3. Clean spacers and combs with distilled water. Fix the spacers to the lateral ends with a paper glue stick.

3.5.3. Casting the Gel

1. In a glass beaker add 80 ml of 5.6% denaturing clear gel, 640 μl of 10% APS (prepared fresh), and 64 μl of TEMED. Swirl carefully to avoid air bubbles. Pour the liquid gel onto the bottom third of the unnotched plate in horizontal position. Set the notched plate at a 90° angle onto the spacers against the unnotched one. Slowly lower the treated side of the notched glass plate avoiding any air enclosure. Line up the corners of both plates, insert comb spacer with the teeth facing outside, and fix both plates together with clamps (*see* **Note 4**).
2. Let the gel polymerize in horizontal position for at least 1 h, then remove clamps and wash of excessive gel with deionized water.

3.5.4. Gel Electrophoresis

1. Prepare the gel electrophoresis station with 150 ml upper buffer (0.5x TBE) and 300 ml lower buffer (1× TBE). Insert shark tooth comb between the glass plates so the teeth gently touch the gel surface. Rinse out any remaining gel pieces in the loading wells with the upper running buffer using a syringe with needle.
2. Transfer 4 μl from each DD-PCR reaction into a new thin-wall 0.2 ml tube. Add 1.5 μl of fdd dye. Prepare a TMR-labeled DNA-standard marker and the no-RT controls (Section 3.2) the same way. Place tubes uncapped into the thermocycler leaving the cover open and heat for 3 min at 95°C. Touch-spin tubes and cool in the PCR cooler.
3. Load the entire remaining content of each tube (~4 μl) slowly into the wells, duplicates adjacent to each other. Run the gel for 5 h at 2,700 V, 100 W, 50°C. Minimize light exposure of the gel.

3.6. Scanning and Band Recovery

1. After electrophoresis has completed, thoroughly clean the glass plate with 95% ethanol or streak-free glass cleaner.
2. Scan the gel in the FMBIOIII with the following parameters: 532 nm laser, 505 nm filter, 100 μm resolution, 90% sensitivity, two scan line repetitions. Visually analyze the scanned image for differential gene expression. Samples should have

at least an estimated twofold intensity difference while duplicates should appear identical (**Fig. 14.1**). The original scan of the fluorescent of the DD-PCR should be saved in the tagged image file format (TIFF). For alignment and recovery of particular bands of interest the following protocol should be followed:

3. Separate the two glass plates using a plastic specula from a bottom corner. Dry the gel on the unnotched plate for 20 min at 50 °C with the drying program of the automated sequencer.
4. Carefully rinse the gel with deionized water to remove the crystallized urea and dry again for 10 min. Repeat rinsing and drying steps at least three times or until no urea crystallization is visible on the dry gel. Remaining urea can interfere with the subsequent PCR reaction.
5. With a graphics editor software program (Adobe Photoshop) overlay the fluorescent gel image with an image of the grid editor (virtual grid) using the orientation marks on the glass plate (1A, H19). Set the virtual grids' opacity to about 40% and line up the marks of both images. Mark bands of interest on the appropriate position of the physical grid with a permanent marker (**Fig. 14.2**; *see* **Note 5**).
6. Place the physical grid with marked band positions face down. Place the unnotched glass with the gel facing up on top. Carefully align the targets on glass and grid and fix grid and plate together with clamps.
7. Prepare a 0.5 ml microcentrifuge tube with 50 µl nuclease-free TE for each band to be excised from the gel.
8. With a clean blade cut along the perimeter of the marked band on the grid. Pipette 2 µl nuclease-free water atop the cutting, scoop up the band and place it in the prepared tubes with TE. Change the blade after each cutting or flame with EtOH.

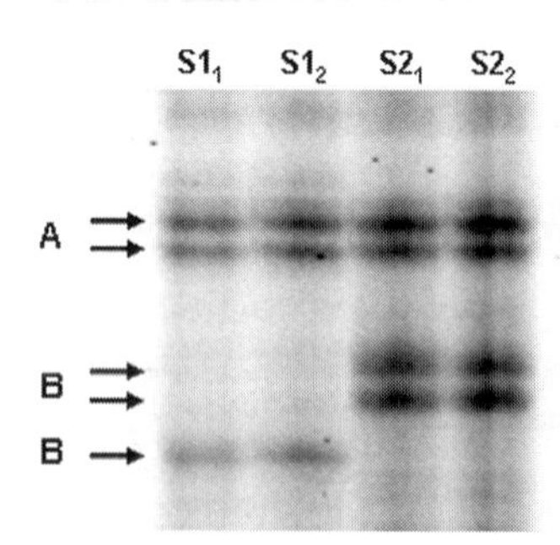

Fig. 14.1. Section of a DD-PCR gel with fluorescent amplicons from two PBMC samples (S1 and S2) each run in duplicates. Arrows indicate bands with equal intensity **(A)** or differential expression **(B)**.

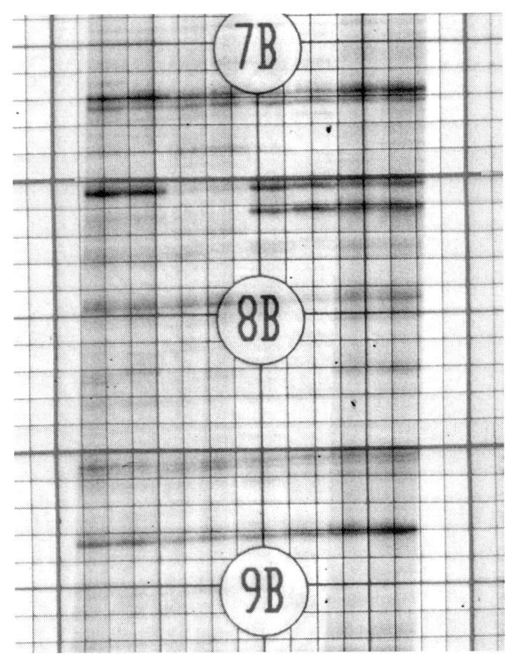

Fig. 14.2. Image of a fluorescent DD-PCR gel overlaid with the grid. The band's location can be transferred onto the physical grid to marks its position on the gel once it is lined up with the glass plate.

9. Incubate gel slices in closed tubes for 60 min at 37°C. Proceed directly to re-amplification (**Section 3.7**) or store tubes at –20°C.
10. Rescan the gel on the glass plate to verify the successful removal of desired bands.

3.7. Re-amplification

1. Use Expand High Fidelity PCR System (Roche Applied Science) for re-amplification. Prepare core mix for each 50 μl PCR reaction: 28.6 μl H_2O, 5 μl 10× PCR, 1 μl dNTP mix (10 mM), 5 μl M13 primer (20 μM), 5 μl 2 μM T7 (20 μM) primer, 0.4 μ *Taq* enzyme mix.
2. Add 45 μl core mix to 5 μl of eluted gel product (Section 3.6.) in a 0.2 ml thin-wall PCR tube and amplify under the following cycle conditions: 2 min at 95°C, ten cycles of 15 s at 92°C, 30 s at 50°C, 2 min at 72°C, 15 cycles of 15 s at 92°C, 15 s at 60°C, 2 min at 72°C, one final step of 7 min at 72°C.
3. Check successful re-amplification by gel electrophoresis. Run 10 μl of each product in a 1.2% agarose gel (**Fig. 14.3**). Direct sequencing will only be successful if a single sharp band is visible.
4. Purify the remaining 40 μl of each PCR reaction with Centricon YM-100 spin columns (Millipore) according to the manufacturers' instructions. The recovered volume should be

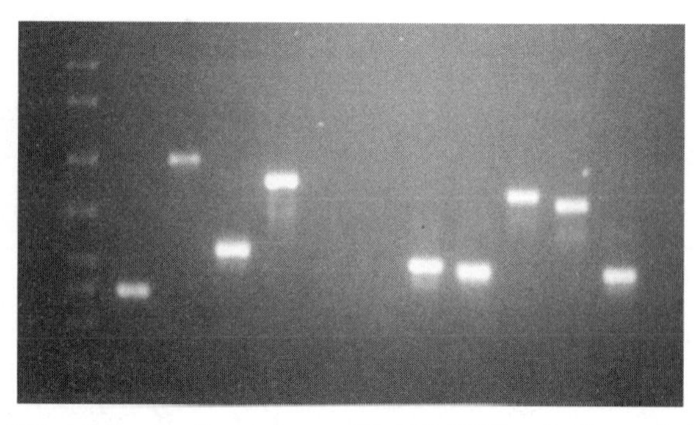

Fig. 14.3. Different size amplicons of reamplified DD-PCR bands. The products are visualized by agarose gel electrophoresis to assess quality and correct molecular weight prior to sequencing.

between 14 and 20 μl and should be kept at –20°C (*see* **Note 7**).

3.8. Direct Sequencing and Gene Identification

The purified reamplified PCR products can be sequenced directly using the M13 primer contained in the 5′-end of the ARPs (**Table 14.2**). Typically, 2 μl of each reamplified band could be end-labeled with BigDye terminator and sequenced with an ABI 3100 Genetic Analyzer using the standard protocol from the manufacturer. The obtained nucleotide sequences were submitted to BLAST for gene identification and annotated subsequently with appropriate software. Sequence quality, i.e., length and number of ambiguous bases can vary widely, and determines its usefulness for gene identification. As a guideline, nucleotide sequence should have at least 100 base pairs with less than 5% ambiguous bases. However, usable BLAST results may be obtained with lower quality sequences.

4. Notes

1. Methods in **Sections 3.3**, **3.4,** and **3.5.4** use components from the HIEROGLYPH mRNA Profile Kit and the fluoroDD TMR-fluorescent anchored primer adapter kit for differential display analysis from Genomyx Corp. (Foster City, CA). Each of the HIEROGLYPH kits contains ten 3′-oligo(dT)-anchoring primers with a set of four kit-specific 5′-arbitrary primers. These kits are no longer available but its components can easily be replaced: All sequences of AP and ARP oligonucleotides are listed in **Tables 14.1** and **14.2**; one set of APs needs to be 5′-end-labeled with TMR. A TMR-labeled molecular weight DNA marker is useful for electrophoresis in the

Table 14.2
Arbitrary primers *(6)*

ARP primer	Sequence (5′ to 3′)
M13-ARP1	5′ ACAATTTCACACAGGACGACTCCAAG 3′
M13-ARP2	5′ ACAATTTCACACAGGAGCTAGCATGG 3′
M13-ARP3	5′ ACAATTTCACACAGGAGACCATTGCA 3′
M13-ARP4	5′ ACAATTTCACACAGGAGCTAGCAGAC 3′
M13-ARP5	5′ ACAATTTCACACAGGAATGGTAGTCT 3′
M13-ARP6	5′ ACAATTTCACACAGGATACAACGAGG 3′
M13-ARP7	5′ ACAATTTCACACAGGATGGATTGGTC 3′
M13-ARP8	5′ ACAATTTCACACAGGATGGTAAAGGG 3′
M13-ARP9	5′ ACAATTTCACACAGGATAAGACTAGC 3′
M13-ARP10	5′ ACAATTTCACACAGGAGATCTCAGAC 3′
M13-ARP11	5′ ACAATTTCACACAGGAACGCTAGTGT 3′
M13-ARP12	5′ ACAATTTCACACAGGAGGTACTAAGG 3′
M13-ARP13	5′ ACAATTTCACACAGGAGTTGCACCAT 3′
M13-ARP14	5′ ACAATTTCACACAGGATCCATGACTC 3′
M13-ARP15	5′ ACAATTTCACACAGGACTTTCTACCC 3′
M13-ARP16	5′ ACAATTTCACACAGGATCGGTCATAG 3′
M13-ARP17	5′ ACAATTTCACACAGGACTGCTAGGTA 3′
M13-ARP18	5′ ACAATTTCACACAGGATGATGCTACC 3′
M13-ARP18	5′ ACAATTTCACACAGGATTTTGGCTCC 3′
M13-ARP20	5′ ACAATTTCACACAGGATCGATACAGG 3′

The sequence of the M13 primer sequence (for re-amplification (**Section 3.7**) and direct sequencing (**Section 3.8**)) is underlined.

HR gel (ideally 0.3–1.0 kb in 50 bp increments). FluoroDD loading dye contains blue dextran and formamide.

2. A fourfold dilution of the cDNA template PCR was found optimal for detection of small differences in gene expression by DD-PCR (**Section 3.4.**). While higher amounts of cDNA masked small expression differences, lower amounts of cDNA input in PCR failed to detect low abundant mRNAs *(3)*.
3. Save portions of the DD-PCR reactions (**Section 3.4.**) to rerun the gel in case of technical failure. Products (covered with aluminum foil) can be stored at 4°C for several weeks.
4. Casting a large gel without air enclosures is somewhat tricky. If gel contains air bubbles, raise the notched plate up until the

bubble is "free". Sometimes, it helps to slightly "wobble" the plate up and down.

5. The physical grid is a fine matrix of approximately 2 mm × 2 mm printed on a clear plastic sheet of the same dimensions as the glass plate. Each sector of 16 × 10 such squares is highlighted and labeled by an alphanumeric system – A1 to H20 (**Fig. 14.2**). The physical grid and a scanned image (virtual grid) of it are used to mark the location of selected DD-PCR bands for excision from the glass plate (**Section 3.6**). Alternatively, a 1:1 printout of the gel image could also be used in a similar way to locate bands of interest.
6. Removing the excised bands from the gel (**Section 3.6**): lay small amount of water exactly within the cutting edges, let the gel beneath soften, push the gel piece laterally with a small pipette tip and then lift it from the glass plate.
7. Save an aliquot of the reamplified bands (**Section 3.7**) as a template for subsequent real-time PCR optimization if validation of the DD-PCR results is desired.
8. It is possible that reamplified bands may represent more than one sequence. Such fragment mixture might require cloning to allow proper sequencing. The T7 primer does not yield good sequence data with this system.

References

1. Liang, P., Pardee, A. B. (1992) Differential display of eucaryotic messenger RNA by means of the polymerase chain reaction. *Science* 257, 967–971.
2. Liang, P., Pardee, A. B. (1997) Differential display. A general Protocol. *Methods Mol Biol* 85, 3–11.
3. Ranamukhaarachchi, D. G., Rajeevan, M. S., Vernon, S. D., Unger, E. R. (2002) Modifying differential display polymerase chain reaction to detect relative changes in gene expression profiles. *Anal Biochem* 306, 343–346.
4. Steinau, M., Unger, E. R., Vernon, S. D., Jones, J. F., Rajeevan, M. S. (2004) Differential-display of peripheral blood for biomarker discovery in chronic fatigue syndrome. *J Mol Med* 82, 750–755.
5. Rajeevan, M. S., Ranamukhaarachchi, D. G., Vernon, S. D., Unger, E. R. (2001) Use of real-time quantitative PCR to validate the results of cDNA Array and differential display PCR technologies. *Methods* 25, 443–451.
6. Genomyx Corporation (1997). FluoroDD TMR-fluorescent anchored primer adaptor kit for the HIEROGLYPH mRNA Profile Kit System for Differential Display Analysis. Application Manual. Foster City, CA.

Chapter 15

cDNA Amplification by SMART-PCR and Suppression Subtractive Hybridization (SSH)-PCR

Andrew Hillmann, Eimear Dunne, and Dermot Kenny

Abstract

The comparison of two RNA populations that differ from the effects of a single-independent variable, such as a drug treatment or a specific genetic defect, can identify differences in the abundance of specific transcripts that vary in a population-dependent manner. There are a variety of methods for identifying differentially expressed genes, including microarray, SAGE, qRT-PCR, and DDGE. This protocol describes a potentially less sensitive yet relatively easy and cost-effective alternative that does not require prior knowledge of the transcriptomes under investigation and is particularly applicable when minimal levels of starting material, RNA, are available. RNA input can often be a limiting factor when analyzing RNA from, for example, rigorously purified blood cells. This protocol describes the use of SMART-PCR to amplify cDNA from sub-microgram levels of RNA. The amplified cDNA populations under comparison are then subjected to suppression subtractive hybridization (SSH-PCR), a technique that couples subtractive hybridization with suppression PCR to selectively amplify fragments of differentially expressed genes. The final products are cDNA populations enriched for significantly over-represented transcripts in either of the two input RNA preparations. These cDNA populations may then be cloned to make subtracted cDNA libraries and/or used as probes to screen subtracted cDNA, global cDNA, or genomic DNA libraries.

Key words: Differential expression, cDNA amplification, SMART-PCR, subtractive hybridization, suppressive PCR, subtracted cDNA.

1. Introduction

Switching mechanism at RNA termini (SMART)-PCR and suppression subtractive hybridization (SSH)-PCR are independent but compatible methods, commercially developed, and marketed by Clontech Laboratories, Inc. SMART-PCR *(1)* – amplification of cDNA begins with conversion of mRNA to cDNA using MMLV- RT, mutated in the RNase H domain, in the

Peter Bugert (ed.), *DNA and RNA Profiling in Human Blood: Methods and Protocols, vol. 496*

DOI 10.1007/978-1-59745-553-4_15 Springerprotocols.com

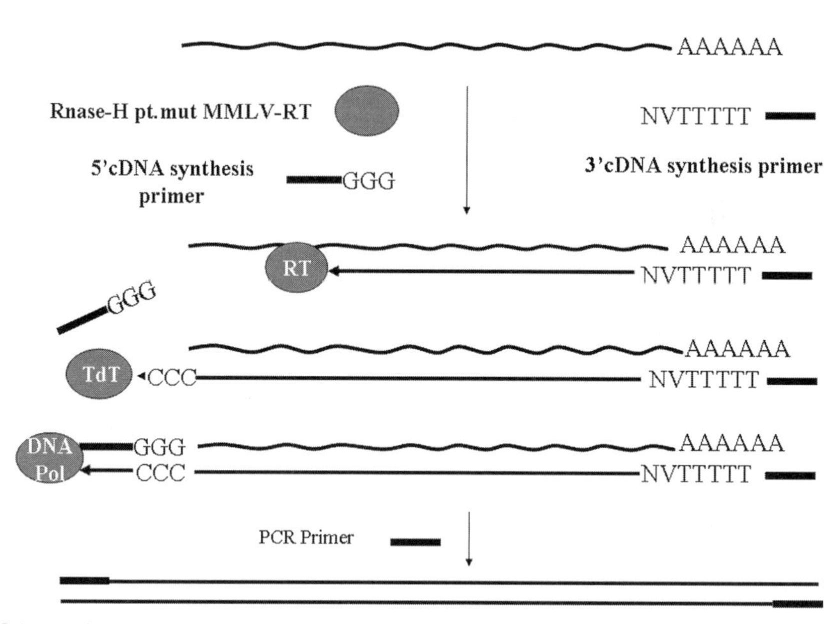

Fig. 15.1. Scheme of double-strand cDNA synthesis by SMART-PCR.

first-strand synthesis reaction (**Fig. 15.1**). This enzyme possesses RT activity, terminal deoxynucleotidyl transferase (TdT) activity, and DNA-dependent DNA polymerase activity. During first-strand synthesis, an anchored oligo-d$(T)_{30}$VN primer (3′ cDNA synthesis primer) anneals to the 5′-end of poly(A) tails, i.e., near the 3′-end of the RNA Pol II transcribed sequences. The 5′-end of this 3′ cDNA synthesis primer contains a defined sequence that serves as a PCR primer target site in the subsequent amplification step. When the first strand has been extended to the end of an mRNA template, the TdT activity of the reverse transcriptase (RT) adds several untemplated deoxycytosine (d(C)) residues to the 3′-end of first-strand cDNAs (in a fashion analogous to the addition of single untemplated d(A) residues to the end of PCR products generated with *Taq* DNA polymerase). The 5′ cDNA synthesis primer contains three (ribo) guanosine residues at its 3′ end that anneal to these untemplated d(C) "tails" and then serve as a template for the DNA-dependent DNA polymerase activity of the MMLV-RT. This SMART allows for MMLV-RT-mediated 3′-extension of all first-strand cDNA products to include a defined, contiguous sequence copied from the 5′ cDNA synthesis primer. The terminal transferase and template switching activities occur during the first-strand cDNA synthesis. Regions of sequence in the 5′ and the 3′ cDNA synthesis primers are identical, allowing for uniform amplification of all first-strand cDNAs with a single-PCR primer *(2)*.

SSH-PCR relies on the kinetics of DNA hybridization *(3)*. The efficiency of identifying a particular differentially expressed gene is a function of both its initial abundance in the transcriptome and its relative difference with that of the population to which it is being compared *(4)*. cDNAs prepared from the two mRNA populations being compared are first digested with a four base cutting restriction enzyme to generate short blunt-ended fragments which are more favorable to the suppressive PCR effect. One cDNA population, believed to contain an over-representation of some unknown genes, is denoted the "tester". The tester cDNA is split into two aliquots and each is ligated to a different adaptor (**Fig. 15.2**).

The other cDNA population is denoted the "driver" and is mixed with the two adaptor-ligated tester cDNAs separately. The two mixtures are heat denatured and then allowed to re-anneal (**Fig. 15.3**). In these complexes, the driver is in excess and hybridizations do not go to completion. Sequences over-represented in the tester cDNA pool remain single stranded and sequences common to both tester and driver re-anneal as they nucleate more frequently due to their higher relative concentrations. In a second hybridization step (**Fig. 15.4**), the single-stranded tester sequences remaining in the two first-hybridization

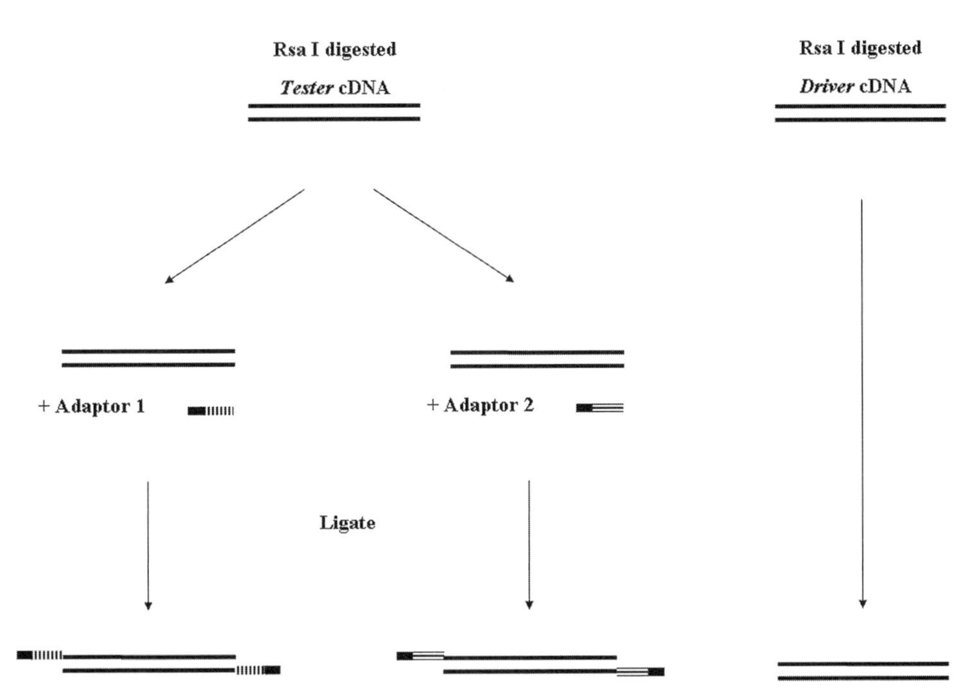

Fig. 15.2. Scheme of adapter ligation.

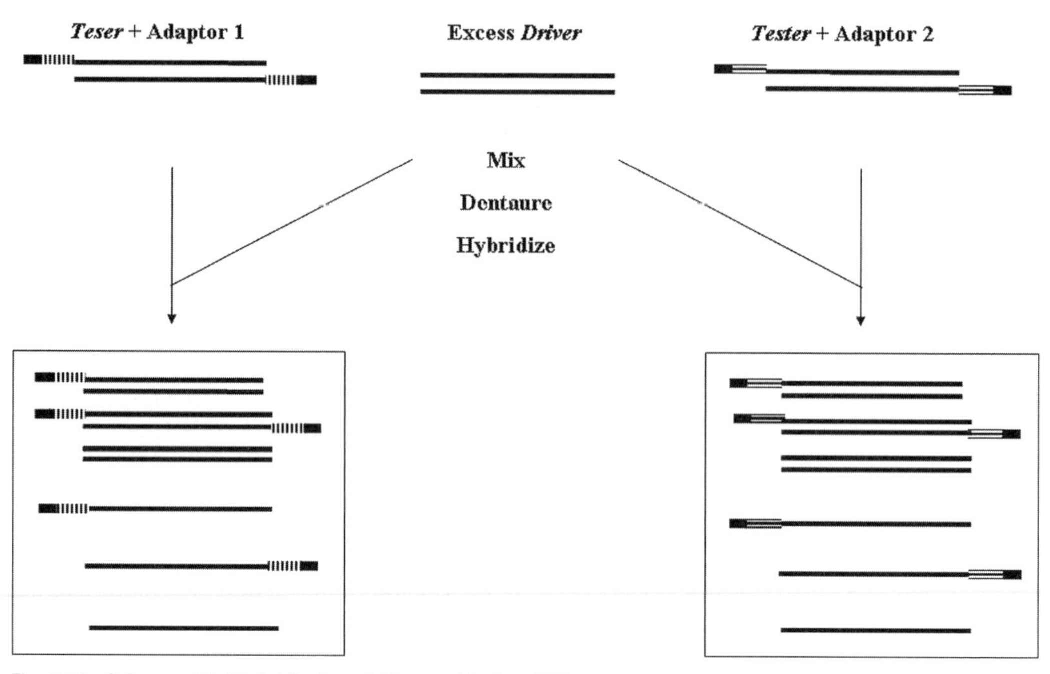

Fig. 15.3. Scheme of first hybridization of driver and tester cDNA.

mixes, which bear different adaptor sequences at their ends, are allowed to fully re-anneal in the presence of additional denatured driver cDNA. In this second hybridization, there is further subtraction of sequences common to both tester and driver cDNAs. As these hybridizations go to completion, single-stranded fragments unique to the tester cDNA form hybrid molecules with different adaptor sequences at either end (*starred* in **Fig. 15.4**).

Two successive PCR reactions are then performed to enrich for the differentially expressed sequences in the subtracted cDNA populations *(5)*. The first-PCR reaction (**Fig. 15.5**) relies on the sequence-suppressive PCR effect to prevent amplification of hybrid molecules with identical adaptors at either end, formed during subtractive hybridization. These fragments form strong secondary structures that prevent primer annealing and extension during thermal cycling (**Fig. 15.5**). Double-stranded molecules containing only one adaptor-ligated strand (arising from adaptor-ligated sequences annealing to driver cDNA) are linearly amplified. Re-annealed driver cDNAs are not amplified. Exponential amplification is favored only from molecules with different adaptor sequences at either end, arising from sequences present only in the tester cDNA population. The first-PCR products are diluted and used as input for the second-PCR reaction.

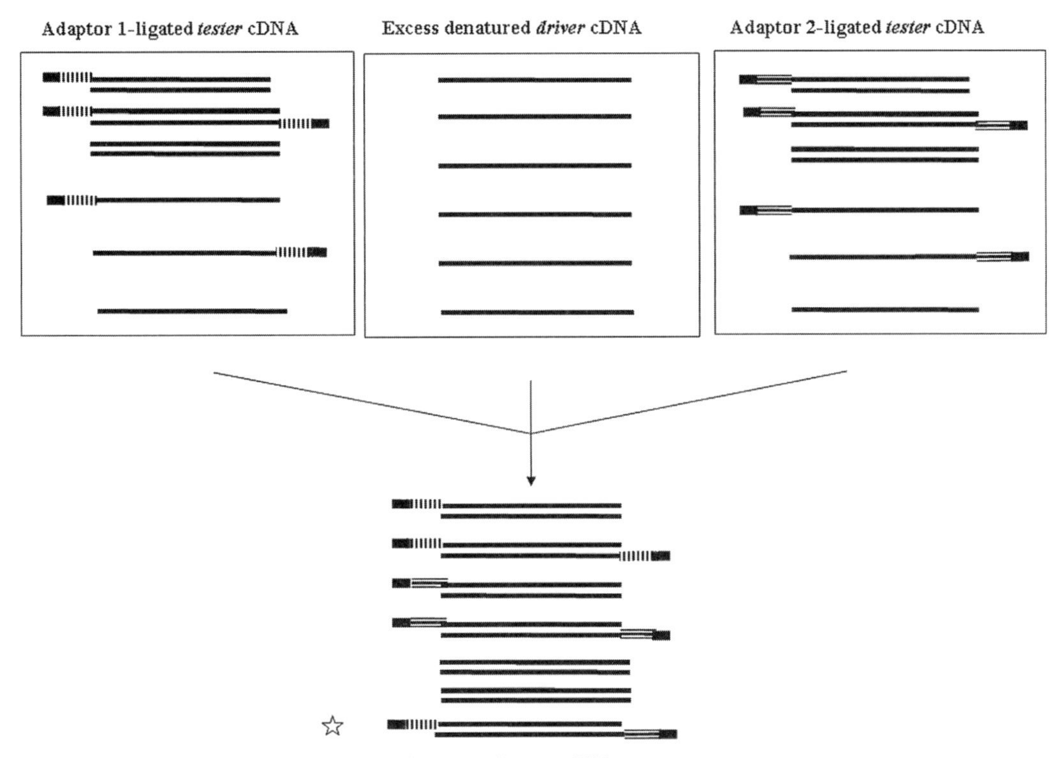

Fig. 15.4. Scheme of second hybridization of driver and tester cDNA.

The use of nested, adaptor-targeted primers in the second-PCR reaction further increases the fraction of differentially expressed sequences. Because the subtractive hybridizations and suppression PCRs enrich for genes over-represented in the tester cDNAs, the method is performed in two parallel processes. One of the two cDNA populations is arbitrarily designated as tester and subtracted with the other as driver in a "forward" subtraction. In parallel, the tester and driver designations are switched and a "reverse" subtraction is performed to enrich for genes over-represented in the other sample. Control reactions are included for both forward and reverse subtractions that result in unsubtracted cDNA pools.

The final products of the entire procedure are thus forward and reverse subtracted and unsubtracted cDNA populations. The subtracted cDNAs can be cloned into plasmid vectors to produce subtracted cDNA libraries *(6)*. The subtracted cDNAs also provide excellent probes for differential screening of these or any relevant gene library, especially if the unsubtracted cDNAs are used as control probes. Clontech Laboratories, Inc. offer kits for subtracted cDNA library construction and screening, as well as for both SMART-PCR and SSH-PCR (PCR-select). These kits also

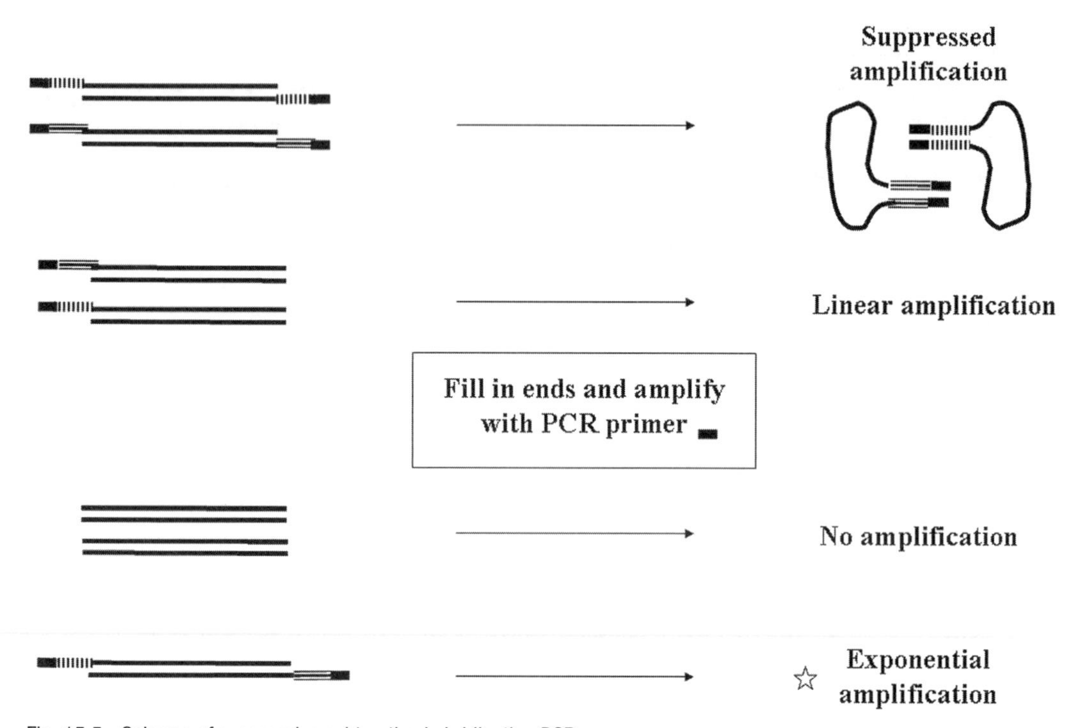

Fig. 15.5. Scheme of suppressive subtractive hybridization PCR.

provide positive control materials which are not described here. The reader will find it useful to refer to the Clontech User Manuals for additional information on amplifying and analyzing subtracted cDNAs prepared from limited amounts of cellular RNA *(7, 8)*. Ji et al. *(4)* describe a theoretically and experimentally tested model of the method which is also recommended for reading when considering this method for a particular application.

2. Materials

All materials should be stored at –20°C unless otherwise stated. Working stocks should be kept on ice unless otherwise stated.

2.1. Preparation and Amplification of cDNA

1. MMLV-reverse transcriptase, RNase H Minus. (e.g., PowerScript from Clontech or Superscript II from InVitrogen).
2. First-strand buffer (5×) (supplied with RT).
3. 3′ cDNA synthesis primer (N=G, A, T, C. V=G, A or C): 5′-AAGCAGTGGTATCAACGCAGA**GTACT**$_{(30)}$VN

4. 5′ cDNA synthesis primer (store at –80°C). 5′-AAGCAGTGGTATCAACGCAGA**GTAC**GCrGrGrG (*see* **Note 1**).
5. PCR primer (see also **Section 2.3.7**): 5′-AAGCAGTGGTATCAACGCAGAGT (*Rsa*I sites are in **BOLD** and identical sequences are UNDERLINED).
6. Thermo-stable DNA polymerase mix (e.g., from Clontech, Advantage 2 PCR kit) (*see* **Note 2**).
7. 10× PCR buffer (supplied with the polymerase).
8. STE buffer: 10 mM NaCl, 10 mM Tris–HCl (pH 7.8), 0.1 mM EDTA. Store at room temperature.
9. CHROMA SPIN 1000 purification columns (Clontech). Store at room temperature.

2.2. Restriction Digest of cDNA

1. *Rsa*I (10 Units/μl).
2. 10× *Rsa*I buffer (supplied with the enzyme, e.g., from New England Biolabs).
3. Silica slurry purification resin and associated buffers (e.g., Clontech NucleoTrap, Qiagen Qiaex II, or equivalent). Store components according to manufacturers' specification.

2.3. Adaptor Ligation

1. T4 DNA ligase (400 U/μl) (containing ATP) (e.g., from BRL, Clontech or InVitrogen).
2. T4 DNA Ligase buffer, 5× (supplied with the ligase).
3. Adapter 1: (10 pmol/μl) 5′-CTAATACGACTCACTATAGGGCTCGAGCGGCCGCCCGGGCAGGT-3′ GGCCCGTCCA-5′
4. Adapter 2R: (10 pmol/μl) 5′-CTAATACGACTCACTATAGGGCAGCGTGGTCGCGGCCGAGGT-3′ GCCGGCTCCA-5′
   Underlined bases match the PCR primer sequence.
   Double underlined bases match the nested (secondary) PCR primer sequences (see **Section 2.5**, **steps 2** and **3**).
5. 5′ G3PDH primer (10 pmol/μl): 5′-ACCACAGTCCATGCCATCAC
6. 3′ G3PDH primer (10 pmol/μl): 5′-TCCACCACCCTGTTGCTGTA These primers will amplify human, mouse, and rat G3PDH.
7. PCR primer (see also **Section 2.5, step 1**): 5′-CTAATACGACTCACTATAGGGC

2.4. Subtractive Hybridizations

1. Hybridization buffer, 4×: 50 mM Hepes (pH 8.3), 0.5 M NaCl, 0.02 mM EDTA (pH 8.0), 10% (w/v) PEG 8000.
2. Dilution buffer: 20 mM Hepes (pH 6.6), 20 mM NaCl, 0.2 mM EDTA (pH 8.0).

2.5. Suppressive and Nested PCR Primers

1. PCR primer: 5′-CTAATACGACTCACTATAGGGC
2. Nested (secondary) primer 1: 5′-TCGAGCGGCCGCCCGGGCAGGT
3. Nested primer 2R: 5′-AGCGTGGTCGCGGCCGAGGT

2.6. General Reagents

General reagents should be stored and used at room temperature unless otherwise stated.

1. 10 mM dNTP's (10 mM each, dGTP, dATP, dTTP, dCTP). Store at –20°C. Keep working stocks on ice.
2. Dithiothreitol (DTT), 20 mM
3. EDTA/glycogen mix: 0.2 M EDTA, 1 mg/ml glycogen.
4. Phenol:chloroform:isoamyl alcohol (25:24:1) (*see* **Note 3**). Store at 4°C. Keep working stock at room temperature.
5. *n*-Butanol.
6. Ammonium acetate, 4 M (pH 7.0).
7. Absolute ethanol. Store at –20°C.
8. EtOH, 80% (80 parts absolute ethanol, 20 parts sterile, deionized water).
9. TE buffer: 10 mM Tris–HCl (pH 7.8), 1 mM EDTA.
10. EDTA, 0.5 M (pH 8.0).
11. Sterile, deionized H_2O.
12. Agarose gel electrophoresis system.
13. Molecular weight markers.

3. Methods

3.1. Amplification of cDNA

3.1.1. First-Strand cDNA Synthesis

1. For each RNA sample, combine the following reagents in a clean, nuclease-free 0.5 ml reaction tube: 1–3 μl RNA sample (50–1,000 ng of total RNA; *see* **Note 4**), 1 μl 3′ cDNA synthesis primer (12 pmol/μl), 1 μl 5′ cDNA synthesis primer (12 pmol/μl), x μl nuclease-free H_2O (5 μl total volume).
2. Mix well and centrifuge briefly to collect reaction mix at the bottom of the tube.
3. Incubate at 70°C for 2 min.
4. Place reactions on ice for 2 min.
5. For each sample, prepare the following RT mix: 2 μl 5× RT buffer, 1 μl DTT (20 mM), 1 μl dNTP mix (10 mM of each dNTP), 1 μl MMLV-reverse transcriptase.
6. Add 5 μl RT mix (for a 10 μl total volume) to each sample, mix well and centrifuge briefly.
7. Incubate the tubes at 42°C for 1 h in an air incubator. (*see* **Note 5**)
8. Terminate the reactions by adding 40 μl of TE and heat denature at 70°C for 10 min.
9. Each sample should now contain 50 μl of first-strand cDNA.

3.1.2. cDNA Amplification by SMART-PCR

Over-cycling of the cDNA amplification reactions can lead to distortion of the relative representation of cDNAs in the amplified preparations. Therefore, it is necessary to determine the optimal cycle number for each sample that generates sufficient amplified cDNA without saturating the PCR reaction. Three identical PCR reactions are set up for each sample – one optimization reaction and two preparative reactions. All reactions are subjected to 15 cycles of PCR, after which the two preparative tubes are temporarily transferred to 4°C. An aliquot is removed from the third (optimization) tube which is then subjected to further cycles of PCR. Additional aliquots are removed from the optimization tubes after every three cycles. These aliquots, collected at 15, 18, 21, and 24 cycles, are then analyzed by electrophoresis. The two preparative tubes held at 4°C are then returned to the thermal cycler and subjected to the remainder of the optimal cycle number. The optimal cycle number is defined as one less than the number of cycles required to achieve the PCR saturation point.

The amount of first-strand cDNA input used for amplification is dependent on the amount of RNA used in the first-strand synthesis reaction. Use the suggested volumes of first-strand cDNA below:

RNA input in first-strand synthesis reaction (ng)	Volume first-strand cDNA used in PCR reactions (μl)
50–100	10
100–250	4
250–500	2
500–1,000	1

1. Dispense the appropriate volume of each cDNA (from the table above) into a clean PCR tube and adjust to 10 μl with H_2O.
2. Prepare enough of the following PCR mix for three PCR reactions for each cDNA sample: Per reaction 74 μl high quality H_2O, 10 μl 10× PCR buffer, 2 μl dNTP mix (10 mM of each dNTP), 2 μl cDNA PCR primer (12 pmol/μl), 2 μl *Taq* polymerase mix.
3. Mix well and centrifuge briefly to collect the mix at the bottom of the tube.
4. Program your thermal cycler as follows and allow it to reach 95°C: 95°C for 2 min (to denature the template and the hot start antibody); 24 cycles with 95°C for 15 s, 65°C for 30 s, and 68°C for 6 min.
5. Add 90 μl of the PCR mix to each 10 μl cDNA sample, mix well and centrifuge briefly.

6. Place all tubes (three identical reactions for each sample) in the preheated thermal cycler.
7. Perform 15 cycles of PCR on all samples.
8. At the end of the 15th cycle, remove the two preparative reaction tubes of each sample and place at 4°C.
9. Remove a 15 μl aliquot from each optimization reaction, place it in a fresh tube and store at 4°C. (*see* **Note 6**).
10. Return the optimization reaction tubes to the thermal cycler and continue the cycling program.
11. At the end of 18th cycle, remove another 15 μl aliquot from each optimization reaction tube and place at 4°C.
12. At the end of the 21st cycle, remove another 15 μl aliquot from each optimization reaction tube and place at 4°C.
13. At the end of the 24th cycle, remove another 15 μl aliquot from each optimization reaction tube and place at 4°C.
14. You should now have 15 μl aliquots of each PCR reaction cycled for 15, 18, 21, and 24 cycles. Electrophorese these aliquots on a 1.2% agarose gel using a 1 kb ladder as a marker.
15. Determine the optimal number of cycles for cDNA amplification, corresponding to one cycle less than the saturation point of the PCR reaction (**Fig. 15.6**).
16. Retrieve the preparative PCR reaction tubes from 4°C storage and subject these to the remaining number of

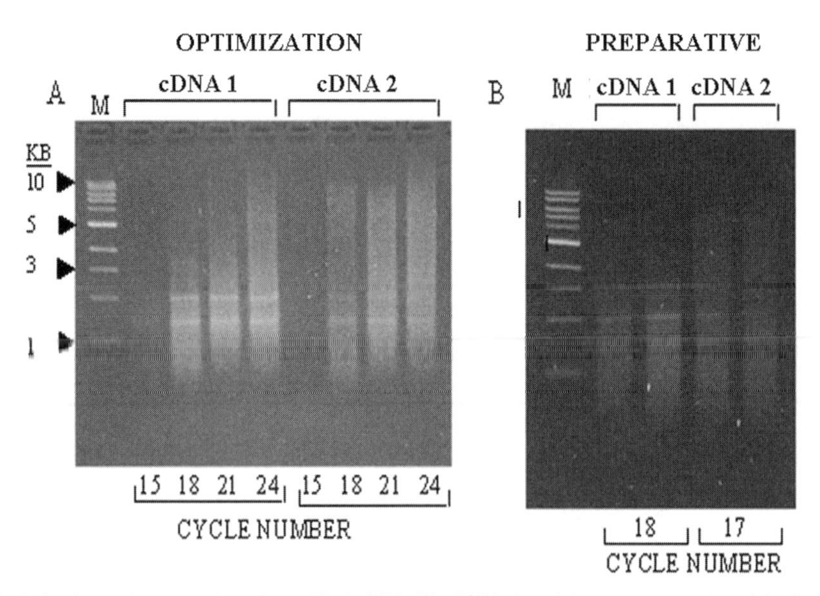

Fig. 15.6. Optimization and preparation of amplified cDNA. (**A**) cDNA's 1 and 2 were prepared as described in **Section 3.1.1** and used as PCR input. SMART-PCR was performed for the indicated number of cycles according to the optimization protocol in **Section 3.1.2** and electrophoresed on a 1.2% agarose gel. Amplified cDNA accumulates with increasing cycle number until saturation, estimated as cycle 18 and 17 for cDNA 1 and 2, respectively. (**B**) Two preparative reaction mixes for each cDNA were then subjected to the optimal number of cycles and again analyzed on a 1.2% agarose gel. M = DNA molecular weight marker.

cycles required to reach the optimal cycle number for each sample.

17. Add 2 μl of 0.5 M EDTA to each tube.
18. Combine the contents of both preparative tubes for each amplified cDNA.
19. Confirm that the PCR reactions were successful by electrophoresing 10–15 μl on a 1.2% agarose gel (**Fig. 15.6B**).

You should now have optimally amplified, double-stranded cDNA from the two mRNA populations being compared. These samples will now be purified and digested prior to linker ligation and subtractive hybridization.

3.1.3. Purification of Amplified cDNA

1. Set aside a 5 μl aliquot of each sample at –20°C as an "input" control. (see step 18)
2. Extract the amplified cDNA products with an equal volume (~190 μl) of phenol/chloroform/isoamyl alcohol (25:24:1). Vortex well.
3. Centrifuge for 2 min at full speed.
4. Transfer the upper aqueous phase to a clean tube.
5. Extract the aqueous phase with two to three volumes of *n*-butanol to concentrate the cDNAs. Vortex well.
6. Centrifuge for 1 min at full speed.
7. The cDNA remains in the lower aqueous phase and should be approximately 50 μl. Transfer the aqueous phase to a fresh tube (*see* **Note** 7). Discard the upper organic phase.
8. Prepare a chroma spin 1000 purification column for each sample by gently but thoroughly resuspending the column matrix.
9. Remove the top cap and bottom cap – in that order – and allow the buffer to drain out by gravity.
10. Equilibrate the column with 1.5 ml room temperature STE.
11. Apply the cDNA sample to the centre of the bed matrix and allow it to enter the column.
12. Add 25 μl of STE and allow the column to flow by gravity until it stops.
13. Add 150 μl STE and allow the column to flow by gravity until it stops.
14. Place a clean 1.5 ml microfuge tube under the column and apply 320 μl STE.
15. Collect the entire flow through as a single fraction. This contains the cDNA.
16. Place a clean 1.5 ml microfuge tube under the column and apply 75 μl STE.
17. Collect the entire flow through as another single fraction. This may contain additional cDNA.
18. On a 1.2% agarose gel, analyse the 5 μl input control (from step 1) and a 10 μl aliquot of the two eluted column fractions. Use a 1 kb ladder as a marker.

19. The first (320 μl) fraction should contain the majority of recovered cDNA. If the second eluate (75 μl) contains a significant amount of cDNA, combine it with the first (320 μl) eluate only if the first fraction contains less than approximately 30% of the total input.

3.2. RsaI Digestion

The purified cDNAs will now be digested with *Rsa*I. This generates blunt ends for linker ligation and produces smaller fragments better suited to subtractive hybridization.

1. Set aside a 10 μl aliquot of each purified cDNA sample at –20°C as an "undigested" control (for step 4).
2. Add an appropriate volume of 10× *Rsa*I buffer to bring the mixture to 1× *Rsa*I buffer.
3. Add 15–20 Units of *Rsa*I, mix well and incubate at 37°C for 3 h (*see* **Note 8**).
4. On a 1.2% agarose gel, analyse the 10 μl undigested controls and 10 μl of the digestion products (**Fig. 15.7**).
5. When digestion is complete, inactivate the enzyme by adding 5–10 μl 0.5 M EDTA (pH 8.0).
6. Purify the digestion products using a silica slurry resin (see **Section 2.2, step 3**) according to the manufacturer's recommendations (*see* **Note 9**).

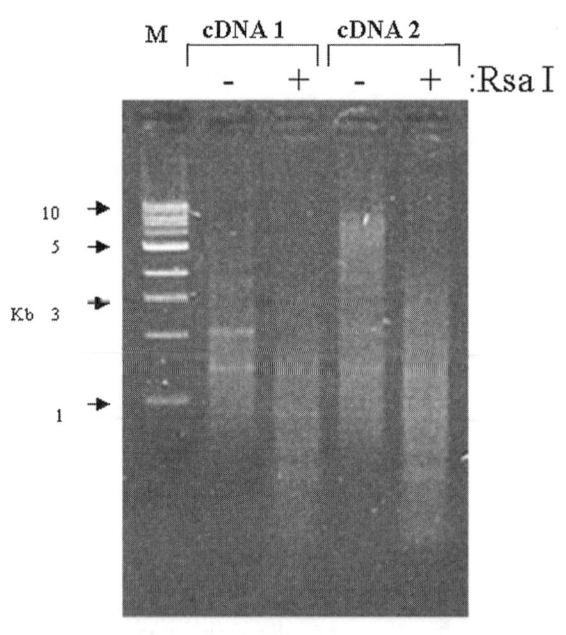

Fig. 15.7. Restriction digestion of amplified cDNAs. The amplified and purified cDNA preparations were digested with the four-base cutter *Rsa*I as described in **Section 3.1** and analyzed alongside undigested cDNA on a 1.2% agarose gel. Complete digestion is confirmed when no further decrease in molecular weight of the digestion products is observed with increased time of digestion.

7. Transfer the eluted, purified cDNA from the final pellet of silica resin to a fresh tube (*see* **Note 10**).
8. Add 1/2 vol. NH_4OAc to the final elution volume containing the silica purified, *RsaI* digested cDNAs and mix.
9. Add 2.5 volumes EtOH to each sample and mix to precipitate the digested cDNA.
10. Centrifuge at full speed (>12,000 ×*g*) for 20 min at room temperature and discard the supernatants.
11. Rinse the pellets with 500 μl 80% EtOH, centrifuge at full speed (>12,000 ×*g*) for 5 min at room temperature and discard the supernatants.
12. Air-dry the pellets at room temperature for 10–15 min. Do not use a speed vacuum.
13. Resuspend the pellets in 6.7 μl of distilled water.
14. Use 1.2 μl to determine the concentration of the purified digested cDNA's by absorbance at 260 nm. You should recover between 2 and 6 μg *Rsa*I digested cDNA after purification. If the concentration is >300 ng/μl, dilute to 300 ng/μl.
15. Store the remaining 5.5 μl of each *Rsa*I digested and purified cDNA at –20°C.

3.3. Preparation of Tester cDNAs

3.3.1. Ligation of Adaptors

Label the prepared cDNAs as number 1- and number 2-. An aliquot of each digested cDNA preparation, 1- and 2-, will be ligated to two separate adaptors; 1 and 2R, making a total of four separate ligation reactions. The ligation reactions are labeled 1-1, 1-2R and 2-1, 2-2R. Two additional ligation reactions, labeled 1-c and 2-c, will be prepared by mixing aliquots of reactions 1-1 with 1-2R and 2-1 with 2-2R. Reaction products 1-c and 2-c will later serve as unsubtracted controls.

1. Prepare a ligation mix sufficient for four reactions: Per reaction 3 μl H_2O, 2 μl 5× ligation buffer, 1 μl T4 DNA ligase.
2. Dilute 1 μl of each digested and purified cDNA from **step 13** in **Section 3.2** with 5 μl distilled water.
3. Set up the following ligation reactions:

Tube	1-1	1-2R	2-1	2-2R
Diluted cDNA 1- (μl):	2	2	–	–
Diluted cDNA 2- (μl):	–	–	2	2
Adaptor 1 (μl):	2	–	2	–
Adaptor 2R (μl):	–	2	–	2
Ligation mix (μl):	6	6	6	6
	10	10	10	10

4. Mix well and centrifuge briefly.
5. Transfer 2 μl from tube 1-1 and 2 μl from tube 1-2R to a fresh tube labeled 1-c and mix.
6. Transfer 2 μl from tube 2-1 and 2 μl from tube 2-2R to a fresh tube labeled 2-c and mix.
7. Centrifuge all tubes briefly.
8. Incubate all six ligation reactions overnight at 16°C.
9. Stop the reactions by adding 1 μl EDTA/glycogen mix to each tube.
10. Denature the ligase by heating all tubes at 72°C for 5 min.
11. Centrifuge all tubes briefly. These are your adaptor-ligated tester cDNAs to be used in step 1 of **Section 3.3.2**.
12. Transfer 1 μl from tube 1-c to a fresh tube containing 1 ml of distilled H_2O. Label this tube "forward unsubtracted 1-c".
13. Transfer 1 μl from tube 2-c to a fresh tube containing 1 ml of distilled H_2O. Label this tube "reverse unsubtracted 2-c".
14. Store all tubes at –20°C.

3.3.2. Ligation Efficiency Assay

It is important to determine if the ligation reactions were efficient by using a simple PCR assay. The two adaptors used in **Section 3.3.1** share a region of common sequence. PCR reactions using a primer pair comprised of a G3PDH-specific primer and the adaptor-targeted PCR primer generate a specific product if the adaptor sequence is present (the 'test' reaction). This product is compared to PCR reaction products generated from a primer pair comprised of two G3PDH targeted primers (control reactions). If the efficiency of the test reaction is similar to that of the control, it indicates efficient ligation of the adaptors to the digested cDNAs.

1. Prepare 1:200 dilutions of each of the four ligation products, 1-1, 1-2R, 2-1. and 2-1R.
2. For each diluted sample, set up two PCR reactions (eight reactions in total)

Reaction:	1 (Test) (μl)	2 (Control) (μl)
Diluted sample: (1-1, 1-2R, 2-1 or 2-2R)	1	1
3′ G3PDH primer (10 pmol/μl)	1	1
5′ G3PDH primer (10 pmol/μl)	–	1
Adaptor-targeted PCR primer (10 pmol/μl)	1	–
	3	3

3. Prepare a PCR mix sufficient for eight reactions: Per reaction 18.5 μl H_2O, 2.5 μl 10× PCR buffer, 0.5 μl dNTP mix, 0.5 μl polymerase mix.

4. Add 22 μl of the PCR mix to each reaction tube from step 2.
5. Subject all tubes to the following thermal profile: 75°C for 5 min (this step fills in the ends of the adaptors to produce fully double-stranded fragments), 95°C for 2 min (to fully inactivate the hot start antibody), 25 cycles with 95°C for 30 s, 65°C for 30 s, 68°C for 2.5 min (*see* **Note 11**).
6. Analyze the PCR products alongside a molecular weight marker on a 2% agarose gel. For rat or mouse samples, you should observe a ~1.2 kb band in the test reactions. For human samples, you should observe a ~0.75 kb band (*see* **Note 12**). For other species, you will need to design specific primers. Successful ligations will result in nearly equal band intensity from both the test and control reactions. If the band intensity of the test reaction products are less than approximately 40% that of the control, you should repeat the procedure starting from the cDNA amplification step (**Section 3.1.2**).

3.4. Subtractive Hybridization

Two subtractions will be performed; a forward and a reverse subtraction. For each subtraction, the adaptor-ligated products are designated the tester cDNAs and the unligated cDNAs (with no adaptors) are designated driver cDNAs. Adaptor-ligated products 1-1 and 1-2R tester cDNAs will be independently hybridized to an excess of driver cDNA 2 in two primary, forward subtraction hybridizations (one for each adaptor-ligated preparation, 1-1 and 1-2R). Likewise, adaptor-ligated products 2-1 and 2-2R tester cDNAs will be independently hybridized to excess driver cDNA 1 in two primary reverse subtraction hybridizations.

3.4.1. Primary Hybridization Reaction

1. Set up the following first-hybridization reactions using the adaptor-ligated tester cDNAs from **step 11** of **Section 3.3.1** and *Rsa*I digested and purified driver cDNAs from **step 15** of **Section 3.2**:

	Forward		Reverse	
First-hybridization number:	*(1)*	*(2)*	*(3)*	*(4)*
*Rsa*I digested driver cDNA 1 (μl)	–	–	1.5	1.5
*Rsa*I digested driver cDNA 2 (μl)	1.5	1.5	–	–
Adaptor ligation cDNA tester 1-1 (μl)	1.5	–	–	–
Adaptor-ligated cDNA tester 1-2R (μl)	–	1.5	–	–
Adaptor ligation cDNA tester 2-1 (μl)	–	–	1.5	
Adaptor-ligated cDNA tester 2-2R (μl)	–	–	–	1.5
4× Hybridization buffer (μl)	1.0	1.0	1.0	1.0
	4.0	4.0	4.0	4.0

2. Heat denature the hybridization reactions in a thermal cycler at 98°C for 1.5 min.
3. Hybridize the denatured cDNAs in a thermal cycler at 68°C for 8 h. Hybridization times over 12 h will allow differentially expressed cDNAs to re-anneal.
4. After hybridization, centrifuge briefly to collect the reaction mixes at the bottom of the tube and return to 68°C briefly while preparing the secondary hybridization reaction mix.

3.4.2. Secondary Hybridization Reactions

1. Prepare fresh denatured drivers for the second hybridization as follows:

	Forward	**Reverse**
*Rsa*I digested driver cDNA 1 (μl)	–	1
*Rsa*I digested driver cDNA 2 (μl)	1	–
4× Hybridization buffer (μl)	1	1
Sterile H_2O (μl)	2	2
	4	4

2. Heat denature these driver cDNA solutions in a separate thermal cycler or heat block (*see* **Note 13**). The first-hybridization products (1) and (2) will now be mixed with freshly denatured driver cDNA 2 in the forward subtraction. Likewise, the first-hybridization products (3) and (4) will be mixed with freshly denatured driver cDNA 1 in the reverse subtraction.
3. Remove 1 μl of freshly denatured driver cDNA 2 and place it on the inside upper wall of the tube containing first-hybridization number (1). Do not let the solutions mix at this point. Set the tube down horizontally on the thermal cyclers heat block momentarily.
4. Working quickly, transfer the contents of first-hybridization number (2) to the tube containing first hybridization number (1) and the freshly denatured driver cDNA 2. As you dispense the mixture, push the denatured driver cDNA down the inside surface of the tube so that all three solutions are mixed simultaneously and return the mixture to the thermal cycler at 68°C. This is your forward subtractive hybridization. Label this tube forward subtracted cDNA.
5. Repeat this process with 1 μl freshly denatured driver cDNA 1 and first hybridization (3) and (4) and return the mixture to the thermal cycler at 68°C. This is your reverse subtractive hybridization. Label this tube reverse subtracted cDNA.
6. Incubate the forward and reverse subtractive hybridization mixes at 68°C overnight. These hybridizations go to completion.

7. When the second hybridization is complete, add 200 μl of dilution buffer to each subtraction mix.
8. Return the diluted hybridization products to the 68°C thermal cycler for 7 min.
9. Store the forward and reverse subtraction mixes at –20°C.

3.5. Suppressive and Nested PCR

The unsubtracted cDNAs prepared in **Section 3.3.1** (**steps 12** and **13**) are used as controls the next steps of suppressive and nested PCRs.

3.5.1. Primary (Suppressive) PCR

1. Prepare enough of the following PCR mix for four reactions: Per reaction 19.5 μl H_2O, 2.5 μl 10× PCR buffer, 0.5 μl 10 mM dNTP mix, 1.0 μl PCR primer (10 pmol/μl), 0.5 μl 50× DNA polymerase mix.
2. Aliquot 1 μl of each of the following into separate PCR reaction tubes, labeled s1, s2, s3, and s4 (s = "suppressive"):
   s1) forward subtracted cDNA (from step 9 of **Section 3.4.2**)
   s2) forward unsubtracted 1-c (from **step 12** of **Section 3.3.1**)
   s3) reverse subtracted cDNA (from **step 9** of **Section 3.4.2**)
   s4) reverse unsubtracted 2-c (from **step 13** of **Section 3.3.1**)
3. Add 24.0 μl of the PCR mix to each tube (to make 25 μl reaction volumes), mix and centrifuge briefly to collect contents of each at the bottom of the tube.
4. Program your thermal cycler as follows and subject all reaction mixes to this PCR program: 75°C for 5 min (this step extends the adaptors to create a complementary target for the PCR primer), 27 cycles with 94°C for 30 s, 66°C for 30 s, and 72°C for 1.5 min.
5. These are your primary (suppressive) PCR products, s1–s4. When cycling is complete, transfer 8 μl of each reaction product to a fresh, appropriately labeled tube, and store at –20°C for later gel analysis.

3.5.2. Secondary (Nested) PCR

1. Prepare 1:10 dilutions of each primary PCR product s1–s4 with water (e.g., 3 μl PCR product and 27 μl water).
2. Aliquot 1 μl of each diluted primary PCR product into separate PCR reaction tubes labeled n1, n2, n3, and n4 (n = "nested").
3. Prepare enough of the following PCR mix for four reactions: Per reaction 18.5 μl H_2O, 2.5 μl 10× PCR buffer, 0.5 μl 10 mM dNTP mix, 1.0 μl nested PCR primer 1 (10 pmol/μl), 1.0 μl nested PCR primer 2R (10 pmol/μl), 0.5 μl 50× DNA polymerase mix.
4. Add 24.0 μl of the PCR mix to each tube (to make 25 μl reaction volumes), mix and centrifuge briefly to collect contents of each at the bottom of the tube.

5. Subject all reaction mixes to the following PCR program: 12 cycles with 94°C for 30 s, 68°C for 30 s, and 72°C for 1.5 min.
6. When cycling is complete, transfer 8 μl of each reaction product to an appropriately labeled tube for gel analysis.
7. The remaining 17 μl of the secondary (nested) PCR products are your final products. They comprise:
   n1) forward subtracted cDNA
   n2) forward unsubtracted cDNA
   n3) reverse subtracted cDNA
   n4) reverse unsubtracted cDNA
8. Electrophorese the 8 μl aliquots of each suppressive and nested PCR product, s1–s4 from **step 5** of **Section 3.5.1**, and n1–n4 from step 6, respectively, on a 2% agarose gel. Both primary and secondary PCR products should appear as a smear, with or without discrete bands (**Fig. 15.8**).
9. The final products (n1–n4) can now be cloned to make subtracted cDNA libraries and/or used as probes to screen subtracted cDNA, global cDNA, or genomic DNA libraries.

3.5.3. Assay for Subtraction Efficiency

1. In fresh tubes, prepare 1:10 dilutions of each secondary PCR product n1–n4 with water (e.g., add 3 μl PCR product to 27 μl water as before).

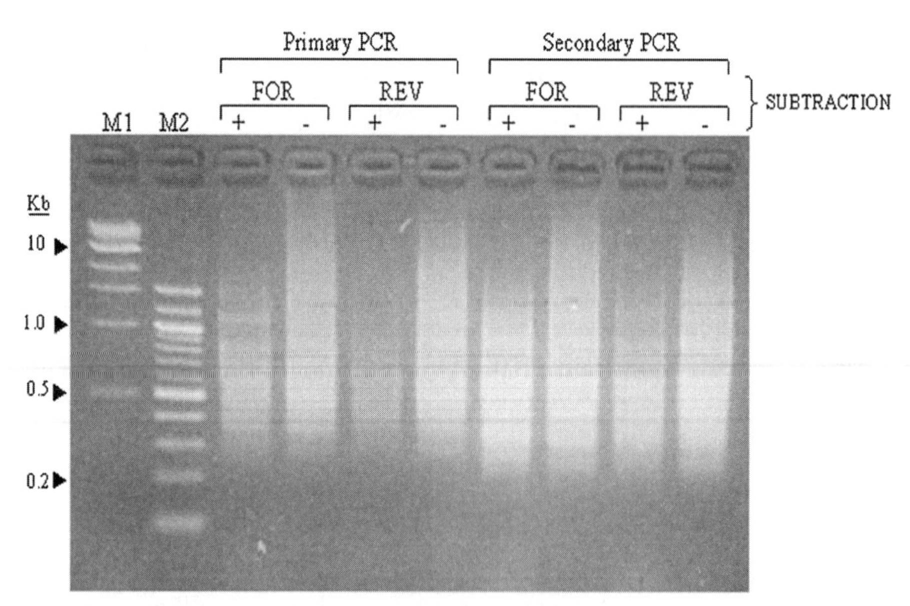

Fig. 15.8. Amplification of subtracted and unsubtracted cDNAs. Primary (suppressive) and secondary (nested) PCR reactions were performed as described in **Sections 3.5.1** and **3.5.2**, respectively, and analyzed on a 2% agarose gel. The input cDNA's were from both forward and reverse subtractions as indicated. Lanes labeled "+" are subtracted and lanes labeled "–" are unsubtracted. The four secondary PCR products (*right*) are the final forward and reverse subtracted and unsubtracted products (referred to as samples n1–n4 in **Section 3.5.2**, **step 7**). M1 and M2 are high and low molecular weight DNA markers, respectively. Kb=kilobase pairs.

2. Place 1.0 μl of each diluted secondary PCR product in a fresh PCR tube.
3. Prepare enough of the following PCR mix for four reactions: Per reaction 22.4 μl H_2O, 3.0 μl 10× PCR buffer, 0.6 μl 10 mM dNTP mix, 1.2 μl G3PDH 5′ primer (10 pmol/μl), 1.2 μl G3PDH 3′ primer (10 pmol/μl), 0.6 μl 50× DNA polymerase mix.
4. Add 29 μl of the PCR mix to each tube (to make 30 μl reaction volumes), mix and centrifuge briefly to collect contents of each at the bottom of the tube.
5. Program your thermal cycler as follows: 33 cycles with 94°C for 30 s, 60°C for 30 s, and 68°C for 1.5 min.
6. Place the reaction mixes in the thermal cycler and run the first 18 cycles.
7. After the 18th cycle, remove 5 μl from each tube and place in separate, fresh tubes.
8. Continue the program for an additional five cycles (for a total of 23 cycles) and again transfer 5 μl from each reaction to fresh tubes.
9. Continue the program for an additional five cycles (for a total of 28 cycles) and again transfer 5 μl from each reaction to fresh tubes.
10. Continue the program for the final five cycles (for a total of 33 cycles) and again transfer 5 μl from each reaction to fresh tubes (for a total of 33 cycles).

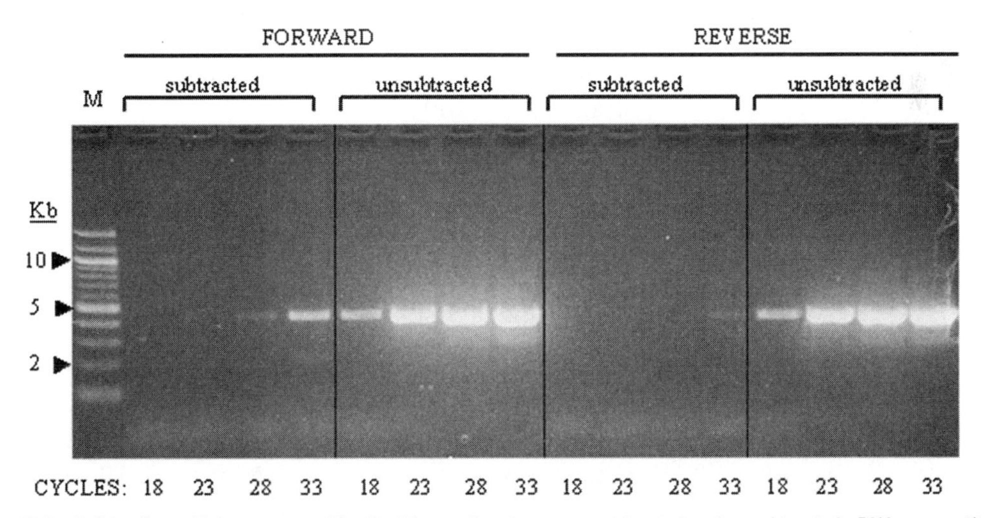

Fig. 15.9. Subtraction efficiency assay. The final forward and reverse subtracted and unsubtracted cDNA preparations were each diluted 1:10 and used as inputs for the G3PDH-specific PCR reaction described in **Section 3.5.3**, **step 7**. Equal aliquots of these reactions were removed during the extension step of the indicated cycle number and analyzed on a 2% agarose gel. Note, for example, in the forward subtraction, 33 cycles were required to achieve G3PDH product yield similar to 18 cycles from the unsubtracted cDNA; a difference of 15 logarithmic cycles. The efficiency of subtraction is indicated by the increase in cycle number needed to generate a similar level of PCR product. (reproduced from Ref. *(6)* with permission from Blackwell Publishing).

11. Electrophorese each of the 5 μl aliquots collected at 18, 23, 28, and 33 cycles on a 2% agarose gel.

    Efficient subtraction is indicated by depletion of G3PDH sequences from the subtracted cDNAs for both the forward and reverse subtractions (**Fig. 15.9**). Unsubtracted cDNAs should show an increase in G3PDH product accumulation with increasing cycle number, to the point of a plateau.

4. Notes

1. This oligonucleotide is available as part of the SMART-PCR cDNA synthesis kit from Clontech. Details of their modifications are proprietary. We successfully amplify cDNA with an independently synthesized oligonucleotide, however, efficiencies in subtractive hybridization and suppressive PCR have not been determined.
2. The *Taq* polymerase used for cDNA amplification should be nuclease deficient, formulated specifically for amplification of long fragments, and should include the Kellog antibody for automatic hot start.
3. Phenol:chloroform:isoamyl alcohol should be prepared with fresh (<2 weeks) phenol saturated with 50 mM Tris–HCl (pH 7.5), 150 mM NaCl, 1 mM EDTA. Phenol appearing yellow from oxidation should not be used.
4. RNA quality is perhaps the most critical parameter for successful comparisons of two RNA populations. Determine RNA quality by, for example, an Agilent bioanalyzer or Northern blot, and determine RNA concentrations carefully by absorbance at 260 nm.
5. To minimize evaporation and condensation in the tube, which can significantly reduce cDNA yields, use a pre-warmed air incubator as opposed to a heat block.
6. It is important that aliquots are removed from each sample while at the 68°C extension step, before the thermal cycler begins to ramp up to the denaturing temperature. Denatured samples will not migrate correctly during electrophoresis.
7. Re-extract the aqueous phase with *n*-butanol or back-extract the organic phase with H_2O, if necessary, to reach an aqueous volume of approximately 50 μl containing your cDNA.
8. Digestion must go to completion. Aliquots (~10 μl) can be removed at several time points and analyzed on an agarose gel. No further decrease in molecular weight with time indicates complete digestion.
9. Alternatively, samples can be extracted with phenol:chloroform:isoamyl alcohol (25:24:1) and ethanol precipitated, although this may lead to decreased adaptor ligation efficiency.

10. After transferring to a fresh tube, re-centrifuge the eluate at full speed for 1 min and transfer to yet another fresh tube to eliminate possible carry over of silica resin. Alternatively, eluates may be passed through a 0.45 μm filter.
11. If G3PDH is highly abundant in your cDNA populations, reduce the cycle number to avoid saturation of the PCR reactions.
12. If a band of ~1.2 kb is generated from a human G3PDH PCR product, this indicates incomplete digestion because human G3PDH contains an *Rsa*I site.
13. It is practical to have this heat block or thermal cycler near the thermal cycler used for the hybridizations in order to avoid premature renaturation of the freshly denatured driver cDNA's.

References

1. Chenchik, A., Zhu, Y. Y., Diatchenko, L., et al. (1998) Generation and use of high-quality cDNA from small amounts of total RNA by SMART PCR, in Gene Cloning and Analysis by RT-PCR. *BioTechniques Books*, MA. 305–319.
2. Chenchik, A., Moqadam, F., Siebert, P. (1996) A new method for full-length cDNA cloning by PCR, in A Laboratory Guide to RNA: Isolation, Analysis, and Synthesis. Wiley-Liss, Inc. 273–321.
3. Wetmur, J. G., Davidson, N. (1968) Kinetics of renaturation of DNA. *J Mol Biol* 31, 349–370.
4. Ji, W., et al. (2002) Efficacy of SSH PCR in isolating differentially expressed genes. *BMC Genomics* 3, 12.
5. Diatchenko, L., et al. (1996) Suppression subtractive hybridization: a method for generating differentially regulated or tissue-specific cDNA probes and libraries. *Proc Natl Acad Sci USA* 93, 6025–6030.
6. Hillmann, A. G., et al. (2006) Comparative RNA expression analyses from small-scale, single-donor platelet samples. J Thromb Haemost 4, 349–356.
7. CLONTECH Laboratories, P.A., CA, USA, PCR-Select cDNA Subtraction kit Kit User Manual. 2007.
8. CLONTECH Laboratories, P.A., CA, USA, SMART PCR cDNA SYNTHESIS Kit User Manual. 2007.

Chapter 16

Transcript Profiling of Human Platelets Using Microarray and Serial Analysis of Gene Expression (SAGE)

Dmitri V. Gnatenko, John J. Dunn, John Schwedes, and Wadie F. Bahou

Abstract

Platelets are anucleated cells that are generated from megakaryocytes via thrombopoiesis. They lack genomic DNA but have a pool of individual mRNA transcripts. Taken together, these mRNAs constitute a platelet transcriptome. Platelets have a unique and reproducible transcript profile, which includes ~1,600–3,000 individual transcripts. In this chapter, we will focus on platelet purification and on transcript profiling using an Affymetrix microarray platform and serial analysis of gene expression (SAGE). Platelet purification is described in detail. Large-scale platelet purification schema is designed to purify platelets from apheresis platelet bags (~3–5 × 10^{11} platelets/bag). Modification of this schema – small-scale platelet purification – is designed to isolate platelets from 20 ml of peripheral blood. This chapter provides detailed protocols for microarray and SAGE transcript profiling. We also discuss peculiarities of platelet purification, RNA isolation, and transcript profiling.

Key words: Platelet RNA, microarray, serial analysis of gene expression.

1. Introduction

Human blood platelets are generated from precursor bone marrow megakaryocytes. They play critical roles in normal hemostatic processes and pathologic conditions such as thrombosis, vascular remodeling, inflammation, and wound repair. Platelets are anucleate and lack nuclear DNA, although they retain megakaryocyte-derived mRNAs *(1)*. Platelets also contain rough endoplasmic reticulum and polyribosomes, and preserve the ability for protein biosynthesis from cytoplasmic mRNA *(2)*. The entire pool of individual platelet mRNAs constitutes a platelet transcriptome. Historically, platelet transcriptome has been regarded as invariant and gradually declining with cell senescence.

Peter Bugert (ed.), *DNA and RNA Profiling in Human Blood: Methods and Protocols, vol. 496*

DOI 10.1007/978-1-59745-553-4_16 Springerprotocols.com

Indeed, younger platelets contain larger amounts of mRNA with a greater intrinsic capacity for protein biosynthesis, as determined using fluorescent nucleic acid dyes such as thiazole orange *(3)*. This traditional platelet/mRNA paradigm was challenged recently when a signal-dependent pre-mRNA splicing was identified in platelets *(4)*. Signal-dependent splicing provides a mechanism for altering the repertoire of translatable mRNAs in response to cellular activation/stimulation. Furthermore, platelets possess essential components of a functional spliceosome and selected unspliced pre-mRNAs. Platelet spliceosomes have a unique ability to splice pre-mRNA in the cytoplasm (as opposed to the typical nuclear location) – a capability not described in any other mammalian cell.

Modern high-throughput technologies such as microarray and serial analysis of gene expression (SAGE) allow sensitive and reliable transcript profiling and have been commonly used for such studies with a wide variety of cell types. To date, however, only a few studies have focused on similar analysis of the human platelet transcriptome. One difficulty in working with platelets is that they contain far fewer transcripts than most nucleated cells, ranging from ~1,600 to 3,000 mRNAs *(5–9)*. In particular, platelets express much less transcripts than leukocytes *(10, 11)* which can easily contaminate the platelet fraction derived from whole blood. To date, three important conclusions can be made from platelet transcript profiling. First, platelets have much fewer individual mRNA transcripts than most nucleated cells, presumably due to the lack of ongoing transcription. Second, platelets have a unique and reproducible transcript profile, distinctive from transcript profiles of any other cell types. Third, this profile is different for normal platelets and for platelets of patients with platelet-associated diseases, in particular of patients with essential thrombocythemia.

Taken together, these studies demonstrate that platelet transcript profiling can be applied to dissect molecular mechanisms of platelet function and to better understand and possibly aid in diagnosing certain platelet-associated diseases. In this chapter we will discuss protocols for platelet purification and transcript profiling, paying specific attention to the critical details and potential limitations of these protocols.

2. Materials

2.1. Platelet Purification

For preparation of all buffers and solutions, deionized and sterilized water (ddH_2O) should be used unless otherwise specified. Reagents and solutions are stored according to manufacturer's instructions. For all procedures involved in RNA isolation, SAGE, and microarray analysis, ddH_2O should be treated with diethyl pyrocarbonate (DEPC), and all glass- and plastic-ware should be DEPC-treated to minimize RNA losses. Use pipette tips with aerosol filters, wear gloves and if possible perform all manipulations with RNA in a "RNAse-free" hood.

2.1.1. Platelet Purification from Platelet Apheresis Bags

1. Sodium ethylenediaminetetraacetate (EDTA) stock solution, 0.5 M, pH 8.0.
2. Sodium citrate, 4% (w/v).
3. Phosphate-buffered saline (PBS), pH 7.4: 137 mM NaCl, 2.7 mM KCl, 4.3 mM Na_2PO_4, 1.4 mM KH_2PO_4.
4. HEPES-buffered modified Tyrodes buffer (HBMT): 10 mM HEPES pH 7.4; 150 mM NaCl; 2.5 mM KCl; 0.3 mM NaH_2PO_4; 12 mM $NaHCO_3$; 0.2% BSA; 0.1% glucose; 2 mM EDTA.
5. Triton X-100.
6. Prostaglandin E (PGE) (Sigma, St Louis, MO, USA): 1 mg/ml stock solution in 95% ethanol, 2.8 mM). Solution is light-sensitive, keep the tube wrapped in foil at –20°C.
7. Aspirin (Sigma).
8. Aqueous chromatography column (5.0 cm × 60 cm; Spectrum Chromatography, Houston, TX, USA).
9. BioGel A50 M (BioRad) or Sepharose CL-4B (Amersham).
10. Mouse anti-CD45 antibody, conjugated to magnetic beads (Miltenyi Biotec Inc., Auburn, CA, USA).
11. MidiMacs starting kit (Miltenyi Biotec) which includes MidiMacs magnetic columns (Miltenyi Biotec) and a magnetic stand.
12. BlueMax™ polypropylene conical tubes, 50 ml (Falcon; BD, Franklin Lakes, NJ USA).
13. Pore size leukocyte membrane filters, 5 μm.

2.1.2. Characterization of the Efficiency of Platelet Purification

1. FITC-conjugated anti-CD41b (BD Pharmingen).
2. PE-conjugated anti-glycophorin (BD Pharmingen).
3. PERCP-conjugated anti-CD45 (BD Pharmingen).
4. Formalin solution (1%) in PBS.
5. Leukocyte contamination of platelet preparation can also be addressed at RNA level by reverse transcription–polymerase chain reaction (RT–PCR) using primers specific to leukocyte common antigen CD45 (NCBI accession number Y00638):
   Forward, 5′-GCTCAGAATGGACAAGTA-3′
   Reverse, 5′-CACACCCATACACACATACA-3′

2.1.3. Platelet Purification from 20 ml of Peripheral Blood

1. Prepare in advance: 20-ml disposable syringe; 19-gauge needle; two columns for platelet gel-filtration (commercially available 60-ml disposable syringes is a simple, inexpensive, and durable columns for platelet gel filtration).
2. Place a circular piece of nylon mesh with 85 micron pore size (Small Parts, Inc.) at the bottom.

3. Place the rubber ring from the plunger of the syringe on top of the mesh to hold it in place.
4. Load suspension of Sepharose 2B beads (Amersham) in HBMT buffer until bead volume reaches 50 ml.
5. Equilibrate the column by passing through 150 ml of HBMT buffer, pH 7.45.
6. Seal top and bottom of the column and keep it at room temperature until use.

2.2. Platelet RNA Isolation

1. Trizol® reagent, (Invitrogen).
2. Chloroform.
3. Isopropanol.
4. Ethanol (100%).
5. "Glyco-blue" (Ambion).
6. SuperRNAseIn^{TM} RNAse inhibitor, 20 U/μl (Ambion).
7. DEPC (Sigma). Treat 1.5- and 0.5-ml Eppendorf tubes with DEPC, autoclave, and store them in a closed beaker.
8. For RNA isolation from platelets isolated from an apheresis bag, treat four 30-ml Corex glass centrifuge tubes with DEPC, autoclave, and store in aluminum foil until use. Use glass pipettes to measure organic solvents.
9. For RNA isolation, use only DEPC-treated (or RNAse-free) plastic- and glass-ware. Prepare DEPC-treated H_2O to dissolve RNA pellets.

2.3. Microarray Profiling

For preparation of all buffers and solutions, deionized and sterilized DEPC-treated water (ddH_2O) should be used unless specified otherwise. Store reagents and solutions according to manufacturer's instructions or as described in Current Protocols in Molecular Biology.

1. $T7dT_{24}$ primer (Proligo).
2. SuperScript II reverse transcriptase (Invitrogen).
3. DEPC-treated H_2O.
4. dNTP mix (10 mM) prepared from individual 100 mM dATP, dCTP, dGTP, dTTP (Invitrogen).
5. *E.coli* DNA ligase (10 U/μl; Invitrogen).
6. DNA polymerase I (10 U/μl; Invitrogen).
7. RNAse H (1–4 U/μl; Invitrogen).
8. T4 DNA polymerase (5 U/μl; Invitrogen).
9. EDTA (0.5 M) in DEPC-treated H_2O.
10. Phaselock tube (Eppendorf).
11. Phenol, chloroform, NH_4O-acetate, absolute ethanol.
12. BioArray High Yield kit (Enzo Biochem).
13. RNeasy mini kit (Qiagen).
14. Dnase I solution (RNase-free DNase set; Qiagen).
15. Fragmentation buffer: 200 mM Tris–acetate, pH 8.1; 500 mM KOAc; 150 mM MgOAc.
16. Control oligonucleotide B2, 20× eukaryotic hybridization controls, 2× MES hybridization buffer all from Affymetrix.

17. BSA (50 mg/ml; Invitrogen).
18. Herring sperm DNA (10 μg/μl; Promega).

2.4. Serial Analysis of Gene Expression

All buffers should be prepared using DEPC-treated ddH_2O.

1. Glycogen, 5 μg/μl (Invitrogen).
2. Phenol:chloroform:isoamyl alcohol.
3. SuperRNAse-In, 20 U/μl (Invitrogen).
4. GlycoBlue (Ambion).
5. Spermidine $(HCl)_3$ (Sigma).
6. *E.coli* DNA ligase, *E.coli* DNA polymerase, *E.coli* RNAse H (SuperScript Choice System for cDNA synthesis; Invitrogen).
7. *E.coli* exonuclease I, 20 U/μl (NEB).
8. TEN/BSA buffer: 10 mM Tris–HCl, pH 8.0; 1 mM EDTA; 1 M NaCl; 1% BSA.
9. *Mme*I restriction buffer (10×): 100 mM HEPES, pH 8.0; 25 mM K-acetate, pH 8.0; 50 mM Mg-acetate, pH 8.0; 20 mM DTT.
10. *Mme*I restriction enzyme (NEB).
11. *Nla*III restriction enzyme (NEB) (Note: this enzyme should be aliquoted and then stored at –80°C).
12. SAM (100×): 4 mM SAM (*S*-adenosylomethionine hydrochloride).
13. Ammonium acetate (7.5 M).
14. *Prepare in advance:* Oligo$(dT)_{25}$-conjugated magnetic beads (Dynal).
15. Dynabeads mRNA DIRECT kit (includes Dynabeads Oligo$(dT)_{25}$ magnetic beads, lysis/binding buffer, washing buffers A and B; Dynal).
16. Magnetic Particle Concentrator MPC-S (Invitrogen or Dynal).
17. Primers and cassettes:
    cassette A primers:
    5′-TTTGGATTTGCTGGTCGAGTACAACTAGGCTTAAT CCGACATG-3′
    5′-pGTCGGATTAAGCCTAGTTGTACTCGACCAGCAAA TCC-3′amino modified
    cassette B primers:
    5′-pTTCATGGCGGAGACGTCCGCCACTAGTGTCGCA ACTGACTA-3′
    5′-TAGTCAGTTGCGACACTAGTGGCGGACGTCTCCG CCATGAA**NN**-3′
18. Primers for PCR amplification:
    Forward, 5′- biotinylated, corresponding to a portion of cassette A top strand: 5′-Biotin-GGATTTGCTGGTCGAG TACA-3′
    Reverse, non-biotinylated, corresponds to a portion of cassette B bottom strand: 5′-TAGTCAGTTGCGACACT AGTGGC-3′

3. Methods

3.1. Platelet Purification

Initially, we have developed a large-scale platelet purification schema to isolate platelets starting from an apheresis bag (~3–5 × 10^{11} platelets) *(6)*. This schema includes four consecutive steps, each based on a different principle of cell separation. Combining the different separation techniques in one purification scheme allows for highly efficient platelet purification (*see* **Note 1**). The resulting product should contain only 5–10 leukocytes per 1 × 10^5 platelets. These steps include *(1)* centrifugation; *(2)* gel-filtration; *(3)* filtration; and *(4)* $CD45^+$ immunodepletion. Combined, these steps complement each other to achieve the highest purity of platelets while minimizing platelet losses. Large-scale schema allows isolation of ~ 50–100 μg of total RNA from highly purified platelets. This amount is sufficient for direct high-throughput transcript profiling by microarray and SAGE technology. The major drawbacks of this schema are that: (i) it is time- and labor-consuming; (ii) it cannot be used for screening large groups of patients, and (iii) it requires researcher access to properly stored fresh apheresis bags and to sophisticated laboratory equipment.

To adapt platelet purification for processing large number of samples, the schema may be modified to isolate platelets from 20 ml of peripheral blood. The modified procedure is simple, reliable and can be used for screening a large number of patients, and yet this schema is efficient in removing contaminating cells. Its main drawback is that it does not provide enough total RNA for direct transcript profiling (**Fig. 16.1**). Platelet purification from 20 ml of peripheral blood using the platelet-rich plasma followed by gel-filtration yields ~2 × 10^9 platelets total. Yield of total platelet RNA varies between 1.0 and 2.5 μg, whereas accurate and reproducible transcript profiling using Affymetrix technology requires ~ 5 μg of total RNA. Although this amount is insufficient for microarray or SAGE analyses, it can be used for the analysis of expression of selected transcript using quantitative real-time RT-PCR (Q-PCR) technique.

Alternatively, total platelet RNA can be converted to cDNA, amplified by PCR or in vitro transcription and then used for transcript profiling. Although amplification can introduce some variance in original transcript ratios, it has been successfully applied for transcript profiling (reviewed at Lockhart et al. *(12)*). In our experience, in vitro amplification is efficient and reproducible. Starting with 60 ng of total platelet RNA and using Ovation Aminoallyl RNA amplification and labeling system (NuGENE), 4–5 μg of labeled cDNA can usually be generated. This amount is sufficient for transcript profiling using custom-made

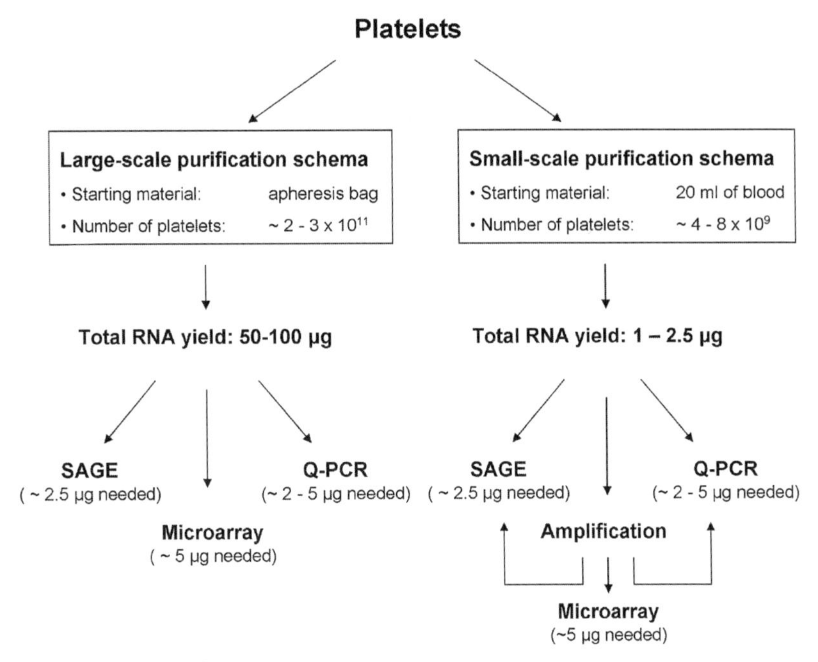

Fig. 16.1. Overview of platelet purification schema.

oligonucleotide platelet-specific microarray chips. Moreover, hybridization of amplified and non-amplified platelet RNA to this chip reveals excellent concordance (**Fig. 16.2**). Potentially, this cDNA can also be hybridized to standard Affymetrix chips.

3.1.1. Platelet Purification from Platelet Apheresis Bags

The volume of liquid and platelet content of each apheresis bag varies depending on many factors, including donor, techniques used for apheresis, and other variables. To standardize the purification schema, the same starting volume (150 ml) should be used

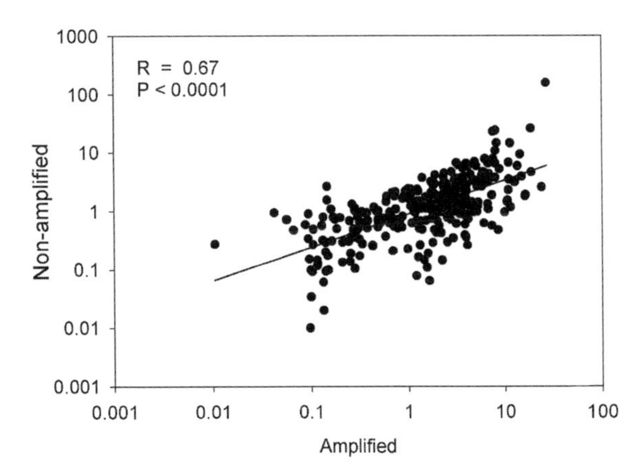

Fig. 16.2. Correlation between amplified and non-amplified samples of total platelet RNA, analyzed by hybridization to a customized platelet-specific microarray chip (R=0.67, P<0.0001).

for every purification. This is an optimal volume for efficient gel-filtration on the column of given size (5 × 60 cm, 1,000 ml support). The remainder of the platelets from the apheresis bag can either be discarded or used for other studies.

For platelet isolation from an apheresis bag, prepare a column for platelet gel-filtration in advance:

1. For large-scale platelet purification a large aqueous chromatography column is packed with BioGel A50M or Sepharose CL-4B in accordance with manufacturer's instructions.
2. Remove air bubbles from the bead slurry by applying a vacuum for 5–7 min before loading into the column. Gel-filtration medium of 1,000 ml bed is used to purify platelets from one apheresis bag.
3. Wash the settled bed in the column by passing through 500 ml of HBMT buffer, containing 5 mM EDTA, 0.05 mM aspirin, and 0.1 μM PGE.
4. Assemble a chromatography system by connecting a pump proximal to the column and an optical unit to read eluate absorbance at 280 nm. Make sure the system is hermetically closed and connected to a pump providing a liquid flow rate from 60 to 600 ml/h. If possible, use a fraction collector to harvest the eluate and a suitable UV recorder to monitor and record the platelet's (A_{280}) elution peak.
   Start with one fresh platelet apheresis bag:
5. Take 150 ml of platelet concentrate (typically 2–3 × 10^{11} platelets), place it into a plastic container and add the following inhibitors of platelet activation: 1.5 ml of 0.5 M EDTA (final concentration = 5 mM); 600 μl of 12.5 mM aspirin (final concentration = 0.05 mM); 7 μl of PGE (final concentration = 0.1 μM), mix well by gentle stirring. These additions are very important in order to prevent platelet aggregation throughout the entire procedure since aggregation complicates platelet purification and dramatically reduces the amount and quality of the resulting platelet RNA.
6. At this point, take a 10 μl sample (Sample 1) for platelet count and flow cytometry analysis.
7. Divide the 150 ml of platelet concentrate equally among three 50-ml falcon tubes and centrifuge at 140*g* for 15 min at room temperature using a swinging bucket rotor centrifuge.
8. Once centrifugation is complete, gently take upper 90% (45 ml) of platelet concentrate, leaving ~ 3–5 ml at the bottom of each tube. This centrifugation efficiently removes a large fraction of contaminating leukocytes.
9. Combine platelet concentrate from three tubes in another plastic container, mix well and take a 10 μl sample (Sample 2) for the analysis. This procedure takes about 20–25 min.
   Start with gel-filtration:

10. Open the chromatography column and remove upper part of HBMT buffer, leaving just enough liquid to cover the beads.
11. Let remaining buffer enter the gel by gravity force and load platelet concentrate (~140 ml) without allowing the bed to dry out.
12. Let platelet concentrate completely enter the gel. If necessary, use the pump.
13. Immediately after the platelets enter the gel, wash column walls with 10 ml of HBMT buffer and let it enter the gel.
14. Then add 50–70 ml of buffer to the top of the gel, close the column and turn on the pump and optical unit to measure the platelet elution profile.
15. Maintain a liquid flow rate at 400–600 ml/h. Monitor movement of platelets through the gel and the liquid flow rate. If platelets become activated, they will clot the column, i.e., liquid flow decreases and buffer starts to accumulate above the gel. If the procedure goes normally, the quiescent platelets will elute from the column as a single wide peak.
16. Stop collecting platelets when the optical density reaches approximately one-third of its maximal height to avoid contamination with plasma proteins, which elute somewhat later than the platelet peak. The volume of the pooled platelet fractions after the column varies between 150 and 200 ml.
17. Combine all the gel filtered platelets in one plastic container, add aspirin to final concentration 0.05 mM and PGE to 0.1 μM, mix well and take another 10 μl sample (Sample 3) for analysis. It takes between 90 and 120 min to complete chromatography from the moment the platelet concentrate is loaded onto the gel. This procedure transfers platelets from plasma into HBMT buffer, removes cell aggregates and further reduces leukocyte contamination.

    Proceed with leukocyte filtration:
18. Assemble a filtration unit by placing a 5 μm pore size leukocyte membrane filter and securing it in its holder.
19. Pre-wet the membrane with HBMT buffer, but do not apply vacuum.
20. Load platelet concentrate and let it pass through the membrane by gravity flow. This procedure takes from 5 to 10 min, the volume of platelet concentrate does not change, and platelet losses are minimal.
21. Once platelet concentrate passes through the membrane, take a 10 μl aliquot (Sample 4) for analysis. This step is designed to further reduce the amount of contaminating leukocytes in the platelet preparation.

    Proceed with $CD45^{+}$cell immunodepletion (*see* **Note 2**):
22. Place platelet concentrate in corresponding number of 50 ml falcon tubes.

23. Harvest platelets by centrifugation using a swinging bucket rotor at 1,500*g* for 15 min at room temperature.
24. Discard supernatant and re-suspend platelets gently but thoroughly by adding 10 ml of HBMT buffer to the first tube and serially transferring it to the next tube until all platelets are re-suspended.
25. Gently pipette platelet concentrate to break any visible clumps or aggregates.
26. Bring the total volume to 20 ml and add 120 μl of mouse anti-CD45 antibody conjugated to magnetic beads.
27. Incubate at room temperature with mixing on a low-speed rotating platform for 45 min.
28. While platelets are incubating, prepare two magnetic columns. Place them on magnets attached to magnetic stands, load 5 ml of HBMT buffer and let the buffer pass through columns by gravity. Do not use a plunger to push buffer through the columns.
29. Once incubation of the platelet fraction with antibody is complete, apply 3–4 ml of the mixture onto the column and let it enter the magnetic mesh. Continue adding platelet concentrate as purified platelets elute from the bottom of the columns.
30. Use plungers and 5 ml of HBMT buffer to displace the entire sample.
31. Combine platelet and HBMT wash fractions and take 10 μl of the material (Sample 5) for analysis.
32. Collect platelets by centrifugation at 1,500*g* for 15 min at room temperature in 50-ml falcon tube. Usually the volume of a platelet pellet is about 5 ml. At this point, the platelet pellet can be frozen at –80°C or used immediately for total RNA isolation.

3.1.2. Characterization of the Efficiency of Platelet Purification

The yield and efficiency of platelet purification should be documented at each step. To analyze platelet yield, obtain platelet count for samples 1–5 and calculate platelet number at each step of purification. Typical, overall platelet yield is ~ 80%. Purification efficiency can be analyzed by flow cytometry.

1. Incubate aliquots containing 2×10^6 platelets from sample 1–5 with saturating concentrations of FITC-conjugated anti-CD41, PE-conjugated anti-glycophorin, and PERCP-conjugated anti-CD45 for 15 min in the dark at 25°C.
2. Wash them once with PBS, and fix in 1× PBS/1% formalin.
3. Analyze samples using a flow cytometer; quantify the number of CD45- and glycophorin-positive events in each sample.
4. Characterize efficiency of platelet purification by expressing it as the number of CD45-positive events (leukocytes) per

100,000 CD41-positive events (platelets). In our experience, platelet purity varied from 3 to 10 leukocytes per 1×10^5 platelets.

Leukocyte contamination of platelet preparation can also be addressed by RT-PCR using a variety of primers specific to leukocyte-expressed transcripts. Primers specific to leukocyte common antigen CD45 (NCBI accession number Y00638) have been successfully used to evaluate efficiency of leukocyte removal *(6, 9)*.

3.1.3. Platelet Purification from 20 ml of Peripheral Blood

1. Prepare in advance: 20-ml disposable syringe; 19-gauge needle; two columns for platelet gel-filtration (commercially available 60-ml disposable syringes is a simple, inexpensive and durable columns for platelet gel filtration).
2. Place a circular piece of nylon mesh with 85 micron pore size at the bottom.
3. Place the rubber ring from the plunger of the syringe on top of the mesh to hold it in place.
4. Load suspension of Sepharose 2B beads in HBMT buffer until bead volume reaches 50 ml.
5. Equilibrate the column by passing through 150 ml of HBMT buffer, pH 7.45.
6. Seal top and bottom of the column and keep it at room temperature until use.
7. Collect blood sample by venipuncture using 19-gauge needle attached to 20-ml syringe.
8. Divide it in two 10-ml aliquots, place each into a plastic container and add 1 ml of 100 mM EDTA to each aliquot as an anti-coagulant.
9. Place 1 μl of blood into a small Eppendorf tube containing 20 μl of 500 mM EDTA for complete blood count (CBC).
10. Generate platelet-rich plasma (PRP) by centrifugation of blood sample for 3.5 min at 700*g* at room temperature.
11. Gently take upper 9/10 of PRP (*see* **Note 3**). The total volume of PRP from 20 ml of blood typically varies between 4 and 8 ml.
12. Gently apply 2–3 ml PRP to each of two Sepharose 2B columns and allow platelets to enter the columns by gravity flow (*see* **Note 4**).
13. Once platelets enter, fill the columns with HBMT buffer to the top. Continue adding buffer during gel filtration to avoid drying of the beads.
14. As platelets elute they have a visible cloudiness that can be detected by eye; pool these fractions. The combined volume from two columns should be from 6 to 11 ml.
15. Harvest the platelets by centrifugation for 10 min at 2,000*g*.
16. Remove as much supernatant as possible, being careful not to disturb the soft platelet pellet.

17. At this point platelets can either be stored at –80°C or immediately used for RNA isolation. With reference to RNA instability, immediate RNA isolation may be a better solution.

3.2. Platelet RNA Isolation

3.2.1. Total RNA Isolation from Apheresis Platelet Bag

1. Thaw platelet pellet from the previous purification step by incubating falcon tube in room temperature water for 3–5 min.
2. Add 10 ml of Trizol reagent using a plastic pipette. If platelet pellet was not frozen, add 10 ml of Trizol directly to the pellet.
3. Use the plastic pipette to break up any clumps which tend to accumulate at the bottom of the tube in the dissolved pellet. It may take several minutes to completely disintegrate large clumps.
4. Add another 10 ml of Trizol to falcon tube, close it tightly and shake well by hand.
5. Transfer lysate into two 30-ml DEPC-treated Corex tubes, aliquoting equal volumes of platelet lysate per tube.
6. To each Corex tube, add another 10 ml of Trizol reagent. Finally, platelet pellet from an apheresis bag should be dissolved in 40 ml of Trizol, divided between two 30 ml Corex tubes. In our experience, we have found that for a 5-ml platelet pellet, the optimal amount of Trizol to be added is 40 ml (*see* **Note 5**).
7. Incubate tubes at room temperature for 5 min.
8. To each tube, add 4 ml of chloroform using a glass pipette. Mix well and incubate for 3 min at room temperature. (Note: parafilm dissolves in chloroform. If necessary seal tops with Dura Seal Laboratory Stretch film from Diversified Biotech, Boston.).
9. Centrifuge at 12,000*g*, 4°C for 15 min.
10. Gently take upper nine-tenths of clear supernatant layer and transfer to two new DEPC-treated Corex tubes; avoid touching protein interphase.
11. To each tube, add 2 μl of GlycoBlue as an indicator of RNA precipitation. Mix well.
12. Add 10 ml of cold (0°C) isopropanol. Close tube tightly with parafilm or Dura Seal, shake well by hand. At this point, tubes can be stored overnight at –20°C.
13. Precipitate total RNA by centrifugation at 12,000*g* at 4°C for 30 min.
14. Gently remove and discard supernatant.
15. Wash platelet RNA pellet – add 5 ml of 75% ethanol per Corex tube, wash tube walls using a pipette, and centrifuge at 12,000*g* for 15 min at 4°C.
16. Repeat washing step one more time.

17. After final wash, remove supernatant and air-dry RNA pellet for 3–5 min. Do not let RNA pellet dry completely since it may make re-suspension of RNA difficult.
18. Carefully re-suspend RNA from both Corex tubes in 100 μl of DEPC-treated water.
19. Transfer RNA solution into DEPC-treated microcentrifuge tube, add 1 μl of RNAse inhibitor – RNAseIn and mix well.
20. Wash walls of Corex tubes with another 50–100 μl of DEPC-treated water, combine with RNA solution, and mix well.
21. Incubate RNA solution at 55°C for 5 min to aid solubilization, quickly transfer onto ice.
22. Gently pipette RNA solution and take 5 μl for quantity and quality analyses. At this point, RNA can be aliquoted in smaller portions and stored at –80°C for further analysis.
23. In our experience, 50–100 μg of total RNA can typically be isolated from 150 ml of an apheresis bag. Total RNA yield for patients with essential thrombocythemia can reach up to 260 μg. This amount is more than sufficient for reliable transcript profiling using both microarray and SAGE.
24. To quantify RNA yield, use a suitable spectrophotometer – we use a Biophotometer from Eppendorf. To address quality of RNA, use a BioAnalyzer (Bio-Rad). A typical capillary electrophoresis of total platelet RNA is shown at **Fig. 16.3**. Note similarities between total RNA isolated from human liver and from platelets.
25. Once purified and characterized, platelet RNA with added RNAse inhibitors can be stored at –80°C for several months without detectable degradation.

3.2.2. Total RNA Isolation from Peripheral Blood Samples

1. Add 600 μl of Trizol to the platelet pellet, thoroughly re-suspend platelets and incubate for 3 min. Follow Trizol protocol after this step.
2. Add 120 μl of chloroform, shake tubes vigorously by hand for 15 s and incubate at room temperature for 3 min.
3. Centrifuge the sample at 12,000*g* for 15 min at 2–8°C. Following centrifugation, the mixture separates into a lower red, phenol–chloroform phase, an interphase, and a colorless upper aqueous phase; RNA remains in aqueous phase.
4. Take upper phase, avoiding cross-contamination with interphase material and transfer into fresh DEPC-treated 1.5 ml microcentrifuge tube. At this point the total volume is typically ~700 μl.
5. Add 1 μl of Glyco-blue to the aqueous layer to facilitate RNA precipitation. Mix sample well.
6. To precipitate RNA, add 600 μl of ice-cold isopropyl alcohol, close tube tightly, shake it well by hand and briefly vortex. Incubate sample for 30 min (or overnight) at –20°C.
7. Mix well and centrifuge at 12,000*g* for 15 min at 2–8°C.

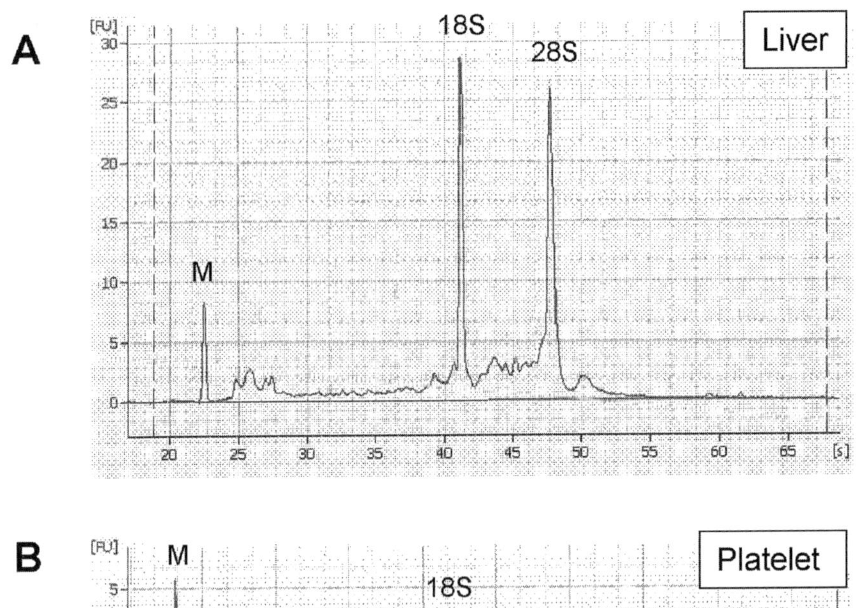

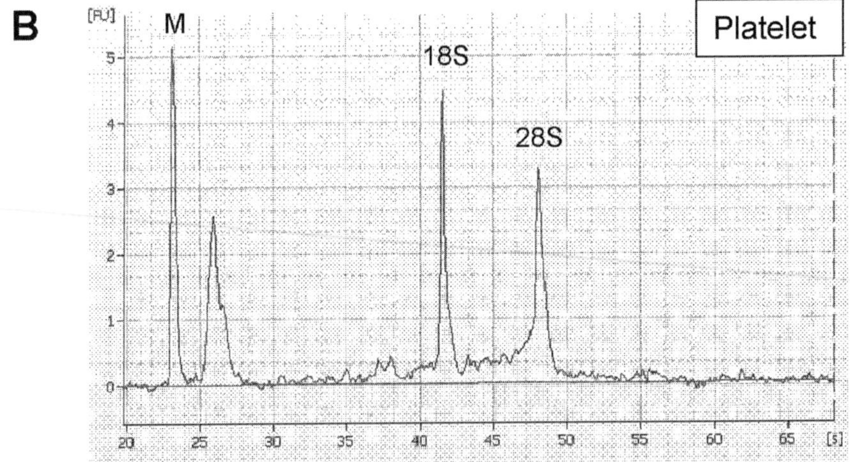

Fig. 16.3. Analysis of total RNA by capillary electrophoresis using BioAnalyzer Model 2,100. Total RNA from liver and from platelets were analyzed under the similar conditions in two separate experiments. (**A**) Adult human male total RNA (Stratagene); (**B**) Total platelet RNA. M – Agilent RNA 6,000 nanomarker. Ribosomal RNA (18S and 28S) peaks are present in both samples.

8. Carefully remove supernatant – at this point RNA pellet will be visible only due to the blue dot formed by Glyco-blue co-precipitation. Be careful not to discard RNA pellet.
9. Wash RNA pellet twice with 1 ml of 75% ice-cold ethanol: add 75% ethanol, vortex briefly, and centrifuge for 5 min at 12,000*g* at 2–8°C. Gently remove the supernatant and repeat the 75% ethanol washing. Exercise extra caution not to discard RNA pellet which at this point is almost invisible. (We routinely save the washes in case of RNA loss).
10. Once washing is complete, gently remove any residual liquid and air-dry pellets for 10–15 min at room temperature.

11. To re-suspend the RNA pellet, add 50 μl of DEPC-treated H_2O containing 1 μl of SuperRNAseIn to inhibit RNAses. Slowly pipette RNA solution up and down using a sterile pipette tip equipped with aerosol filter.
12. Incubate RNA at 53°C for 5 min to aid resolubilization, quickly transfer on ice, incubate for 5 min and briefly centrifuge at 5,000*g* to collect any vapor condensate from the walls of the tube.

3.3. Platelet Transcript Profiling Using Affymetrix Microarray Platform

Development of global mRNA profiling technologies *(13, 14)* (such as microarray and SAGE), coupled with recent completion of the Human Genome project provides novel approaches to study platelet function *(15, 16)*. Microarray transcript profiling uses grids of "probes", specific to known DNA sequences, to capture and quantify RNA transcripts of corresponding genes. To date, many different microarray platforms have been developed, all of them include (1) isolation of RNA from biological sample; (2) copying of RNA into either cRNA or cDNA, simultaneously incorporating either fluorescent nucleotides or a tag that is later used for fluorescent labeling; (3) hybridization to a microchip containing a grid of DNA probes; (4) washing; (5) labeling (if necessary, depending on the protocol); (6) scanning under laser light; and (7) image processing, data extraction, and analysis *(17)*. Hybridization to microarrays and quantitative real-time PCR are "closed-architecture" tools, i.e., these methods rely totally on pre-established sequence information whereas SAGE and similar techniques which do not require any a priori knowledge of the sample's structure or complexity are considered "open-architecture" tools. Each method has its own strengths and weaknesses and their utility is somewhat sample dependent. In our experience, hybridization to microarrays and quantitative real-time PCR are the methods of choice for analyzing platelet transcriptomes.

3.3.1. cDNA Synthesis

The cDNA synthesis is performed using reagents from Invitrogen Inc.

1. Mix 5 μg total RNA with 1 μl (100 pmol) $T7dT_{24}$ primer and DEPC-treated water to a final volume of 12 μl in 1.5 ml tube. Heat to 70°C for 10 min. Quick spin then place on ice.
2. Prepare a master mix with a 10% excess. For each sample mix 4 μl of 5× first strand buffer, 2 μl of 0.1 M DTT (first strand buffer and DTT supplied with Super Script II RT) and 1 μl of 10 mM dNTP mix (2.5 mM each dNTP).
3. Add 7 μl of the master mix to each sample tube of total RNA and primer. Mix and place in a 42°C bath for 2 min.
4. Add 1 μl (200U) of Super ScriptIIRT and incubate at 42°C for 1 h.

5. Prepare a master mix for the second strand synthesis containing: 91 μl DEPC-treated H_2O (Ambion), 30 μl second strand buffer (Invitrogen), 3 μl of 10 mM dNTP, 1 μl (10 U) DNA ligase, 4 μl (40 U) DNA polymerase I, and 1 μl (2 U) RNaseH. Cool a bath to 16°C.
6. Add 130 μl of the second strand master mix to each sample mix and spin and place into the 16°C bath for 2 h.
7. Add 2 μl (10 U) T4 DNA polymerase, incubate at 16°C for 5 min. Add 10 μl of 0.5 M EDTA to each sample. At this point the samples can be frozen at –20°C or purified.

3.3.2. cDNA Purification

1. Label a Phaselock tube for each sample and spin the tubes at 12,000*g* for 25 s. Add 162 μl of phenol/chloroform (21:1) to each sample and mix. Transfer the entire 324 μl of sample to a Phaselock tube and spin for 2 min at 12,000*g*.
2. Carefully transfer the supernatant to a new 1.5 ml tube. Add 0.5 volumes 7.5 M NH_4OAc and 2.5 volumes cold absolute ethanol. Mix and centrifuge for 20 min at > 14,000*g*.
3. Remove the supernatant – a faint, small white pellet should be visible. Wash the pellet with 1 ml of 80% ethanol and centrifuge for 15 min at > 14,000*g*.
4. Remove all of the supernatant and air-dry the pellets for a few minutes.
5. Re-suspend the pellet in 22 μl DEPC-treated H_2O.

3.3.3. IVT Reaction

1. The BioArray High Yield kit is used to generate the cRNA.
2. Mix 18 μl of the master mix and the 22 μl of sample, aliquot into several tubes, and incubate at 37°C overnight.

3.3.4. cRNA Purification

The cRNA is purified using the RNeasy mini kit (Qiagen).

1. Add 60 μl DEPC-treated water and 350 μl buffer RLT (do not add the BME to the buffer RLT) mix.
2. Add 250 μl of 100% EtOH, mix by pipetting several times and transfer to a Qiagen RNeasy column.
3. Centrifuge 18 s at 12,000*g*. Reapply the sample and spin again for 18 s at 12,000*g*.
4. Discard tube and flow-through. Place column in a new tube. Apply 350 μl buffer RW1 to the column. Spin 18 s at 12,000*g*. Discard flow-through.
5. Carefully apply 80 μl DNAse I solution (10 μl DNAse I + 70 μl RDD buffer. Incubate 15 min at room temperature.
6. Apply 350 μl buffer RW1. Centrifuge at 12,000*g* for 18 s.
7. Discard tube and flow-through. Place column in a new tube and apply 500 μl buffer RPE. Centrifuge at 12,000*g* for 18 s. Discard tube and flow-through.
8. Place column in a new tube and apply 500 μl buffer RPE and centrifuge 2 min at 14,000*g*. Discard tube and flow-through

making sure that there is no residual flow-through on the column.

9. Place column in a labeled new microfuge tube. Carefully apply 50 μl DEPC-treated H_2O to the center of the column. Centrifuge 1 min at 12,000*g*. Carefully apply 50 μl DEPC-treated H_2O to the center of the column a second time. Centrifuge 1 min at 12,000*g*. Use of H_2O at 37°C improves cRNA recovery.
10. Add 10 μl of 5 M NH_4Oac and 275 μl of –20°C absolute EtOH. Incubate at –20°C for 30–60 min. Centrifuge at 18,000*g* at 4°C for 25 min. A large white pellet will be visible. Discard supernatant.
11. Wash pellet with ice-cold 70% EtOH, centrifuge at 18,000*g* for 15 min. Remove all supernatant and air-dry pellet.
12. Once all the EtOH as evaporated re-suspend the pellets in 32 μl DEPC-treated H_2O.
13. Measure nucleic acid concentration on a spectrophotometer.

3.3.5. Fragmentation of cRNA

1. In a microfuge tube combine 20 μg cRNA and DEPC-treated H_2O to a final volume of 32 μl.
2. Add 8 μl fragmentation buffer.
3. Incubate at 94°C for 35 min. Be sure to secure the top of the tube so it does not open during the incubation.
4. Place tubes on ice.

3.3.6. Hybridization of Arrays

1. Create a master mix of the hybridization cocktail containing: 15 ug fragmented cRNA, 5 μl control oligonucleotide B2, 15 μl 20× eukaryotic hybridization controls, 3 μl BSA 50 mg/ml, 3 μl sonicated herring sperm DNA 10 mg/ml, 150 μl 2× MES hybridization buffer (prepared as per Affymetrix GeneChip Expression Analysis Technical Manual.).
2. Add DEPC H_2O to a final volume of 300 μl.
3. Pre-hybridize the arrays with 1× MES hybridization buffer (Affymetrix GeneChip Expression Analysis Technical Manual.) at 42°C at 60 rpm in the Affymetrix Hybridization Oven Model 420.
4. Heat hybridization cocktail to 98°C for 5 min.
5. Transfer tubes to 42°C for 5 min.
6. Centrifuge samples at 18,000*g* for 5 min.
7. Remove arrays from hybridization oven and remove the 1× MES hybridization buffer.
8. Load the appropriate volume hybridization cocktail – 80 μl for Affymetrix Test 3 arrays chip or 200 μl for Affymetrix GeneChip. In our studies we used HGU133A microarray chip. Currently it is discontinued and replaced with more advanced microarray chips, which allow better quantification

of gene expression and include probes for more transcripts. For details of hybridization to these chips, refer to the manufacturer's instructions.

9. Place a "tough spot" over the lower loading port as this port leaks sometimes.
10. Place arrays into carriers and into the Affymetrix Hybridization Oven Model 420 set to 42°C at 60 rpm. Hybridize the arrays overnight ~16 h.

3.3.7. Washing and Staining the Arrays

1. Prepare wash buffers as per instructions in the Affymetrix GeneChip Expression Analysis Technical Manual.
2. Prepare 2× Stain 1 and 1× Stain 2 as per Affymetrix GeneChip Expression Analysis Technical Manual.
3. Remove the arrays from the Affymetrix Hybridization Oven, remove the tough spot.
4. Remove the hybridization cocktail and store at –80°C.
5. Load to appropriate volume of "buffer A" – 100 μl for the Test 3 array or 250 μl for the U133A array.
6. Wash and stain the arrays using the Micro v1.1 protocol for the Test 3 array and the EukGE-WS2v4 for the GeneChip U133.
7. When the third wash cycle begins start the scanner so that it completes its warm up cycle.

3.3.8. Scanning and Image Analysis

1. After the washing and staining protocol is complete check the arrays to make sure there are no air bubbles present.
2. Scan the arrays. Once the scan is complete check the grid alignment in the corners and in the center of the array.
3. Analyze the image using Microarray Analysis Suite 5.1 (Affymetrix) utilizing the following parameters scaling: All Probe sets: Target Signal 250, Normalization: User Defined, value 1. All other parameters are set at default values. The percent of genes called present (%*P*) is usually low in platelets (<20%), whereas the 5'/3' ratios is higher. The latter may be the result of partial degradation of platelet mRNA due to platelet aging and/or RNAse activity in the absence of ongoing transcription
4. Further data processing and bioinformatic analysis is performed using GeneSpring software or its analogs using standard microarray data mining algorithms. Description of bioinformatic techniques is beyond the scope of this chapter.

3.4. Serial Analysis of Gene Expression Analysis

SAGE is a powerful mRNA-based method for comprehensive analysis of gene expression patterns *(18,19)*. In the original SAGE procedure, double-stranded cDNA is synthesized from poly $(A)^+$ mRNA by priming first-strand cDNA synthesis with a biotinylated oligo $(dT)_{18}$ primer. The cDNA is then cut with a restriction endonuclease having a 4-bp recognition sequence (typically *Nla*III, recognition sequence CATG, which theoretically results in cleavage on average every 256 bp), and the 3'-terminal cDNA fragments are captured on streptavidin-coated magnetic beads. These fragments are ligated with DNA cassettes containing a recognition sequence for *Bsm*FI, a type IIS restriction endonuclease. Subsequent cleavage with the *Bsm*FI releases short (13–14 bp) but positionally defined sequences, referred to as tags, which are eventually concatenated into arrays and cloned into a plasmid vector for DNA sequencing. The power of the method, as its name implies, is that many tags can be read serially from each clone during the sequencing step which vastly increases throughput. Over 10 million cDNA tags have been analyzed by this method since it was first described, many of which are publicly available at (http://www.ncbi.nlm.nih.gov/SAGE).

In recent years, several new commercially available enzymes which cleave further into the DNA thereby increasing tag length, have superseded *Bsm*F1 since longer tags are particularly useful in characterizing expression patterns in the absence of complete genome sequence data, i.e., from "uncharted transcriptomes" and in designing primers to obtain full-cDNAs from transcripts whose tags are not currently present in RefSeq or similar expression databases. One very useful enzyme is *Mme*I which cleaves 20/18 bases past its non-palindromic (TCCRAC) recognition sequence *(20, 21)*. *Mme*I is now commercially available from NEB which has led to its use in numerous SAGE-type studies; typically referred to as Long SAGE due to the increase in tag length. In addition, detailed protocols are available on the web which explain many of the steps for preparation and analysis of SAGE libraries. A good starting point is the description of the I-SAGE^{TM} Long kit (Catalog no. T5000-03) from Invitrogen (www.invitrogen.com)

3.4.1. Magnetic Beads Preparation and mRNA Capturing

1. Briefly vortex tube with Dynal oligo(dT) magnetic beads.
2. Remove 100 μl of bead suspension, place into clean 1.5 ml siliconized microcentrifuge tube and place tube on the magnetic holder. Beads will collect on the wall of the tube near the magnet.
3. Gently remove supernatant without disturbing the beads.
4. Wash the beads with 400 μl lysis-binding buffer from the Dynal Dynabeads mRNA Direct kit. To do this, remove tube from the magnet, wash beads off the wall using buffer, place the tube back on the magnet and wait for ~5 min to collect the beads.
5. Mix total platelet RNA (3–5 μg) with 200 μl lysis/binding buffer supplemented with 10 μg/ml glycogen.
6. Add RNA solution to the beads, incubate at 60°C for 5 min to help melt the secondary structure, cool slowly to room temperature with occasional mixing (once–twice per minute) to allow the poly (A) tracks in the mRNA to hybridize to the oligo dT.

7. Collect the beads by placing the tube back on the magnetic holder, carefully remove the supernatant, and store it at 4°C.
8. Wash beads twice using 400 μl wash buffer A, supplemented with 20 μg/ml glycogen.
9. Wash beads three times with 400 μl wash buffer B, supplemented with 20 μg/ml glycogen. Move beads to a fresh tube after the first wash.
10. Wash beads two times with 400 μl RT first strand buffer supplemented with 20 μg/ml glycogen and 2 μl SuperRNAseIn.

3.4.2. First Strand cDNA Synthesis

1. Re-suspend beads in 25 μl RT first strand buffer supplemented with 20 μg/ml glycogen and 2 μl SuperRNAseIn.
2. Incubate at 42°C for 2 min followed by incubation at 37°C for 2 min.
3. In a separate tube, prepare the following pre-mix for the reverse transcription reaction: 9 μl DEPC-treated water, 1 μl SuperRNAsin, 5 μl RT first strand buffer, 2.5 μl dNTP (10 mM each), 5 μl 0.1 M DTT, 2.5 μl SuperScript II reverse transcriptase.
4. Scrape beads off sides if necessary, add the pre-mix to the beads. Gently pipette up and down and incubate at 37°C for 1 h with gentle mixing.
5. Heat to 60°C for 3 min, incubate at 37°C for 2 min, and then add an additional 2 μl of reverse transcriptase.
6. Incubate at 37°C for an additional hour. Repeat one more time. Total volume of reverse transcriptase used for three cycles of cDNA synthesis equals 6.5 μl.
7. Collect beads and carefully remove RT first strand buffer.

3.4.3. Second Strand synthesis

1. In a separate tube, prepare a pre-mix: 254 μl DEPC-treated water, 70 μl 5× second strand buffer, 8 μl dNTP (10 mM each), 2.5 μl *Escherichia coli* DNA ligase, 10 μl *E.coli* DNA polymerase, 2.5 μl *E.coli* RNAse H, 3.5 μl glycogen.
2. Add pre-mix to the beads, incubate at 16°C overnight with gentle mixing for first 3 h.
3. Collect beads and wash six times with TEN/BSA buffer, using 500 μl per wash. For the first wash re-suspend in TEN/BSA, heat to 75°C for 10 min, and, slowly cool to room temperature.
4. Wash six more times, however, during the subsequent washes heating is not needed. After the fourth wash, transfer the beads with attached cDNA to a clean tube.

3.4.4. Digestion with NlaIII

1. Wash beads three times with 200 μl 1× New England Biolabs #4 buffer (optimal for *Nla*III).

2. Re-suspend beads in 200 μl 1× New England Biolabs #4 buffer, supplemented with 1× BSA and 4 mM spermidine $(HCl)_3$.
3. Add 2 μl *Nla*III (∼10 U/μl).
4. Mix by pipetting up and down. Incubate at 37°C for 2 h.
5. Add an additional 2 μl *Nla*III and incubate for an additional 2 h at 37°C.
6. Capture beads, carefully remove as much supernatant as possible.
7. Wash beads with an additional 100 μl 1× NEB #4 buffer, pool and then precipitate the released DNA by adding 30 μl 2.5 M Na acetate, pH 5.2, and 1 μl glycogen. Mix well by vortexing.
8. Add 500 μl of cold 95% ethanol, vortex briefly, and incubate at –20°C for 2 h.
9. Centrifuge for 30 min at 12,000*g* in microcentrifuge.
10. Air-dry under vacuum for 3–5 min. Store beads in TEN/BSA buffer at 4°C.

3.4.5. Ligation of First Linker Cassette (Cassette A) Containing Restriction Site for MmeI Tagging Enzyme

The schema of *Mme*I restriction, cassette A and its ligation to cDNA is shown in **Fig. 16.4.**

1. Capture beads using magnetic holder, wash three times in 200 μl LoTE (10 mM Tris–HCl, pH 8.0; 0.1 mM EDTA), one time – in 200 μl 1× T4 DNA ligase buffer (Takara).
2. Make the following pre-mix for cassette A ligation: 38 μl LoTE, 5 μl 10× T4 DNA ligase buffer, 4 μl linker cassette A (mix of LS–LT primer and LS–LB primer, 10 pmoles/μl each)

A. Cassette A

```
5'-TTTGGATTTGCTGGTCGAGTACAACTAGGCTTAATCCGACATG-3'
3'-CCTAAACGACCAGCTCATGTTGATCCGAATTAGGCTp-5'
```

B. *MmeI* restriction site

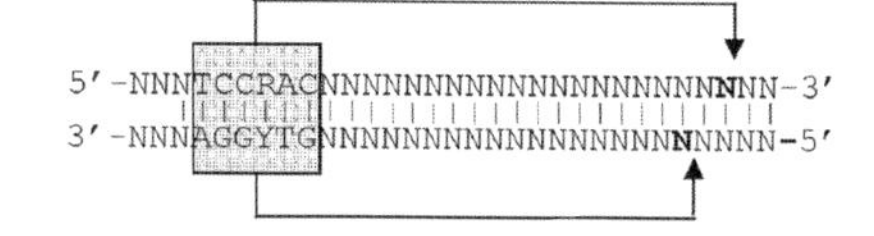

C. Ligation of Cassette A to *NlaIII*-cleaved cDNA/tag

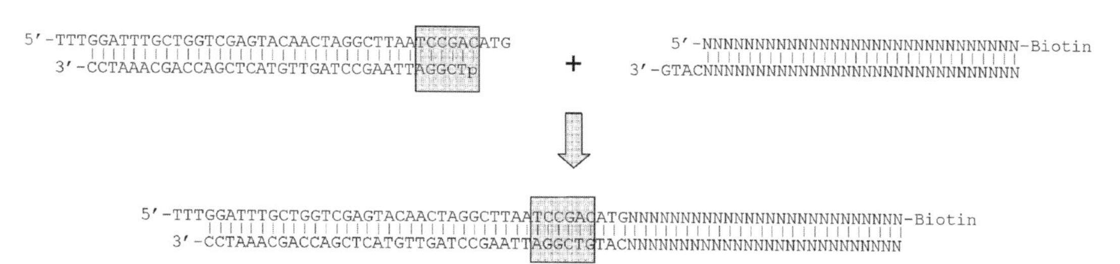

Fig. 16.4. Cassette A, *Mme*I restriction site and ligation of cassette A.

3. Heat pre-mix at 50°C for 2 min, cool to RT on the bench, and incubate at RT for 15 min.
4. Add 3 μl of T4 DNA ligase (~10 U). Mix well by pipetting.
5. Add to the beads and incubate at 16°C for 2 h with occasional gentle mixing.

3.4.6. MmeI Digestion to Generate Cassette A Ligated Beads

1. Wash beads six times with 400 μl TEN.
2. Capture beads, remove wash. Wash beads two times with 200 μl 1× *Mme*I buffer (NEB #4)
3. In a separate tube, make the following pre-mix: 85 μl ddH_2O (total volume will be 100 μl), 10 μl 10× *Mme*I buffer, 1 μl 100× SAM (4 mM – freshly prepare from 32 mM stock provided with the enzyme), 4 μl *Mme*I (~2 U/μl, 8 U total)
4. Incubate reaction at 37°C for 2 h and, mixing occasionally.
5. Collect beads using magnetic holder.
6. After *Mme*I digestion the cDNA/tags will be released into supernatant. Carefully remove as much supernatant as possible and transfer it to a fresh tube.
7. Rinse beads with 100 μl LoTE and combine with first supernatant.
8. To inactivate enzyme, extract supernatant with equal volume of TE-saturated phenol. Add phenol, vortex vigorously for 1 min and spin in microcentrifuge at top speed (~12,000*g*) for approximately 5 min at 4°C.
9. Carefully transfer upper aqueous phase to a fresh tube and extract with an equal volume of chloroform/IAA (24:1 v/v). Vortex briefly and spin in microcentrifuge at top speed for 15 min at 4°C.
10. Gently remove as much upper phase as possible, avoid disturbing the interphase. Place supernatant in a fresh tube.
11. Precipitate the released DNA/tags with ethanol. To 200 μl of supernatant, add 3 μl of GlycoBlue and 133 μl of 7.5 M ammonium acetate. Mix well and add 1 ml of cold 100% ethanol. Vortex and incubate at –70°C for 1 h or at –20°C overnight.
12. Spin at 12,000*g* for 30 min at 4°C. Wash twice with 75% ethanol, dry under vacuum for 5–10 min.

3.4.7. Ligation of Second Linker Cassette (Cassette B) Containing NlaIII Restriction Site

The schema of cassette B and its ligation to cassette A tag is shown in **Fig. 16.5**. *Mme*I cuts 20–21/18–19 bp past its recognition site. Therefore cutting the linkered DNA in step 6, releases the linker and 17–18/15–16 bp immediately 3′ to the *Nla*III site. These sequences become the SAGE tags which are PCR amplified and ligated together to form ≥500 bp long concatemers prior to cloning and DNA sequencing. Because each clone contains multiple tags, sequencing throughput increases accordingly. Using the strategy described below, the two unique nucleotides at the 3′

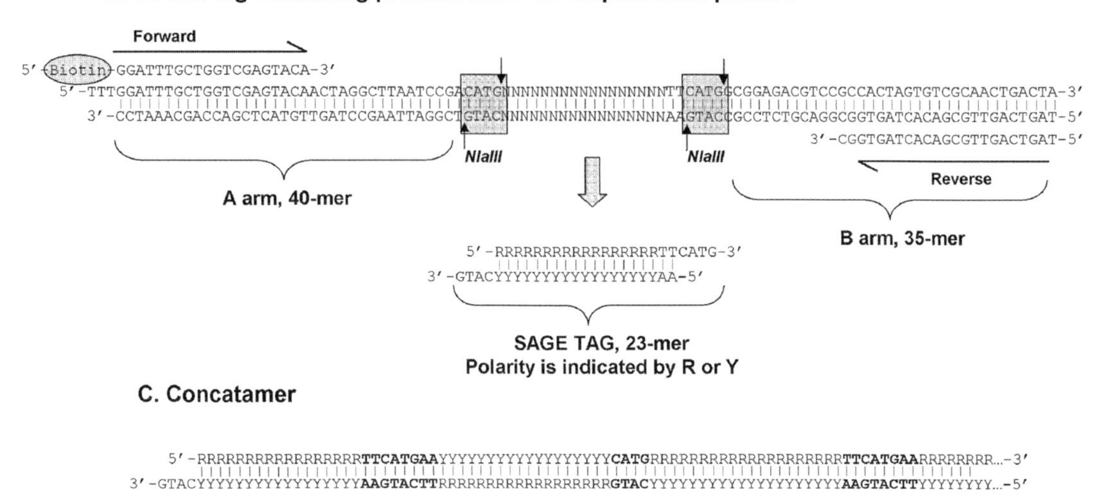

Fig. 16.5. Cassette B, PCR amplification product and concatamerized SAGE tags.

end of each tag are not lost. One strategy for capturing these nucleotides is based on the approach used in the process called TALEST (tandem arrayed ligation of expressed sequence tags) developed by Spinella et al. *(22)* and modified by our laboratory for our original GST protocol (6). It employs ligation with a 16-fold degenerate oligonucleotide (cassette B) to capture all the sequence information in the *Mme*I cleaved DNAs 3′ extensions.

1. Re-suspend DNA pellet (can be visible only as a blue dot because of GlycoBlue) in 29.5 μl TEsl.
2. Add 4 μl 10× T4 DNA ligase buffer and 3.5 μl linker cassette B (mix of LS–RT primer and LS–LB primer, 10 pmol/μl each).
3. Mix well, incubate at room temperature for 15 min, and then add 3 μl of T4 DNA ligase (~10 U).
4. Mix well by pipetting and incubate 2 h to overnight at 16°C.
5. Final product of ligation should be 98-mer with two *Nla*III restriction sites, as shown in **Fig. 16.5**, panel B.

3.4.8. PCR Amplification of cDNA Tags Using Biotinylated Primers

For this PCR, use biotinylated forward and reverse primers that corresponds to a portion of cassette A upper strand and to a portion of cassette B bottom strand (see **Fig. 16.4**).

1. Place clean microcentrifuge tube on ice and make PCR mix as follows: 371 μl ddH$_2$O, 50 μl 10× Promega buffer, 20 μl 50 mM MgSO$_4$, 15 μl dNTP (10 mM each), 20 μl 10

μM forward biotinyl-primer, 20 μl 10 μM reverse primer, 2.0 μl platinum *Taq* polymerase, 2.0 μl cDNA product from step 3.4.7. Store remaining cDNA product at –80°C for further analysis, if needed.

2. Mix well by brief vortexing, spin to remove traces of solution from walls of the tube.
3. Divide 500 μl PCR mix into ten small thin-wall PCR tubes, 50 μl per tube. Keep tubes on ice.
4. Run PCR using the following cycle parameters:
5. 95°C, 2 min 30 cycles with 95°C, 30 s;
   58°C, 30 s; 72°C, 30 s 72°C,
   4 min 10°C - hold

3.4.9. Optimizing PCR by Linear Amplification to Resolve Heteroduplexes (LARHD)

In our experience, standard PCR as above usually gives the expected ~94 bp product plus variable amounts of material migrating as a more slowly migrating diffuse band. As discussed in detail in Ref (GST paper), this material probably represents amplicons with perfectly base-paired cassette regions flanking SAGE tags with some miss paired bases. These products are then converted to completely duplex forms by a process, we termed LARHD.

1. Pool solution after PCR from ten tubes into one tube and mix well. Total volume should be 500 μl.
2. In a fresh tube, make the following pre-mix, total volume 1,250 μl: 682.5 μl ddH_2O, 125 μl 10× Promega buffer, 50 μl 50 mM MgSO_4, 37.5 μl dNTP (10 mM each), 50 μl 10 μM forward biotinyl-primer, 50 μl 10 μM reverse primer, 250 μl first round amplified tags from previous step (1/2 of the total volume), 5.0 μl platinum *Taq* Hi Fi polymerase.
3. Divide pre-mix into smaller thin-wall PCR tubes, keeping volume 50–100 μl per tube.
4. Run LARHD reaction (one cycle only): 95°C, 2 min 30 s; 58°C, 30 s; 72°C, 5 min; 10°C, hold.
5. Pool PCR products, saving 10 μl for gel analysis if desired. Products should be 94 bp in length, as shown in **Fig. 16.5.**

3.4.10. Exonuclease Digestion of Primers

1. To PCR products, add, 6.25 μl *E. coli* exonuclease I, 10 U/μl to 1,250 μl of LARHD products.
2. Incubate at 37°C for 30 min, remove and store 5 μl for gel analysis.
3. Perform phenol/chloroform purification as described in **Section 3.4.6.**
4. Wash lower (phenol) phase with a small volume of TEsl.
5. Pool and then precipitate in as many tubes as needed. One tube should contain: 270 μl sample, 30 μl 3 M Na acetate, pH 6.0, and 750 μl 100% ethanol (–20°C)
6. Place at –20°C for 20 min.

7. Spin down amplicons at 4°C, 30 min, 12,000*g*.
8. Redissolve in 0.3 M Na acetate, re-precipitate in one tube with 2.5 volumes of cold 100% ethanol. Place at –20°C for 20 min.
9. Centrifuge at 12,000*g* for 30 min at 4°C, remove supernatant, and carefully wash with 1 ml of cold 70% ethanol.
10. Spin down amplicons and air-dry.

3.4.11. NlaIII Digestion to Release SAGE Tags

1. Dissolve amplicons in 50 μl of 1× NEB buffer #4, supplemented with 1× BSA, and 4 mM spermidine HCl.
2. Add 4 μl of *Nla*III, mix well by pipetting, and incubate for 2 h at 37°C.
3. Add another 4 μl of *Nla*III and incubate for additional 2 h. (*see* **Fig.** 16.5 for schema of this restriction). This step releases the monomeric tags from the flanking cassettes used for PCR amplification.
4. Bring total volume to 200 μl with 1× NEB buffer #4 and mix well. Keep tube on ice.
5. Conduct phenol/chloroform purification as described in **Section 3.4.6** and precipitate with ethanol.

3.4.12. Gel-Purification of 23-mer SAGE Tags

In our hands, removing the biotinylated cassettes by capturing on streptavidin beads usually does not leave pure tags in the unbound fraction. Therefore we recommend loading the entire sample on a 10–12% non-denaturing polyacrylamide gel. It is important to electrophorese the sample slowly (100 V) until the bromophenol blue dye front reaches 1–2 cm from the bottom so that the tags are not heat denatured.

1. After electrophoresis, stain the gel in 0.5 μg/ml ethidium bromide plus 50 mM NaCl and visualize the bands under long wave length UV light. You should see a sharp 23–24 band below the cassette arms.
2. Using a clean razor blade, excise the tag band from each lane of the gel, place on a clean piece of parafilm, and then cut each gel piece into smaller pieces and place them into one or more 0.5-ml microcentrifuge tubes as needed that have had a hole pierced into the bottom by a 18-gauge needle.
3. Place each 0.5-ml microcentrifuge tube in a sterile 1.5-ml nonstick microcentrifuge tube and centrifuge at maximum speed for 2–3 min. The excised gel pieces will pass through the hole and be broken into much smaller pieces and be collected in the 1.5-ml microcentrifuge tube.
4. To each tube add 200–400 μl of LoTEsl plus 2.5 M ammonium acetate making sure that enough buffer is used to completely submerge the gel pieces.

5. Incubate at 37°C for several hours with occasional vortexing to elute the DNA.
6. Remove the eluates and pass them though a clean spin filter to remove any particles of polyacrylamide.
7. Combine the eluants and precipitate with ethanol. From this point, the procedure follows the standard SAGE methods for tag concatemerization, cloning into pZERO, and sequence analysis. The caveat, of course, is that our protocol generates runs of monomeric tags (**Fig. 16.5**), not the standard ditags that is the norm with other SAGE procedures.

4. Notes

1. All procedures of platelet purification should be performed at room temperature unless specified otherwise to avoid platelet activation. In our experience, platelet activation can be prevented during purification by using three inhibitors – EDTA or citrate, PGE, and aspirin. If platelets were activated during purification, the yield of total RNA is minimal and its quality is not good for microarray analysis, as estimated by hybridization to the test chip in Affymetrix protocol.
2. At this step, magnetic separation columns are used to capture any remaining $CD45^+$ cells (leukocytes) by positive selection. The amount of time needed for this step is somewhat unpredictable, but may take from 30 min to 2 h. Buffer flow may dramatically decrease or platelets may start to clog the columns. Although this procedure helps to remove traces of leukocyte contamination, it may be omitted with minimal effect on final platelet purity.
3. Avoid taking too much PRP for gel filtration since this may result in leukocyte contamination of the sample.
4. In small-scale platelet purification schema, platelets if necessary can be concentrated before gel filtration by centrifugation at 2,000*g* for 10 min and re suspended 1–2 ml of the running buffer (HBMT) (**Section 3.3.2, step 1**). Add inhibitors to avoid platelet activation (see **Section 3.1.1**)
5. Trizol protocol for total RNA isolation allows simultaneous isolation of a protein fraction. This may be useful if further studies include immunodetection of protein of interest or validation of transcript profiling at the protein level. If necessary, harvest protein pellet before guanidine step exactly as described in Trizol reagent protocol and store it at –80°C for further analysis. Protein pellets can be stored for up to 6 months at –80°C without detectable degradation and be analyzed by the Western blotting.

Acknowledgments

This work was supported by NIH/NHLBI grants R21 HL076457 (D.V.G.) and HL086376 (W.F.B); Department of Defense grant MP048005 and a Targeted Research Award (D.V.G.) from Stony Brook University. Studies at the Brookhaven National Laboratory were supported by a Laboratory Directed Research and Development award (J.J.D.) and by the Offices of Biological and Environmental Research, and of Basic Energy Sciences (Division of Energy Biosciences) of the US Department of Energy.

References

1. Newman, P. J., Gorski, J., White, G. C., 2nd, et al. (1988) Enzymatic amplification of platelet-specific messenger RNA using the polymerase chain reaction. *J Clin Invest* 82, 739–743.
2. Kieffer, N., Guichard, J., Farcet, J. P., et al. (1987) Biosynthesis of major platelet proteins in human blood platelets. *Eur J Biochem* 164, 189–195.
3. Rinder, H., Schuster, J., Rinder, C., et al. (1998) Correlation of thrombosis with increased platelet turnover in thrombocytosis. *Blood* 91, 1288–1294.
4. Denis, M. M., Tolley, N. D., Bunting, M., et al. (2005) Escaping the nuclear confines: signal-dependent pre-mRNA splicing in anucleate platelets. *Cell* 122, 379–391.
5. Bugert, P., Dugrillon, A., Gunaydin, A., et al. (2003) Messenger RNA profiling of human platelets by microarray hybridization. *Thromb Haemost* 90, 738–748.
6. Gnatenko, D. V., Dunn, J. J., McCorkle, S. R., et al. (2003) Transcript profiling of human platelets using microarray and serial analysis of gene expression. *Blood* 101, 2285–2293.
7. McRedmond, J. P., Park, S. D., Reilly, D. F., et al. (2004) Integration of proteomics and genomics in platelets: a profile of platelet proteins and platelet-specific genes. *Mol Cell Proteomics* 3, 133–144.
8. Gnatenko, D. V., Cupit, L. D., Huang, E. C., et al. (2005) Platelets express steroidogenic 17beta-hydroxysteroid dehydrogenases. Distinct profiles predict the essential thrombocythemic phenotype. *Thromb Haemost* 94, 412–421.
9. Hillmann, A. G., Harmon, S., Park, S. D., et al. (2006) Comparative RNA expression analyses from small-scale, single-donor platelet samples. *J Thromb Haemost* 4, 349–356.
10. Shim, M. H., Hoover, A., Blake, N., et al. (2004) Gene expression profile of primary human CD34+CD38lo cells differentiating along the megakaryocyte lineage. *Exp Hematol* 32, 638–648.
11. Tenedini, E., Fagioli, M. E., Vianelli, N., et al. (2004) Gene expression profiling of normal and malignant CD34-derived megakaryocytic cells. *Blood* 104, 3126–3135.
12. Lockhart, D. J., Winzeler, E. A. (2000) Genomics, gene expression and DNA arrays. *Nature* 405, 827–836.
13. Hoheisel, J. D. (2006) Microarray technology: beyond transcript profiling and genotype analysis. *Nat Rev Genet* 7, 200–210.
14. Sausville, E. A., Holbeck, S. L. (2004) Transcription profiling of gene expression in drug discovery and development: the NCI experience. *Eur J Cancer* 40, 2544–2549.
15. Gnatenko, D. V., Perrotta, P. L., Bahou, W. F. (2006) Proteomic approaches to dissect platelet function: one-half of the story. *Blood* 108, 3983–3991.
16. Gnatenko, D. V., Bahou, W. F. (2006) Recent advances in platelet transcriptomics. *Transfus Med Hemother* 33, 217–226.
17. Butte, A. (2002) The use and analysis of microarray data. *Nat Rev Drug Discov* 1, 951–960.
18. Velculescu, V., Zhang, L., Vogelstein, B., et al. (1995) Serial analysis of gene expression. *Science* 270, 484–487.
19. Velculescu, V., Zhang, L., Zhou, W. et al. (1997) Characterization of the yeast transcriptome. *Cell* 88, 243–251.
20. Boyd, A. C., Charles, I. G., Keyte, J. W., et al. (1986) Isolation and computer-aided characterization of MmeI, a type II restriction endonuclease from Methylophilus methylotrophus. *Nucleic Acids Res* 14, 5255–5274.

21. Tucholski, J., Skowron, P. M., Podhajska, A. J. (1995) MmeI, a class-IIS restriction endonuclease: purification and characterization. *Gene* 157, 87–92.

22. Spinella, D. G., Bernardino, A. K., Redding, A. C., et al. (1999) Tandem arrayed ligation of expressed sequence tags (TALEST): a new method for generating global gene expression profiles. *Nucleic Acids Res* 27, e22.

Chapter 17

Genome-Wide Platelet RNA Profiling in Clinical Samples

Angelika Schedel and Nina Rolf

Abstract

Human blood platelets are anucleate cells that contain minute amounts of translational active mRNA. Investigation of the gene expression profile by microarray analysis has become an excellent tool for better understanding of normal and pathological platelet function. Its use, however, is often limited by the low yield of megakaryocytic-derived mRNA, the possible contamination with leukocytes during platelet preparation and the small platelet volume in clinical settings, especially in pediatric patients. In this chapter, we present a protocol for the isolation of leukocyte-depleted platelet samples in clinical settings and an optimized procedure for transcript profiling, using the Agilent oligo microarray technology. In particular, we discuss the special aspects of platelet purification when working with blood sample volumes not exceeding 3–5 ml, which is typical in pediatric patients and we furthermore provide detailed information for transcript profiling of extremely small amounts of platelet RNA.

Key words: Platelet disorders, pediatric patients, thrombocytopenic patients, leukocyte depletion, platelet isolation, platelet RNA profiling.

1. Introduction

Although circulating platelets are anucleate cells and therefore devoid of genomic DNA, they do contain minute amounts of residual plasmatic mRNA upon separation from the megakaryocytic cytoplasm. During this process only a vestigial amount of megakaryocytic mRNA is integrated into the developing platelet which has hampered gene expression analysis in platelets for many years. Platelet mRNA is not very stable and degrades in the peripheral circulation within 24–48 h *(1)*. Flow cytometric methods could prove the existence of mRNA within platelets after incubation with the DNA stain thiazole orange *(2)* confirming that the highest mRNA-content is found in young platelets.

Peter Bugert (ed.), *DNA and RNA Profiling in Human Blood: Methods and Protocols, vol. 496*

DOI 10.1007/978-1-59745-553-4_17 Springerprotocols.com

Identification of platelet-specific mRNA is often limited due to possible contamination of platelet preparations with leukocytes. Since the amount of mRNA in single leukocytes is approximately 100,000 times higher than in platelets, a small number of leukocytes can significantly distort the mRNA profile of platelet gene expression. It is calculated that a single platelet contains approximately 0.002 fg mRNA (based on our own experience). Therefore, large volumes of leukocyte-depleted blood (up to 500 ml) or platelet apheresates were often employed for gene-profiling studies. However, drawing such quantities of blood in a clinical setting is unrealistic, particularly in pediatric patients where blood samples often cannot exceed 3–5 ml, or when donors are thrombocytopenic.

Growing interest in isolating residual amounts of RNA from platelets in order to perform multiple genetic studies, such as RNA profiling *(3, 4)* and real-time PCR *(5)* called for a protocol that can reliably be applied to smaller blood samples (3–5 ml) possibly with limited platelet numbers when collected in thrombocytopenic patients.

Thus, in order to guarantee reliable and platelet-genome specific results it is mandatory to first isolate highly purified platelet yields before analysis of platelet-specific gene expression is feasible. Our protocol for the isolation of leukocyte-depleted platelet samples successfully copes with all three major problems such as extremely minute amounts of mRNA per platelet, contamination of platelet preparations with leukocyte as well as small blood sample volumes in a clinical setting *(6)*. Thus, a high-platelet yield (40%) with a degree of leukocyte depletion of 0–0.1/μl platelet isolate is feasible. It is estimated that $3–5 \times 10^8$ platelets, corresponding to 5 ml citrated blood with an initial platelet count of 100,000/μl, are at least required for safely performing further platelet genome studies.

The expected amount of platelet RNA obtained from $3–5 \times 10^8$ platelets can be as little as 50–100 ng. For direct transcript profiling at least 5 μg of total RNA is required. Therefore a method that amplifies the starting mRNA, without alteration of the gene expression profile, is required. Several RNA amplification strategies are currently under use, they include reverse transcription, which is followed either by exponential PCR amplification *(7–9)* or by T7-based linear in vitro transcription *(10–12)*.

We and our co-workers investigated the reliability and precision of RNA amplification, using the SMART – (switching mechanism at the 5' end of RNA templates) amplification method *(13)*. In six independent microarray experiments, original and SMART-amplified platelet RNA was profiled across 9850 genes and revealed a strong linear relationship. In addition, we successfully applied the SMART protocol to clinical samples *(6)*. Even so, the SMART protocol is a reliable method for RNA amplification, it remains a time consuming procedure.

In order to decrease the hands on time, we introduced several modifications to the existing T7-protocol from Agilent Technologies. We will illustrate and discuss a detailed protocol for platelet purification and RNA isolation from blood samples not exceeding 3–5 ml. With the proposed procedure platelet RNA profiling is feasible and easier to perform for studies on pediatric and/or thrombocytopenic patients in clinical settings.

2. Materials

2.1. Isolation of Leukocyte-Depleted Platelet Samples

1. Plastic aspiration tubes with one-tenth volume of 0.106 M trisodium citrate solution (Sarstedt, Germany).
2. Purecell® PL High-Efficiency Leukocyte Reduction Filter for Platelet Transfusion (PALL Medical, USA).
3. Record positive, luer negative adapter (Söllner GmbH, Deggendorf, Germany).
4. Sterile 10 ml syringe.
5. Propidium iodide.
6. Cell counter.
7. Flow cytometer.
8. Microcentrifuge (room temperature).

2.2. Total RNA Extraction

1. Trizol Reagent (cat. No. 15596-018; Invitrogen).
2. Liquid nitrogen.
3. Chloroform.
4. Isopropyl alcohol.
5. Ethanol (70%) diluted in DEPC-treated water.
6. DEPC-treated water.
7. Refrigerated microcentrifuge (4°C).

2.3. cRNA Synthesis and Fluorescent Labeling

1. DEPC-treated water.
2. Agilent low RNA input fluorescent linear amplification kit containing: T7 promoter primer; cDNA-mastermix: 5× first strand buffer, 0.1 M DTT, 10 mM dNTP mix, MMLV-RT, RNaseOUT.
3. Cyanine 3-CTP (10 mM), (PerkinElmer/NEN Life Science).
4. Cyanine 5-CTP (10 mM), (PerkinElmer/NEN Life Science).
5. Transcription master mix: 4×transcription buffer, 0.1 M DTT, NTP mix, 50% PEG, RNaseOUT, Inorganic pyrophosphatase, T7 RNA polymerase.
6. Water baths (40°C, 65°C, 80°C).
7. Ice and ice bucket.

2.4. Purification and Quantification of cRNA Products

1. RNeasy mini Kit (Qiagen).
2. DEPC-treated water.
3. Ethanol (96–98%) diluted in DEPC-treated water.
4. Microcentrifuge.
5. UV-Spectrophotometer.

2.5. Microarray Hybridization, Scanning, and Analysis

1. Agilent G2545A Hybridization Oven (Agilent Technologies).
2. Agilent microarray hybridization chamber kit (Agilent Technologies) containing chamber base, chamber cover, clamp assembly, tweezers, gasket slide, 1 microarray/slide format (G4112A), glass cover slip, 25× fragmentation buffer and 2× hybridization buffer.
3. Slide rack.
4. Horizontal shaker.
5. Water bath (60°C).
6. DEPC-treated water.
7. Gen expression wash buffer 1 and 2 (Agilent Technologies).
8. Stabilization and drying solution (Agilent Technologies).
9. Microarray Scanner including data analysis software.

3. Methods

Although the following platelet isolation protocol is simple and straightforward, we suggest to proceed slowly with gentle strength for obtaining highest platelet yield and complete leukocyte depletion.

3.1. Isolation of Leukocyte-Depleted Platelet Samples

1. Collect venous blood into plastic aspiration tubes with one-tenth volume of 0.106 M trisodium citrate solution (*see* **Note 1**).
2. Prepare the set-up for the platelet isolation system:
   a. Remove re-adjustable plunger from 10 ml syringe, place and adjust syringe upside down onto laboratory stand in a slightly tilted position so that the opening of the syringe is in the lowest position allowing for best output of fluid.
   b. Cut off filter system approximately 3 cm above the filter, discard collecting bag, turn around the clamp on the tube and tighten.
   c. Adjust tube via adapter onto the syringe opening and place a falcon tube under filter for collection of platelet isolate.
3. Determine platelet count and white blood cell count by using an automated cell counter.

4. For preparation of platelet-rich plasma (PRP), centrifuge citrated blood sample for 15 min at 150*g* (at room temperature, with brake).
5. Aspirate approximately 8/10 of PRP gently and very slowly, beginning from the surface near the tube-wall, neither aspirating leukocytes from the buffy coat nor activating platelet aggregation due to heavy suction. Gently discharge the aspirated PRP from pipette, again lean the tip against the tube-wall into the syringe (*see* **Note 2**).
6. Open the clip and collect platelet isolate into a falcon tube.
7. Once the isolate stops draining through the filter (occasionally it stops inexplicably half-way); slowly adjust the plunger into the syringe and gently press down (*see* **Note 3**). Carefully watch the filter tip for light bubbles and remove filter before bubbly isolate falls into tube in order to avoid leukocyte contamination.
8. In order to validate the degree of leukocyte contamination in the platelet isolate, use approximately 120 μl of highly purified platelet isolate for FACS analysis that sufficiently detects residual leukocytes as rare events after staining with DNA-intercalating dye propidium iodide (*see* **Note 4**). Only proceed with samples not exceeding 0.1 leukocyte/μl platelets.
9. Determine the approximate platelet count in the platelet isolate by using an automated cell counter (*see* **Note 5**).
10. Prepare aliquots from platelet isolate into 2 ml Eppendorf tubes containing approximately 3–5 × 10^8 platelets based on estimated platelet counts by centrifugation at 5,000*g* for 10 min.
11. Carefully remove supernatant (*see* **Note 6**).
12. Store pellets at –80°C or proceed with RNA extraction.

3.2. Total RNA Extraction

Working with RNA requires certain precautions to prevent RNase contamination; all the plastic ware and solutions for RNA extraction should be reserved for RNA work only. Disposable cloves should always be worn and the work should only been done on a clean bench.

1. Add 1 ml of Trizol reagent to the platelet pellet (3–5 × 10^8 cells) and lyse the cells by repetitive pipetting (*see* **Note** 7 and **Note 8**).
2. Incubate samples for 5–10 min at room temperature.
3. Add 0.2 ml of chloroform, shake tubes vigorously for 15 s, and incubate samples at room temperature for 5 min.
4. Centrifuge the samples at 12,000*g* at 4°C for 15 min.
5. Obtain the aqueous phase and transfer it to a fresh tube.
6. Add 0.5 ml of isopropanol to precipitate the RNA, shake carefully and let the tube sit for another 10 min at room temperature (*see* **Note 9**).
7. Centrifuge the samples at 12,000*g* at 4°C for 10 min.

8. Remove the supernatant and wash the RNA pellet in 1 ml of 75% ethanol (*see* **Note 10**).
9. Centrifuge at 7,500*g* at 4°C for 5 min.
10. Discard the wash and air-dry the pellet for 20–30 min (*see* **Note 11**).
11. Dissolve the RNA pellet completely by adding 10.3 μl DEPC-treated H_2O. If the sample is not used immediately for cDNA-synthesis store RNA at –80°C (*see* **Note 12**).

3.3. cDNA Synthesis from Total RNA and Generation of Fluorescent cRNA

For the cDNA synthesis and generation of fluorescent cRNA use the protocol provided by Agilent Technologies (briefly reproduced here) with our modification.

1. Add 1.2 μl of T7 promotor primer to the tube containing the total RNA in 10.3 μl DEPC-treated water. The total reaction volume should be 11.5 μl.
2. Denature the primer and the template by incubating the reaction in a 65°C water bath for 10 min.
3. Place the tube immediately on ice and incubate for 5 min.
4. Prepare the reverse transcription reaction mix in the following order: for each reaction combine 4 μl of 5× first strand buffer, 2 μl of 0.1 M DTT, 1 μl of 10 mM dNTPmix, 1 μl of MMLV-RT, and 0.5 μl of RNaseOUT (*see* **Note 13**).
5. Add 8.5 μl of this mix to each sample tube and incubate samples at 40°C for 4 h.
6. Incubate the tubes in a 65°C water bath for 10 min.
7. Place the tubes on ice and incubate for 5 min.
8. To each sample tube, add either 2.4 μl of cyanine 3-CTP (10 mM) or 2.4 μl of cyanine 5-CTP (10 mM).
9. Prepare the transcription master mix in the following order: for each reaction combine 15.3 μl of DEPC-treated water, 20 μl of 4× transcription buffer, 6 μl of 0.1 M DTT, 8 μl of NTP mix, 6.4 μl of 50% PEG, 0.5 μl of RNaseOUT, 0.6 μl of inorganic phosphatase, and 0.8 μl of T7 RNA polymerase (*see* **Note 14**).
10. Add 57.6 μl of the transcription master mix to each sample.
11. Incubate samples for 15 h in a 40°C water bath.

3.4. Purification and Quantification of cRNA Products

For the purification of the amplified cRNA we recommend using Qiagen's RNeasy mini spin columns. The RNeasy mini kit protocol is reproduced here.

1. Add 20 μl of DEPC-treated water to the cRNA sample, to obtain a total volume of 100 μl.
2. Add 350 μl of buffer RLT and mix thoroughly.
3. Add 250 μl of 96–98% ethanol and mix thoroughly by pipetting. Do not centrifuge.
4. Transfer 700 μl of cRNA sample to an RNeasy mini column in a 2 ml collection tube. Centrifuge the sample for 30 s at 10,000*g*.

5. Transfer the RNeasy column to a new collection tube and wash the sample by adding 500 μl of buffer RPE to the column. Centrifuge the sample for 30 s at 10,000*g*.
6. Wash the sample again by adding 500 μl of buffer RPE to the column. Centrifuge the sample for 60 seconds at 10,000*g*.
7. Elute the cleaned cRNA sample by transferring the RNeasy column to a new 1.5 ml collection tube. Add 30 μl DEPC-treated water directly onto the RNeasy filter membrane. Incubate 60 seconds and centrifuge 30 seconds at 10,000*g*. Save the flow through in the collection tube.
8. Again, add 30 μl DEPC-treated water onto the filter membrane and wait 60 second before centrifuging for 30 seconds at 10,000*g*. The final volume is approximately 60 μl (*see* **Note 15**).
9. Determine the concentration of cRNA in each sample, using an appropriate spectrophotometer (*see* **Note 16**).
10. Store the cRNA sample at –80°C or continue directly with the hybridization protocol.

3.5. Microarray Hybridization

1. Prepare 2× cRNA target solution by mixing 5 μg each of cyanine 3- and cyanine 5-labeled linear amplified cRNA. Add DEPC-treated water to yield a total volume of 240 μl per tube (*see* **Note 17**).
2. To each tube add 10 μl 25× fragmentation buffer and incubate in a 60°C water bath in the dark for 30 min.
3. To terminate the fragmentation reaction, add 250 μl 2× hybridization buffer to reach a final volume of 500 μl per microarray. Try to avoid introducing bubbles while mixing.
4. Without introducing bubbles, place 490 μl of the labeled sample onto the gasket well and flip the slide with the microarray onto it to obtain a sandwich (*see* **Note 18**).
5. Place the hybridization chamber into the hybridization oven and set the rotator to rotate at approximately 6 rpm.
6. Hybridize the slides at 60°C for 22 h.
7. Disassemble the chambers and immediately place the microarray in the slide rack.
8. Move the slide rack in wash solution 1 and place it on a horizontal shaker.
9. Wash the microarray at 60 rpm for 10 min.
10. Move the slide rack into wash buffer 2 and wash for 5 min at 60 rpm.
11. Dry the microarray slide prior to scanning by immersing it into stabilization and drying solution (*see* **Note 19**).
12. Place the microarray in a clean-slide box until scanning.

3.6. Scanning and Analysis

1. Scan the microarray with a high-resolution laser confocal scanner to obtain raw data images with fluorescence signals.

2. Quantitate and analyze the data using computer software for microarrays (*see* **Note 20**).

4. Notes

1. Depending on the patient's platelet count and age draw 3–20 ml blood. In order to reduce formation of platelet aggregates clotting the filter system and thereby leading to significant platelet loss particularly in small sample volumes, blood should be drawn whenever possible without tourniquet, discarding the first 2 ml of blood, while aiming at distracting children from crying (including EMLA® patch application).
2. Where applicable, note initial blood volume as well as removed amount of supernatant for future calculations.
3. It is very important not to apply any force to the filter for the possibility of removing caught leukocytes from the filter into the purified PRP.
4. The FACS method is successfully used in the standardized and quality-controlled evaluation and validation of leukocyte-depleted platelet concentrates *(14)*. It allows for detecting rare leukocyte events up to 0.1/μl platelet isolate. Leukocyte contamination should not exceed 0,1 leukocyte/μl platelet isolate. Since platelets are believed to contain approximately 100,000 times less RNA than leukocytes, the leukocyte–platelet ratio should be recalculated for the RNA level.
5. Note that some microaggregates are usually present. If indicated perform blood smear for optical evaluation.
6. If necessary, perform multiple volume steps into the same tube. Note that the platelet pellet might be invisible during the first volume step.
7. For better cell-lysis, we recommend to shock freeze the homogenized cells in liquid nitrogen and refreeze them in 40°C water bath. Freezing circles can be repeated several times to ensure complete lysis of the platelets. If liquid nitrogen is not available, a homogenizer can be used instead.
8. Washing cells before addition of Trizol should be avoided as this increases the degradation of RNA.
9. RNA samples can be stored at –80°C in isopropanol if not used immediately.
10. The RNA precipitate forms a gel like structure on the side bottom of the tube and is often hardly visible. Care should be taken when removing the supernatant

because the small pellet is easily lost while performing this step.

11. It is important not to let the RNA pellet dry for to long as this will decrease its solubility. On the other hand, remaining ethanol will inhibit enzymes essential for the subsequent cDNA synthesis. The air drying is a crucial step and needs to be monitored carefully.
12. The dry pellet is dissolved in 10.3 μl H_2O, the exact amount necessary for subsequent cDNA synthesis. Due to the low amount of total RNA we abstain from RNA quantification. Based on the initial platelet number of $\sim 5 \times 10^8$we expect to reach a total RNA yield of at least 50 ng (based on our own experience).
13. Prewarm the 5× first strand buffer by incubating the vial in an 80°C water bath for 3–4 min prior to use.
14. Prewarm the 50% PEG solution by incubating the vial in a 40°C water bath for 1 min.
15. When the amount of purified cRNA is expected to be extremely low, we recommend elute only once. The final flow through volume should be approximately 30 μl.
16. The typical cRNA amplification reaction, starting with approximately 50 ng (initial platelet number of 5×10^8) will yield 200–300 ng/μl. We recommend using the Gene Quant Spectrophotometer (Amersham/Bioscience) to minimize the amount of sample consumed by the measurement.
17. The volume of DEPC-treated water will vary with the concentration of the cRNA. If necessary, the 2× target solution can be stored at –80°C up to 1 month.
18. Flip the oligo microarray so that the numeric side is facing up and lower it onto the gasket slide. Gently place the oligo microarray slide against the gasket slide to complete the sandwich. Be sure that the microarray slide and the gasket slide are completely aligned, before you correctly place the chamber cover onto the sandwiched slides and close the chamber. Inspect the sandwiched slides and note the bubble formation. A large mixing bubble should have formed. For more detailed instructions, refer to the Agilent microarray hybridization chamber user guide.
19. The slide should be fully covered with stabilization and drying solution and be removed again very slowly preventing the development of bubbles on top of the microarray side of the slide.
20. Comparative analysis of individuals from two study groups revealed strong correlation (**Fig. 17.1**). This indicates that the protocol for platelet RNA profiling leads to highly reproducible results.

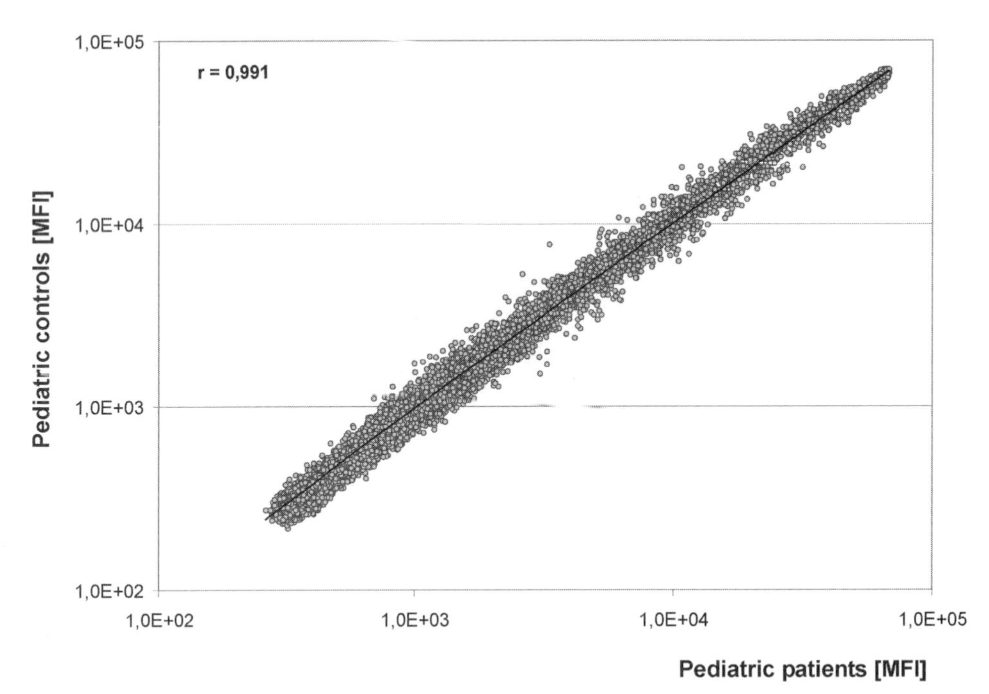

Fig. 17.1. Scatter plot of mean signal intensity values from comparative microarray analysis of more than 40,000 human genes in pediatric patient (*x-axis*) and pediatric control (*y-axis*) platelet RNA. The *line* indicates the linear relationship between the microarray hybridization results from patient and control samples ($r = 0.991$) The mean values of 21 patients and 15 controls are represented.

References

1. Ault, K. A., Knowles, C. (1995) In vivo biotinylation demonstrates that reticulated platelets are the youngest platelets in circulation. *Exp Hematol* 23, 996–1001.
2. Kienast. J., Schmitz, G., (1990) Flow cytometric analysis of thiazole orange uptake by platelets: a diagnostic aid in the evaluation of thrombocytopenic disorders. *Blood* 75, 116–121.
3. Gnatenko, D. V., Dunn, J. J., McCorkle, S. R., et al. (2003) Transcript profiling of human platelets using microarray and serial analysis of gene expression. *Blood* 101, 2285–2293.
4. Bugert, P., Dugrillon, A., Gunaydin, A., et al. (2003) Messenger RNA profiling of human platelets by microarray hybridization. *Thromb Haemost* 90, 738–748.
5. Fink, L., Holschermann, H., Kwapiszewska, G., et al. (2003) Characterization of platelet-specific mRNA by real-time PCR after laser-assisted microdissection. *Thromb Haemost* 90, 749–756.
6. Rolf, N., Knoefler, R., Suttorp, M., et al. (2005) Optimized procedure for platelet RNA profiling from blood samples with limited platelet numbers. *Clin Chem* 51, 1078–1080.
7. Matz, M., Shagin, D., Bogdanova, E., et al. (1999) Amplifikation of cDNA ends based on template switching effect and step-out PCR. *Nucleic Acids Res* 27, 1558–60.
8. Schwabe, H., Stein, U., Walther, W. (2000) High-copy cDNA amplification of minimal total RNA quantities for gene expression analysis. *Mol Biotechnol* 14. 165–172.
9. Zhu, Y. Y., Machleder, E. M., Chenchik, A., et al. (2001) Reverse transcription template switching: a SMART approach for full length cDNA library construction. *Biotechniques* 30(4), 892–897.
10. Phillips, J., Eberwine, J. H. (1996) Antisense RNA amplification: A linear amplification method for analyzing the mRNA population from single living cells. *Methods* 10(3), 283–288.

11. Baugh, L. R., Hill, A. A., Brown, E. L., et al. (2001) Quantitative analysis of mRNA amplification by in vitro transcription. *Nucleic Acids Res* 29(5), E29.
12. Hu, L., Wang, J., Baggerly, K., et al. (2002) Obtaining reliable information from minute amounts of RNA using cDNA microarrays. *BMC Genomics* 3, 16.
13. Rox, M. J., Bugert, P., Müller, J., et al. (2004) Gene expression analysis in platelets from a single donor: Evaluation of a PCR-based amplification technique. *Clin Chem* 50(12), 2271–2278.
14. Schlenke, P. (2005) Leukocyte reduction in blood component supply: the impact of flow cytometry in assessing residual leukocytes. *Transfus Med Hemother* 32, 12–19.

Chapter 18

PCR-Based Amplification of Platelet mRNA Sequences Obtained From Small-Scale Platelet Samples

Jutta M. Rox, Jens Müller, and Bernd Pötzsch

Abstract

Platelet transcriptome studies provide a powerful tool in the analysis of disorders affecting the megakaryocytic-platelet lineage. However, individualised platelet gene expression profiling is hampered by the exceptionally low yield of platelet mRNA. The yield of mRNA transcripts that can be obtained from small-scale platelet preparations is generally not sufficient for standard RNA-labeling reactions used in expression profiling. Furthermore, leukocyte contaminants in platelet preparations are a potential source of 'unwanted' mRNA since they contain several orders of magnitude more mRNA than platelets. To overcome these limitations a strategy that combines leukocyte filtration and a PCR-based amplification technique (SMART^{TM}) has been developed and extensively evaluated.

Key words: Platelets, platelet mRNA, leukodepletion, SMART^{TM} amplification, platelet transcriptome.

1. Introduction

Platelets are anucleate but they contain a significant number of megakaryocyte-derived RNA transcripts *(1)* and maintain functionally intact protein translational capabilities. Posttranscriptional regulation of de novo protein synthesis has been described for Bcl-3, pro-IL-1 beta, and tissue factor *(2–4)*.

However, platelet transcriptome analysis is limited by the exceptionally small yield of mRNA and the risk of leukocyte contamination during platelet preparation. We describe a protocol for PCR-based amplification of mRNA (SMART^{TM}) from single-donor platelets to provide a template for gene expression studies like microarray analysis. The maintenance of the original

Peter Bugert (ed.), *DNA and RNA Profiling in Human Blood: Methods and Protocols, vol. 496*

DOI 10.1007/978-1-59745-553-4_18 Springerprotocols.com

transcript ratios was assessed using quantitative PCR for selected transcripts and microarray analysis *(5)*.

The protocol described makes platelet mRNA isolated from single patients available for gene expression studies aimed at detecting target genes involved in the development of megakaryocytic/platelet disorders.

2. Materials

2.1. Platelet Preparation

1. Blood collection tubes with 0.105 M trisodium citrate stock solution (final concentration of 10% (v/v)), e.g., S-Monovette Coagulation (Sarstedt, Nürnbrecht, Germany).
2. Tyrodes buffer: 139 mM NaCl, 2.9 mM KCl, 1.0 mM Na_2HPO_4, 5 mM HEPES, 10 mM D-Glucose; pH 6.5. Store at 4°C.
3. Leukocyte removal filter (Purecell PL leukocyte removal filter; Pall Biomedizin GmbH, Dreieich, Germany).

2.2. Isolation of Platelet poly A+ RNA

1. Dynabeads® Oligo (dT)25 magnetic beads (Invitrogen, Karlsruhe, Germany). Store at 4°C.
2. Magnet for 1.5 ml tubes, e.g., Dynal MPC™-S (Invitrogen).
3. Gauge needle (25) and 1 ml syringe.
4. Lysis buffer: 0.5 M LiCl, 0.1 M Tris–HCl (pH 8.0), 0.01 M EDTA, 1% LiDS, 5 mM dithiothreitol (DTT). Store at 4°C.
5. Binding buffer: 20 mM Tris–HCl, pH 7.5, 1.0 M LiCl, 2 mM EDTA. Store at 4°C.
6. Washing buffer with LiDS: 10 mM Tris–HCl, pH 7.5, 0.15 M LiCl, 1 mM EDTA, 0.1% LiDS. Store at 4°C.
7. Washing buffer without LiDS: 10 mM Tris–HCl, pH 7.5, 0.15 M LiCl, 1 mM EDTA. Store at 4°C.
8. Deionised H_2O, RNase free. Store at –20°C.

2.3. First-Strand cDNA Synthesis and Amplification

1. SMART™ PCR cDNA Synthesis Kit (Clontech) containing: SMART II™ A oligonucleotide (5′ AAGCAGTGGTATCAA CGCAGAGTACGCGGG-3′, 12 μM; store at –70°C); -3′ SMART™ CDS Primer II A (5′-AAGCAGTGGTATCAA CGCAGAGTACT(30)VN-3′, 12 μM; store at –20°C); 5× First-strand buffer: 250 mM Tris–HCl (pH 8.3), 375 mM KCl, 30 mM $MgCl_2$; store at –20°C; 5′ PCR Primer II A (5′-AAGCAGTGGTATCAACGCAGAGT-3′, 12 μM; store at –20°C); dNTP mix (10 mM each; store at –20°C); dithiothreitol (DTT, 20 mM; store at –20°C); deionised H_2O (store at –20°C).
2. SuperScript™ III reverse transcriptase, 200 U/μl (Invitrogen). Store at –20°C.

3. Advantage 2 PCR Kit (Clontech) containing: deionised H_2O; 10× Advantage 2 PCR buffer; 50× dNTP mix; 50× Advantage 2 polymerase mix; all components stored at –20°C.
4. DNA size marker (e.g., 1-kb DNA ladder). Store at 4°C.

2.4. Purification of cDNA Amplification Products

1. QIAquick PCR Purification Kit (Qiagen Hilden, Germany) containing: QIAquick spin columns, buffer PBI, buffer PE (concentrate, add stated volume of ethanol), collection tubes.
2. Sodium acetate 3 M, pH 5.0.
3. Ethanol (p.a.).

3. Methods

Platelet gene expression analysis faces two main technical obstacles: (i) leukocyte contamination of platelet preparations and (ii) the low level of residual cytoplasmatic RNA in platelets.

It has been estimated that a single platelet contains about 0.2 fg total RNA while a single leukocyte contains about 10^4 as much *(6)*. Thus, a small number of contaminating leukocytes can significantly disturb a gene expression profile obtained from a platelet sample. Many techniques, such as differential centrifugation, filtration , magnetic beads coated with anti-leukocyte antibodies, or laser-assisted microdissection can be used for leukodepletion.

We describe leukofiltration that results in a 1,000-fold reduction of leukocyte mRNA as shown by quantification of leukocyte-specific CD45 transcripts in platelet mRNA obtained before and after one filtration step (**Fig. 18.1**). The extend of leukodepletion that has to be performed depends on the sensitivity of downstream applications.

The minimal platelet number required has to be determined for each application. Platelet loss during sample preparation is high (with this protocol about 70%) and depends mainly on the extent of leukodepletion steps. In our hands, 40 ml whole blood or 2 × 10^9 leukodepleted platelets were sufficient to generate reliably 20–30 μg of amplified platelet cDNA.

3.1. Platelet Sample Preparation from Whole Blood

1. Collect 40 ml peripheral blood from the patient or donor in tubes with a sodium citrate stock solution to final concentration of 10% v/v with the anticoagulant. Avoid platelet activation by gentle aspiration of the blood sample. Mix the blood with the anticoagulant by inverting the tube immediately. Do not cool the sample. Spin the blood samples at 150 × *g* for 20 min at room temperature without brake to obtain platelet-rich plasma (PRP).

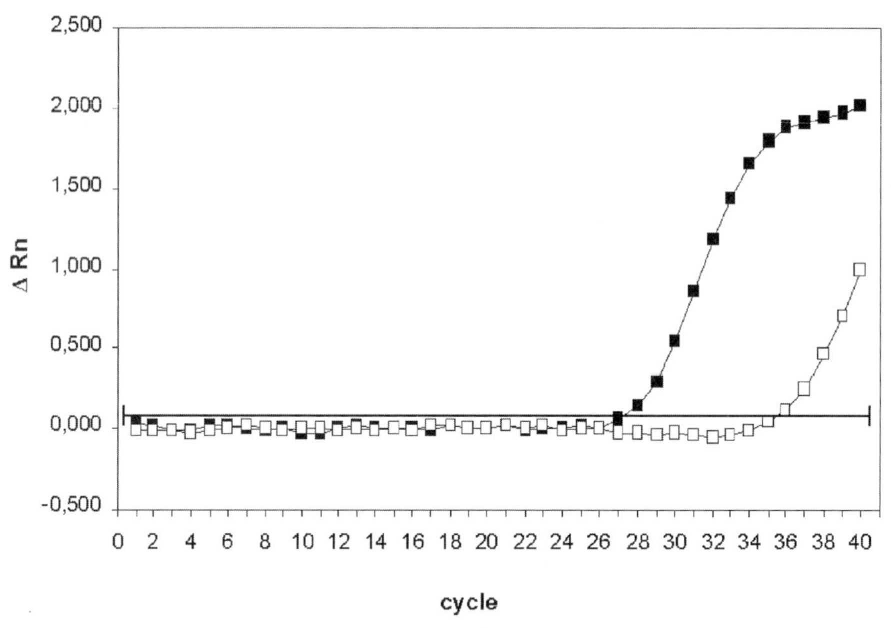

Fig. 18.1. Amplification curves of CD45 real-time RT-PCR. To determine leukocyte filtration efficacy, real-time RT-PCR was performed using oligonucleotide primers and probe specific for CD45 RNA. Platelet mRNA was prepared as described in this protocol, with (open symbols) or without (closed symbols) one leukocyte filtration step. The leukocyte filtration step results in an approximate 10^3 reduction of CD45 RNA as estimated on cycle threshold values (*x*-axis). Data points represent the mean from duplicate wells (reproduced from Ref. *(7)* with permission from S. Karger Medical and Scientific Publishers).

2. Aspirate carefully the upper two thirds of the PRP in a 50 ml tube.
3. Add twice a volume of Tyrodes buffer and mix gently by inverting the tube.
4. Flush the leukocyte depletion filter with 10 ml Tyrodes buffer. Remove the bag connected to the filter.
5. Filter the PRP by gravitation force only. Use a 50 ml tube to collect the filtrated samples. Do not use any pressure.
6. Centrifuge the sample at 1,000 × g for 15 min at room temperature and remove the supernatant.
7. Freeze the platelet pellet in liquid nitrogen and store at –80°C.

3.2. Isolation of Platelet poly A+RNA

1. Wash Dynabeads® Oligo (dT)25 (0.2 ml) in lysis buffer (0.2 ml) using the magnet.
2. Remove leukodepleted platelet pellets from the –80°C freezer and immediately add 1.2 ml of lysis buffer.
3. Use a 25G needle and a syringe to resuspend the cell pellet in the lysis buffer and repeat syringe uptake and repulsion until the solution is no longer viscous. This homogenisation process should not take more than a few minutes.
4. Transfer the lysed material in a microcentrifuge tube and spin for 45 s at g_{max}. Transfer the supernatant to the microcentrifuge tube containing the washed Dynabeads® Oligo (dT)25.

5. The beads are hybridised to poly A+RNA for 4 min at room temperature (*see* **Note 1**).
6. Place the tube in the magnet for 1–5 min (until all beads are collected on the tube wall) and remove all supernatant gently.
7. Wash the beads twice with 1 ml washing buffer containing LiDS and twice with 1 ml washing buffer without LiDS, to remove all of the detergent that can inhibit the following enzymatic reactions.
8. Elute the poly A+ RNA by adding 10 μl H_2O, heat to 85°C, and immediately place the tube on the magnet. Gently and immediately transfer all supernatant containing the eluted mRNA into a chilled 1.5 ml tube and place on ice (*see* **Note 2**).

3.3. First-Strand cDNA Synthesis and Amplification

3.3.1. First-Strand cDNA Synthesis

1. For each sample, combine the following reagents in a 0.2 ml reaction tube to a total volume of 5 μl: 3 μl of the poly A+ RNA sample (*see* **Section 3.2**); 1 μl 3′ SMART™ CDS Primer II A (12 μM); 1 μl SMART™ II A Oligonucleotide (12 μM).
2. Mix contents by pipetting and spin briefly.
3. Incubate the tube at 72°C for 2 min and subsequently cool on ice for 2 min.
4. Centrifuge the tube briefly to collect all contents at the bottom. Then add the following to each reaction tube: 2 μl 5× first-strand buffer; 1 μl DTT (20 mM); 1 μl dNTP mix (10 mM of each dNTP); 1 μl SuperScript™ III reverse transcriptase (200 U/μl).
5. Mix by gently pipetting and spin briefly.
6. Incubate the tubes at 42°C for 1 h, preferably in a hot lid thermal cycler.
7. Place the tube on ice to terminate first-strand synthesis.
8. Generated first-strand cDNA can be directly applied to LD PCR amplification or stored at < –20°C until used.

3.3.2. Amplification of First-Strand cDNA by Long-Distance PCR

1. For each sample, combine the following reagents in a 0.2 ml reaction tube to a total volume of 98 μl or, if appropriate, prepare a corresponding master mixture: 80 μl deionised H_2O; 10 μl 10× Advantage 2 PCR buffer; 2 μl 50× dNTP mix (10 mM of each dNTP); 4 μl 5′ PCR Primer II A (12 μM); 2 μl 50× Advantage 2 polymerase mix.
2. Centrifuge the tube briefly and subsequently, add 2 μl of the generated first-strand cDNA sample (*see* **Section 3.3.1**)
3. Mix contents by gently flicking the tube.
4. Centrifuge the tube and place it in a preheated (95°C) hot lid thermal cycler.
5. Start the amplification reaction using the following temperature profile: 95°C, 1 min; 18 cycles with 95°C, 15 s, 65°C, 30 s and 68°C, 6 min.

6. When cycling is completed, electrophorese 5 μl of each sample on a 1.1% agarose/EtBr gel in 1× TAE buffer. Also run 0.1 μg of a 1-kb DNA ladder.
7. A smear of cDNA should be visible from approximately 0.5–6 kb with an intensity comparable to that of the loaded (0.1 μg) DNA size-marker (*see* **Note 3**).

3.4. Purification of cDNA Amplification Products

1. Depending on desired downstream applications, LD PCR reaction mixtures need to be purified from unused primers and dNTPs in order to allow accurate quantification of generated SMART^{TM} amplification products. The following described application of chromatography using silica-gel membrane spin columns is a fast but also reproducible method for the purification of PCR amplicons.
2. Add 5 volumes of buffer PBI to 1 volume of the PCR sample and mix. Check that the color of the mixture is yellow. If the color of the mixture is orange or violet, add 10 μl of 3 M sodium acetate, pH 5.0, and mix.
3. Apply the sample to a QIAquick spin column placed in a 2 ml collection tube and centrifuge at approximately 18,000 × *g* for 30–60 s. Discard flow-through and place the QIAquick column back into the same tube.
4. To wash, add 0.7 ml buffer PE to the QIAquick column and centrifuge at approximately 18,000 × *g* for 30–60 s. Discard flow-through and place the QIAquick column back into the same tube. Perform the washing step three times, place the column in the emptied collection tube and centrifuge for an additional 1 min to remove any traces of washing buffer.
5. Place the QIAquick column in a clean 1.5 ml microcentrifuge tube, add 50 μl of ddH_20 to the center of the membrane, incubate for 2 min and finally centrifuge at approximately 18,000 × *g* for 30–60 s. Store the RNA at –80°C.

4. Notes

1. Dynabeads® Oligo dT(25) (200 μl) are able to bind about 2 μg poly A+ RNA. For economy, only 200 μl Dynabeads® Oligo dT(25) are sufficient to obtain platelet mRNA from single-donor samples.
2. Dynabeads® Oligo dT(25) can be used up to five times. After each mRNA isolation, suspend the used beads in 0.1 M NaOH (200 μl), heat the mixture to 65°C for 2 min. Place the tube immediately in the magnet. After 30 s discard the supernatant. Wash the beads two times with 0.1 M NaOH (200 μl). Then wash the beads four times with 200 μl storage buffer (250 mM Tris–HCl, pH 7.5, 20 mM EDTA, 0.1% Tween-20, 0.02%

sodium azid). After the last washing step, measure the pH of the supernatant. If the pH is >8.0 perform additional washing steps with storage buffer until the pH is <8.0. Suspend the beads in 200 μl storage buffer and store at 4°C.

3. Although the given protocol had been optimised and evaluated to work with low numbers of platelets gained from individual subjects, optimal amplification conditions may vary depending on used reagent batches and thermal cycler characteristics. It might therefore be necessary to adapt the number of PCR cycles to the given conditions in your laboratory. If a banding pattern is present but appears weak and does not comprise a size distribution of at least 5 kb, perform additional PCR cycles and electrophorese 5 μl samples until the optimum pattern characteristics is obtained. In contrast, an obviously strong smear of cDNA that comprises a size distribution of more than 7 kb is indicative of PCR overcycling. If this is the case, repeat the experiment described in **Section 3.3.2** but initially perform only 15 amplification cycles. Alternatively, run an extra tube for each sample and subject all tubes to 15 cycles. Then, use the extra tubes to determine the optimal number of additional PCR cycles that finally have to be applied to the sample tubes. By monitoring the amplification of various platelet-derived cDNA sequences during LD PCR, we found a linear increase of both low- and high-abundance sequences

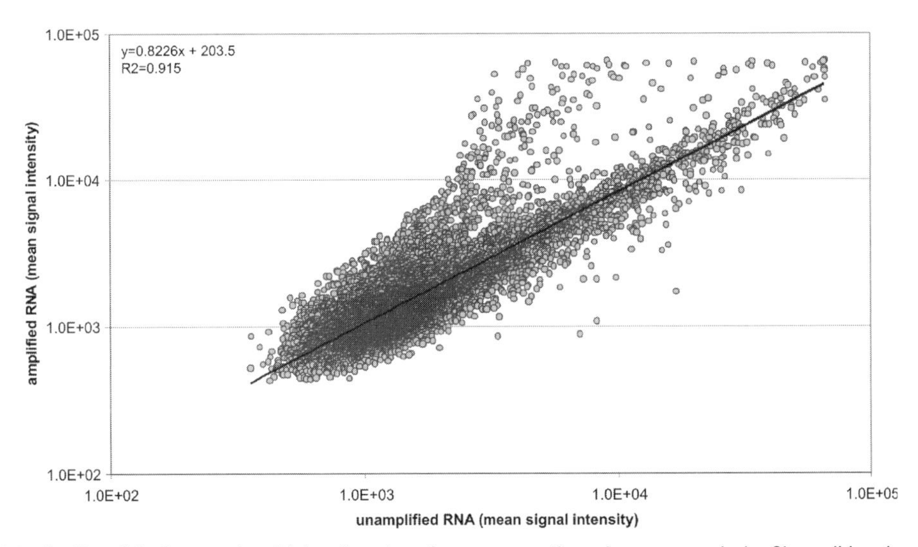

Fig. 18.2. Scatter plot of mean signal intensity values from comparative microarray analysis. Glass slide microarrays, representing 9,850 human genes, were used to characterise the mRNA profile in human platelets. Mean values of hybridisation signals were calculated from six individual experiments for each individual gene and compared between unamplified (*x*-axis) and SMART-amplified samples (*y*-axis), indicated by gray circles in the scatter plot. The line indicates the linear relationship between the microarray hybridisation results from unamplified and amplified samples ($R_2 = 0.915$) calculated by the equation: $y = 0.8226 + 203.5$ (reproduced from Ref. *(5)* with permission from the American Association for Clinical Chemistry).

if low starting concentrations of platelet mRNA were used, as achieved in the single-donor-protocol presented here *(5)*. Accordingly, original platelet mRNA profiles were found to be sufficiently preserved for microarray analysis after a total of 24 amplification cycles (**Fig. 18.2**).

References

1. Newman, P., Gorski, J., White, G., et al. (1988) Enzymatic amplification of platelet-specific messenger RNA using the polymerase chain reaction. *J Clin Invest* 82, 739–743.
2. Lindemann, S., Tolley, N. D., Dixon, D. A., et al. (2001) Activated platelets mediate inflammatory signalling by regulated interleukin 1beta synthesis. *J Cell Biol* 154, 485–490.
3. Pabla, R., Weyrich, A. S., Dixon, D. A., et al. (1999) Integrin-dependet control of translation: engagement of intergrin alpha2bbeta3 regulates synthesis of proteins in activated human platelets. *J Cell Biol* 144, 175–184.
4. Schwertz, H., Tolley, N. D., Foulks, J. M., et al. (2006) Signal-dependent splicing of tissue factor pre-mRNA modulates the thrombogenecity of human platelets. *J Exp Med* 203, 2433–2440.
5. Rox, J. M., Bugert, P., Müller, J., et al. (2004) Gene expression analysis in platelets from a single donor: evalutation of a PCR-based amplification technique. *Clin Chem* 50, 2271–2278.
6. Wicki, A. N., Walz, A., Gerber-Huber, S. N., et al. (1989) Isolation and characterization of human blood platelet mRNA and construction of a cDNA library in lambda gt11. Confirmation of the platelet derivation by identification of GPIb coding mRNA and cloning of a GPIb coding cDNA insert. *Thromb Haemost* 61, 448–453.
7. Rox, J. M., Müller, J., Pötzsch, B. (2006) Platelet Transcriptome Analysis. *Transfus Med Hemother* 33, 177–182.

Chapter 19

MicroRNA Profiling of Megakaryocytes

Ramiro Garzon

Abstract

MiRNAs are non-coding RNAs of 19–25 nucleotides in length that regulate gene expression. Recent work suggests a role for miRNAs during hematopoiesis. Here, we described in detail how to successfully obtain highly pure megakaryocytes precursors from human CD34+ stem cells in culture to perform high-throughput miRNA profiling using a custom microarray platform.

Key words: MicroRNAs, megakaryocytes, hematopoiesis, microarrays, cytokines.

1. Introduction

MiRNAs are non-coding RNAs of 19–25 nucleotides in length that regulate gene expression by inducing translational inhibition or cleavage of their target mRNA through base pairing to partially or fully complementary sites *(1)*. Initially discovered as regulators of mechanisms involved in cell proliferation, cell death and development, stress resistance, and fat metabolism *(1)*, more recently miRNAs have been shown to be involved in hematopoiesis *(2–5)*. We have recently described the miRNA expression during megakaryocytic differentiation of CD34+ cells in vitro *(2)*. Using a novel approach to culture stem cells in vitro we were able to obtain enough numbers of high-quality megakaryocytes precursors to perform phenotypic characterization by flow cytometry and miRNA microarray studies *(2)*. To analyze miRNA expression we used a custom microarray platform that contained 250 miRNAs probes (OSU_CCC version 2) *(6)*. Oligonucleotide probes (about 40-mer) for each miRNA were designed, one containing the mature miRNA sequence and

Peter Bugert (ed.), *DNA and RNA Profiling in Human Blood: Methods and Protocols, vol. 496*

DOI 10.1007/978-1-59745-553-4_19 Springerprotocols.com

the other the miRNA precursor without the mature sequence. Using these distinct sequences, we were able to separately analyze the expression of mature miRNA and pre-miRNA transcripts for the same miRNA. This microarray platform has been recently updated (OSU_CCC version 4) and now includes more than 700 cloned and predicted miRNAs probes. Here, we will discuss in detail, first how to obtain high-quality differentiated megakaryocytes and second how to perform miRNA profiling using these samples.

2. Materials

2.1. In Vitro Culture of CD34 Cells

1. For optimal results, we recommend to use high-quality CD34+ stem cells obtained from "steady state" bone marrow healthy donors (*see* **Note 1**).
2. Alternatively cord blood CD34+ samples could be used. We do not recommend the use of mobilized CD34+ stem cells because the cytokines employed for mobilization (usually G-CSF) will modify the miRNA expression of the stem cells (unpublished data).
3. To culture the CD34+ cells we use StemSpam SFEM (serum-free expansion media) (StemCell technologies, Vancouver, BC) which contains: Isocove- modified Dulbecco's medium supplemented with human transferrin, insulin, human low-density lipoprotein, and glutamine. This media should be stored at –20°C until use. Once in use, store at 4°C.
4. To obtain megakaryocytes precursors from CD34+ cells we use the following cytokines: recombinant human thrombopoietin (rh-TPO) alone (PeproTech, Rocky Hill, NJ) and Cytokine cocktail CC-200, which includes: rh-TPO, rh-interleukin-3 (Il-3), and rh-stem cell factor (SCF) (StemCell technologies).
5. Trizol reagent for RNA extraction (Invitrogen, Carlsbad, CA).

2.2. Microarray Analysis

1. To generate cDNA from total RNA we use the following reagents: random oligonucleotideprimer [3′-(N)8-(A)12-biotin-(A)12-biotin-5′]; 5× first-strand buffer, DTT, 10 mM dNTP mix, and SuperScript II RNaseH$^-$reverse transcriptase (200 units/μl) (Invitrogen).
2. NaOH (0.5 M)/50 mM EDTA solution.
3. Tris–HCI (1 M), pH 7.6 (Sigma).
4. SSPE/formamide (6×): 0.9 M sodium chloride; 60 mM sodium phosphate; 8 mM EDTA; 30% formamide. Adjust pH to 7.4.
5. TNT (0.75×): 0.1 M Tris–HCl, pH 7.5; 0.15 M NaCl; 0.05% Tween 20.

6. Streptavidin-Alexa647 conjugate is required for the detection of the biotin conjugates.
7. Custom microarray contains 385 probes corresponding to 250 mature and precursors human and mouse miRNA genes.All probes on these microarrays are 40-mer oligonucleotidesspotted by contacting technologies and covalently attached toa polymeric matrix *(6)*.

3. Methods

3.1. Differentiation of CD34+ Cells to Megakaryocyte Precursors In Vitro

To obtain highly pure megakaryocyte precursors from CD34+ hematopoietic stem cells in vitro, first we culture steady state bone marrow CD34+ cells with a serum free media supplemented initially only with the specific megakaryocyte cytokine (TPO) for 4 days. Previous work have shown that unilineage CD34+ cultures using only TPO yield a high percentage of megakaryocyte precursors, though at lower numbers *(7)*. Our strategy was to direct the CD34+ differentiation process to the megakaryocyte lineage using only TPO for the first 4 days and then to add other cytokines to stimulate the proliferation of the committed clones to the megakaryocyte lineage. Here we will describe in detail our method.

3.1.1. Day 0

1. Remove 1×10^6cells of high-quality CD34+ cells from frozen storage and quickly thaw in a 37°C waterbath by gently agitating vial.
2. As soon as the ice crystals melt, pipette gently into a 15 ml centrifuge tube containing prewarmed StemSpam medium.
3. Centrifuge the cells at 1,000 rpm for 10 min, pour off the supernatant and resuspend in 10 ml of prewarmed StemSpam media containing only TPO at 100 ng/ml and no antibiotics (*see* **Note 2**).
4. Plate the cells in two wells of a six-well plate (5 ml each well).
5. Put the plate at 37°C with 5% CO_2 in a standard cell incubator for 48 h.

3.1.2. Day 2

1. After 48 h of culture, mix thoroughly the cells by pipetting and split the cells 2:1, resulting in four wells with 2.5 ml each.
2. Add 2.5 ml of prewarmed StemSpam media with TPO (100 ng/ml) to each well. Place again the plate in the incubator for 48 h.

3.1.3. Day 4

1. Combine all the cells from the four wells (total volume 20 ml) into a 50 ml centrifuge tube.
2. Take an aliquot (100 ul) and check the cell number and viability.

3. Centrifuge at 1,100 rpm for 7 min.
4. Remove the old supernatant from the tube and replace with fresh prewarmed StemSpam media with the following cytokine combination; TPO 100 ng/ul, IL-3 (10 ng/ml), and SCF (50 ng/ml).
5. Dilute the cells to a concentration of 0.2×10^6 /ml in 12-well plates (2 ml each well).
6. Place the plates in the incubator at 37°C. At this stage, it is possible to obtain $2–4 \times 10^6$ cells (**Fig. 19.1**). An aliquot of cells (1×10^6) could be used for RNA extraction (*see* **Note 3**).

3.1.4. Day 6, 8, 10, 12, and 14

Cultures should be examined daily, observing the morphology, the color of the medium and the density of the cells.

1. Count cells and adjust to keep a concentration of 0.20×10^6 cells/ml.
2. Harvest cells (at least 1×10^6 cells) to obtain RNA to perform microarray analysis.

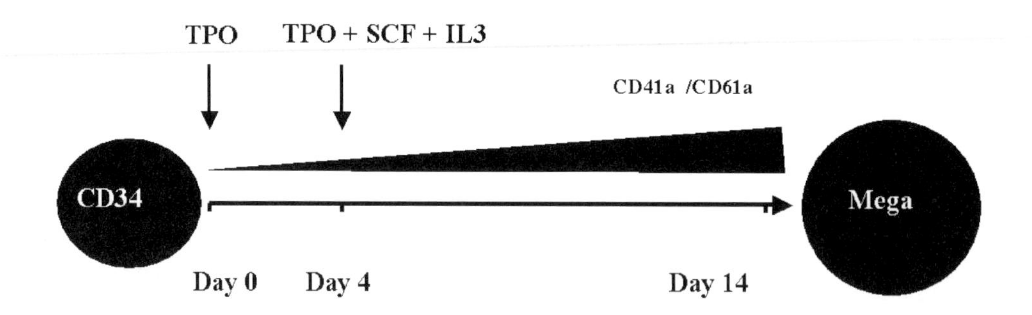

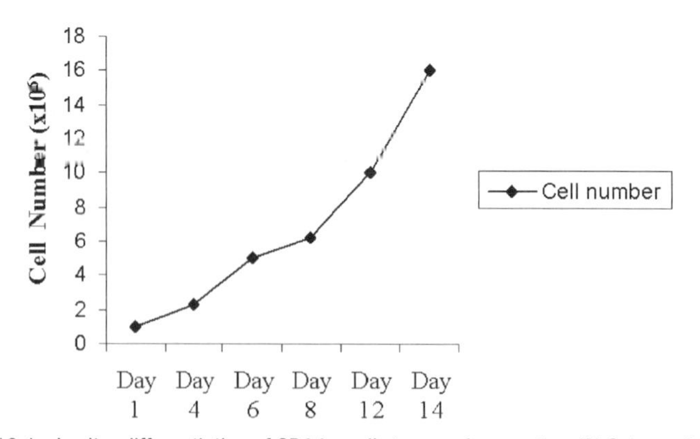

Fig. 19.1. In vitro differentiation of CD34+ cells to megakaryocytes. (**A**) Schematic representation of the culture of CD34+ stem cells in vitro. On top, different cytokines are shown in relationship to the day when they are added to the media. The specific markers for megakaryocytic differentiation CD41a and CD61a increased during the differentiation process more than 85% with respect to CD34+ stem cells. (**B**) Number of differentiated CD34+ cells in culture with the described protocol in a representative experiment. The number of cells is shown as $n \times 10^6$.

3. For characterization of megakaryocyte-specific antigens (CD41a, CD61a, CD42) using FACS analysis, we also recommend to use a minimum of 1×10^6 cells.
4. After 2 weeks in culture, the cells will exhibit high expression of CD41a, CD61a, and CD42 and low expression of CD34 (**Fig. 19.1A**).
5. The total amount of cells obtained after 2 weeks of culture is between 10 and 20×10^6 (**Fig. 19.1B**).

3.2. MiRNA Profiling of Megakaryocyte Precursors

To perform high-throughput miRNA expression analysis of megakaryocyte precursors we used a custom miRNA microarray chip, which is based on a one-channel system (OSU_CCC version 2) *(6)*. The chips contain gene-specific oligonucleotide probes, corresponding to the 250 human mature and precursor miRNAs in quadruplicate spotted by contacting technologies and covalently attached to a polymeric matrix.

3.2.1. Target Preparation

1. For RNA extractions we use Trizol reagent according to the manufacturer protocol.
2. Five micrograms of total RNA is separately added to a reaction mix with a final volume of 12 μl, containing 1 μg of [3′-(N)8-(A)12-biotin-(A)12-biotin-5′] random oligonucleotide primer (*see* **Note 4**).
3. Incubate the mixture for 10 min at 70°C and chill on ice.
4. With the mixture remaining on ice, 4 μl of 5× first-strand buffer, 2 μl of 0.1 M DTT, 1 μl of 10 mM dNTP mix, and 1 μl of SuperScript II RNaseH- reverse transcriptase (200 units/μl) are added to a final volume of 20 μl, and the mixture is incubated for 90 min in a 37°C water bath.
5. After incubation for first-strand cDNA synthesis, 3.5 μl of 0.5 M NaOH/50 mM EDTA was added into 20 μl of first-strand reaction mix and incubated at 65°C for 15 min to denature the RNA/DNA hybrids and degrade RNA templates.
6. Then, 5 μl of 1 M Tris–HCI, pH 7.6 is added to neutralize the reaction mix, and labeled targets are stored at –80°C prior to hybridization.

3.2.2. Array Hybridization

1. Hybridize labeled targets on each miRNA chip in 6× SSPE/formamide at 25°C for 18 h.
2. Wash chips in 0.75× TNT at 37°C for 40 min.
3. Process chips by using direct detection of the biotin-containing transcripts by Streptavidin-Alexa 647 conjugate (*see* **Note 5**).

3.2.3. Array Scanning

Processed slides are scanned using a GenePix Axon 4000B microarray scanner, with the laser set to 635 nm, at fixed PMT setting of 800, and a scan resolution of 10 mm.

4. Notes

1. It is critical to use high-quality CD34+ stem cells. These cells could be obtained commercially from Allcells.com or Stem-cell technologies. Other cheaper alternatives include the use of cord blood-selected CD34+ cells. Failure to use high-quality products will result in decreased numbers and low percentages of differentiated megakaryocytes.
2. Other major issue is bacterial contamination, due to the lack of use of antibiotics in the culture. Extra precautions should be taken to avoid contamination of the cultures.
3. Differentiated cells at early stages of the culture (days 4–8) may need to be sorted for CD41a + CD34- to obtain homogenous populations for microarrays analysis.
4. In the event that there is not enough RNA to obtain cDNA, it is possible to use successfully as little as 2.5 ug of total RNA, according to a previous report from our group *(6)*.
5. Experimental design. When designing experiments the researcher should consider to include proper controls, assure all the samples are hybridized in the same batch (one array can hold up to 100 different samples) and in triplicates whenever possible. The remaining material from each sample used for the microarray should be stored at –80°C to be used for validation of the microarrays using a different technology.

References

1. Bartel, D. (2004) MicroRNAs: genomics, Biogenesis, Mechanism, and Function. *Cell* 116, 281–297.
2. Garzon, R., Pichiorri, F., Palumbo, T., et al. (2006) MicroRNAs fingerprints during human megakaryocytopoiesis. *Proc Natl Acad Sci USA* 103, 5078–5083.
3. Chen, C. Z., Li, L., Lodish, H., et al. (2004) MicroRNAs Modulate Hematopoietic lineage Differentiation. *Science* 303, 83–86.
4. Felli, N., Fontana, L., Pelosi, C., et al. (2005) MicroRNAs 221 and 222 inhibit normal erithropoiesis and erythroleukemic cell growth via kit receptor down-modulation. *Proc Nat Acad Sci USA* 102, 18081–18086.
5. Fazi, F., Rosa, A., Fatica, A., et al. (2005) A mini-circuitry comprising microRNA-223 and transcription factors NFI-A and C/EBPa regulates human granulopoiesis. *Cell* 123, 819–831.
6. Liu, C. G., Calin, G. A., Meloon, B., et al. (2004) An oligonucleotide microchip for genomic-wide microRNA profiling in human and mouse tissues. *Proc Natl Acad Sci USA* 101, 11755–11760.
7. Guerriero, R., Testa, U., Gabbianelli, M., et al. (1995) Unilineage megakaryocytic proliferation and differentiation of purified hematopoietic progenitors in serum-free liquid culture. *Blood* 86, 3725–3736.

Chapter 20

Serial Analysis of Gene Expression Adapted for Downsized Extracts (SAGE/SADE) Analysis in Reticulocytes

Béatrice Bonafoux and Thérèse Commes

Abstract

Reticulocytes are the last stage of erythropoiesis before red blood cells (RBC). Although most of the RBC proteins have been characterized, little is known about expression profile of their mRNA during differentiation. Our goal was to initiate a genomic global solution to provide a transcriptional data base on which it will be possible to follow the erythroid differentiation and study RBC disorders. We used a modified protocol of serial analysis of gene expression (SAGE), the serial analysis of downsized extracts (SADE) which allows identification of genes from small amounts of mRNA and is thus an appropriate method starting from reticulocyte. SAGE does not depend on previous knowledge of genes. The method produces a set of cellular transcribed gene signatures (tags) and provides a comprehensive view of cells phenotypes by computational analyses.

Key words: SAGE, SADE, reticulocyte, blood, mRNA, red blood cells, transcriptome.

1. Introduction

Of the approximately 100,000 genes in the human genome, only a fraction are thought to be active in each type of cell, but there are several thousand different types of cells in the human body and each has a unique pattern of gene expression. SAGE allows us to study thousands of genes simultaneously, to measure their expression, and to quickly identify the genetic differences between normal and tumor cells. SAGE is a technique used by molecular biologists to produce a snapshot of the messenger RNA population in a sample of interest. The original technique was developed by Velculescu at the Oncology Center of Johns Hopkins University, and was published in *Science* in 1995 *(1)*.

Peter Bugert (ed.), *DNA and RNA Profiling in Human Blood: Methods and Protocols, vol. 496*

DOI 10.1007/978-1-59745-553-4_20 Springerprotocols.com

Elalouf *et al.* *(2)* set up a SAGE adaptation for down-sized extracts, enabling a 1,000-fold reduction of the amount of starting material. The potential of this approach was evaluated by studying gene expression in microdissected kidney tubules (50,000 cells). It is concluded that SAGE adaptation for down-sized extracts makes possible the large-scale quantitative gene expression measurements in small biological samples and will help to study the tissue expression and function of genes not evidenced with other high-throughput methods.

For the construction of a SAGE/SADE library of reticulocytes *(3)* we considered the following points: (1) we first choose healthy adult donors of reticulocytes; (2) we purified reticulocytes from whole blood, delicate step like seen in other SAGE library which all contain transcripts from reticulocytes (e.g., HBB); (3) SAGE/SADE method was employed; and (4) sequence analysis was realized by C+tag and Preditag softwares for tag sequence data analysis and tag to gene mapping, respectively *(4, 5)*.

2. Materials

2.1. Reticulocyte Purification

1. VACUETTE® heparin tube (Greiner bio-one, Courtaboeuf).
2. Syringe (10 ml) (Terumo, Guyancourt).
3. α-Cellulose (Sigma-Aldrich).
4. Microcrystalline cellulose (Sigma-Aldrich).
5. Whatman no. 1 (VWR).
6. Hepes buffer (pH 7.4): 154 mM NaCl, 10 mM Hepes, 1 g/L BSA.

2.2. RNA Extraction

1. TRIzol® (Life Technologies).
2. Chloroform.
3. Isopropyl alcohol.
4. Ethanol (70%) (prepared with molecular biology grade water).
5. RNase-free water.
6. Glycogen for molecular biology (Boehringer Mannheim).

2.3. SAGE/SADE Analysis

1. Dynabeads mRNA DIRECT™ kit (Dynal).
2. Magnetic particle concentrator (MPC; Dynal).
3. cDNA synthesis kit for first and second strand cDNA (Invitrogen).
4. *Sau*3AI (4 U/μL) (Promega).
5. TEN: 10 mM Tris–HCl, pH 8; 1 mM EDTA; 1 M NaCl.
6. TE: 10 mM Tris–HCl, pH 7.5; 1 mM EDTA.
7. loTE: 3 mM Tris–HCl, pH 7.5; 0.2 mM EDTA.
8. BSA (bovine serum albumin).
9. Linkers (**Table 20.1**).
10. T4 DNA ligase 2,000,000 cohesive end units/ml (NEB).

Table 20.1
Linkers and PCR primers

Name	Sequence (5′ to 3′)
Linker 1A	TTTTGCCAGGTCACTCAAGTCGGTCATTCATGTCAGCACAGGGAC
Linker 1B	GATCGTCCCTGTGCTGACATGAATGACCGACTTGAGTGACCTGGCA
Linker 2A	TTTTTGCTCAGGCTCAAGGCTCGTCTAATCACAGTCGGAAGGGAC
Linker 2B	GATCGTCCCTTCCGACTGTGATTAGACGAGCCTTGAGCCTGAGCAA
Primer 1	GCCAGGTCACTCAAGTCGGTCATT
Primer 2	TGCTCAGGCTCAAGGCTCGTCTA

11. *Bsm*FI (2000 U/ml) and NEB IV (NEB).
12. Ethanol (75 and 100%).
13. Isopropyl alcohol.
14. PCI: Phenol/chloroform/isoamyl alcohol (25:24:1) (Promega).
15. Glycogen (Boehringer Mannheim).
16. Ammonium acetate (3 M).
17. Blunt-ending T7 DNA polymerase kit (Promega).
18. AmpliTaq Gold (Applied Biosystem).
19. Ampliwax® PCR Gem 100 (Perkin Elmer).
20. Primers 1 and 2: unmodified for first step "12 PCR" and 5′-biotinylated for second step "200 PCR" (**Table 20.1**).
21. Ammonium persulfate (APS) 10%.
22. TEMED.
23. Acrylamide/bis, 19:1 and 37.5:1 (Sigma-Aldrich).
24. Costar® Spin-X® Centrifuge Tube Filters, 0.22 μm pore CA membrane corning.
25. Bromophenol blue loading buffer (BB): 0.125% bromophenol blue, 10% Ficoll 400, 12.5 mM EDTA (filtered on 0.45 μm membrane).
26. SYBR Green® I (Invitrogen).
27. TAE 1×: 40 mM Tris–acetic acid, pH 8.0, 1 mM EDTA.
28. Streptavidin magnetic beads (Promega).
29. W&B buffer (2×): 10 mM Tris–HCl, pH 7.5, 1 mM EDTA, 2 M NaCl.
30. *Mbo*I (20 U/ml) (NEB).

3. Method

3.1. Reticulocyte Purification

One of the difficulties of SAGE library construction is to obtain a purified cell population. This artifact is observed in all SAGE libraries from fresh tissues, most of them containing *HBB* transcripts, the gene of the beta globin chain, one of the most abundant proteins in RBC.

3.1.1. Adult Donors Selection

1. Five men and five women healthy Caucasian adult volunteers gave fresh whole blood, after consent.
2. They had normal blood cell counts, blood smears, hemoglobin electrophoresis, red cell membrane resistance tests, no martial deprivation, and no biological signs of hemolysis (serum haptoglobin, LDH, total serum bilirubin were normal).
3. A volume of 4.5 ml of venous blood in heparin tube was drawn from each donor.

3.1.2. Reticulocytes Purification

The aim of this step is to eliminate leukocytes and reticulated platelets, containing mRNA. The Beutler technique was used for purification *(6)*. The principle of this technique is based on separation of cells by filtration through a cellulose column.

1. Spin each heparin tube of fresh whole blood, 5,000*g* for 5 min, remove plasma and buffy coat.
2. At the same time, prepare the cellulose column: Cut a disc of filter paper Whatman with a diameter equal to that of a 10 ml syringe, put it at the bottom of the syringe.
3. Fill the syringe by the mixture of cellulose (equal volume of α-cellulose and of microcrystalline cellulose suspended in Hepes buffer), maintain the syringe vertically.
4. Filter blood cells through the column, carry out Hepes buffer in order to obtain a final volume equal to the starting volume. Wash cells three times in saline solution.
5. After filtration, a typical reticulocyte count of 50,000 /μL is obtained. Other typical counts are given in **Table 20.2**.

3.2. Total RNA Extraction

We used TRIzol® reagent for the isolation of total RNA from cells, an improvement of the single-step RNA isolation method

Table 20.2
Typical blood cell counts in whole blood (A) and after filtration through cellulose column (B)

Cell type	A	B
Red blood cells (/μL)	4×10^6	4×10^6
Reticulocytes (/μL)	50,000	50,000
Leukocytes (/μL)	6,000	<200
Leukocytes/red blood cells	0.0015 (0.15%)	0.00005 (0.005%)
Leukocytes/reticulocytes	0.12 (12%)	0.004 (0.4%)
Platelets (/μL)	300,000	Undetectable

developed by Chomczynski and Sacchi *(7)* and allowing manufacturer's instructions. After purification by the column, each blood sample was treated separately.

1. Spin the cells at 1,000*g* for 3 min and remove supernatant.
2. Add 10 ml of TRIzol® and vortex the sample.
3. Add 2 ml of chloroform, incubate 2–3 min at room temperature (RT) and spin for 15 min at 12,000*g* at 4°C. Transfer the aqueous phase in a fresh tube.
4. Add 5 ml of isopropyl alcohol and incubate for 5 min at RT.
5. Add 10 μL of glycogen. Vortex and spin for 10 min at 12,000*g*.
6. Gently remove the supernatant, leaving a rest of about 50 μL in the tube.
7. Wash the RNA pellet once with 1 ml of ethanol (75%), spin 10 min at 7,500*g* at 4°C, remove supernatant, air dry and dissolve in 20 μL H_2O RNase-free.
8. Pool total RNA from ten samples. Keep at –80°C (*see* **Note 1** and **2**).

3.3. SAGE/SADE Analysis

A detailed description of the SAGE method is available: *http://www.sagenet.org*. Three principles underlie the SAGE methodology. First, a major principle of SAGE was to reduce sequence tags to their minimal size by using a type IIS restriction enzyme *(1)*. Associated with the positional information, this size (14 pb in conventional method) is usually sufficient for identifying each transcript, considering that 1,048,576 (10^4) different sequences can be distinguished. The strategy used to amplify the whole population of tags preserves the same quantitative distribution as in the initial mRNA population. Second, another characteristic of SAGE was to assemble tags into concatemers before DNA sequence analysis so that multiple tags (20–30 or more) could be read from each sequencing lane. Third, quantification of the number of times a tag is observed provides the expression level of the corresponding transcript.

The main steps of SAGE are as follows:

1. Trap mRNA with magnetic beads and convert mRNA into cDNA.
2. Make a cut in each cDNA with "anchoring enzyme" so that there is a broken end sticking out. Attach a "docking module" or a "linker" to this end; here a new enzyme (called "tagging enzyme") can dock, reach down the molecule, and cut off a short tag. Combine two tag in a unit, a ditag. Make billions of copies of the ditags using PCR.
3. Remove the modules and glue the ditags together into long concatemers. Put concatemers into bacteria and copy them millions of times. Pick the best concatemers and sequence them.

4. Use software to identify how many different cDNAs are there and count them. Match the sequence of each tag to the gene that produces the RNA. The SAGE adaptation for downsized extracts (SAGE/SADE) method was established by Elalouf and collaborateurs *(2)* in order to reduce RNA quantity (*see* **Note 3**).
   The main modifications are the following:
5. Total RNA is extracted from cells and pooled. Then, mRNA is purified with oligo(dT) magnetic beads.
6. "Anchoring" enzyme *Sau*3AI is used instead of *Nla* III. Moreover, *Sau*3AI has no restriction site on *HBA1* and *HBA2* mRNA. This procedure eliminates all mRNA of alpha globin, the most abundant protein known in the reticulocyte.
7. T7 DNA polymerase for blunt-ending cDNA tags.
8. Biotinylated PCR primers are used for ditags amplification and purification.
9. Only one step of PCR amplification of ditags is done in the SAGE method. In SAGE/SADE, a first step of PCR amplification (12 PCRs of 28 cycles) is followed by a purification step on acrylamide gel and a second PCR amplification.
10. Concatemers are cloned into pBluescript II vector, after linearization with *Bam*HI and dephosphorylation.

In the following paragraphs, we describe in detail the steps of the SAGE/SADE method.

3.3.1. mRNA Isolation Using Magnetic Beads

Poly(A) RNAs were isolated from purified total RNA (total volume of 300 μL) through hybridization of 500 μL magnetic beads coupled with oligo-dT, and following manufacturer's instructions. The protocol was adapted to reticulocytes mRNA purification (*see* **Note 4**).

1. Dispense 500 μL of beads in a fresh tube and wash them twice with the binding buffer.
2. Add 200 μL of total RNA, anneal mRNA to the beads by incubating 10 min at RT on a rocking platform.
3. Using the MPC, remove the supernatant and perform the washes (each wash buffer contains 20 μg/ml glycogen): twice with 600 μL washing buffer containing lithium dodecyl sulfate, twice with 300 μL washing buffer alone and twice with 300 μL 1× first strand reaction buffer.

3.3.2. cDNA First and Double Strand Synthesis

1. Prepare first strand reaction buffer: 10.5 μL H_2O, 4 μL 5× first strand buffer, 2 μL 10 mM DTT, 1 μL 100 mM each dNTP, 1 μL oligo-dT primer (500 ng/μL).
2. Resuspend the beads containing mRNA in 20 μL first strand mix and incubate at 42°C for 2 min, then at 37°C for 1 min.
3. Add 2.5 μL of Superscript II reverse transcriptase and incubate at 37°C for 1 h.
4. Chill the first strand reaction on ice for 2 min.

5. Prepare the second strand reagents on ice: 89 μL RNase-free H_2O, 30 μL 5× second strand buffer, 3 μL 10 mM each dNTP, 1 μL *Escherichia coli* DNA ligase, 4 μL *E. coli* DNA polymerase I, 1 μL *E. coli* RNase H, 0.5 μL glycogen (5 μg/ml).
6. Add the mix to the beads and incubate at 16°C for 2 h or over-night.

3.3.3. Digesting with Anchoring Enzyme *Sau*3AI

1. Wash the beads, four times with 200 μL TEN, 1% BSA, twice with 200 μL *Sau*3AI 1× buffer and twice with 400 μL 1× *Sau*3AI buffer.
2. Add to the beads the mix for *Sau*3AI digestion: 173 μL H_2O, 20 μL 10× *Sau*3AI buffer, 2 μL 1% BSA, 5 μL *Sau*3AI (20 U). Incubate 3 h at 37°C, vortex intermittently after 1 and 2 h.
3. Place the tube on MPC: keep the supernatant (5′ end) to check for cDNA synthesis and *Sau*3AI digestion by PCR with appropriated *HBB* primers (5′ and 3′ ends).

3.3.4. Ligating Adapters to the cDNA

1. Wash the bound fraction (3′ end) twice with 75 μL *Sau*3AI buffer: 178 μL H_2O, 20 μL 10× *Sau*3AI buffer, 2 μL 1% BSA and twice with 200 μL TEN, 1% BSA.
2. Resuspend the beads in 200 μL TEN, 1% BSA and divide the beads (cDNA) into two fractions of 100 μL called 1A/B and 2A/B.
3. Prepare the mix of the linkers (1A/B and 2A/B): 29 μL H_2O, 1 μL linker 1A and 1B or 2A and 2B (7 pmol each), 4 μL 10× ligase buffer.
4. Add 34 μL of the mix to each fraction 1A/B and 2A/B.
5. Incubate 5 min at 45°C, put on ice.
6. Add 4 μL 10 mM ATP, 2 μL T4 ligase (5 U/μL) and incubate at 16°C over-night.

3.3.5. Cleaving with "Tagging" Enzyme *Bsm*FI and Blunt-Ending

1. Wash the beads three times with 200 μL TEN, 1% BSA and three times with 200 μL NEB IV, 1% BSA. It is important to wash the beads extensively to eliminate unligated linkers.
2. Prepare the mix for *Bsm*FI digestion : 87 μL H_2O, 1 μL 1% BSA, 10 μL 10× NEB IV buffer, 2 μL *Bsm*FI. Add 100 μL in each tube and incubate 2 h at 65°C, and then 5 min at RT.
3. Collect supernatant which contains tags and wash the beads twice with 75 μL iced TE, 1% BSA.
4. Pool the supernatants and purify each one with 250 μL PCI and 60 μg glycogen, centrifuge for 10 min at 10,000*g* at 4°C.
5. Transfer the aqueous phase to a fresh tube and ethanol precipitate by adding 125 μL of 10 mM ammonium acetate and 1.125 ml of 100% ethanol.

6. Centrifuge for 20 min at 15,000*g* at 4°C, and wash the pellet twice with 400 μL 70% ethanol, vacuum dry and suspend in 10 μL loTE.
7. Prepare and add 15 μL of the following mix: 40 mM Tris–HCl (pH 7.5), 20 mM $MgCl_2$, 50 mM NaCl in each tube and incubate for 2 min at 42°C.
8. Prepare and 25 μL of the following mix: 7.5 μL H_2O, 5.5 μL 0.1 M DTT, 11 μL dNTP mix (2 mM each), 1 μL T7 DNA polymerase. Incubate 10 min at 42°C and pool the fractions.
9. Rinse the two tubes with 150 μL loTE and 20 μg glycogen and add to the pooled reaction (final volume: 250 μL).
10. Purify with 250 μL PCI, precipitate with ethanol 100%, wash twice with ethanol 75%, vacuum dry and suspend in 6 μL loTE.

3.3.6. Create Ditags and PCR Amplification (first step – "12 PCR")

1. Ligation is made with 1 μL of T4 DNA ligase (5 U/μL), 2 μL 5× ligase buffer, 1 μL 10 mM ATP in 10 μL final volume. Incubate at 16°C over-night.
2. Add 90 μL loTE.
3. Perform a semi-quantitative PCR to estimate the optimal number of cycles. We have successfully used standard PCR reactions and *Taq* polymerase from various companies, the annealing temperature for the PCR primers is 58°C. The PCR master mix (Promega) per reaction is as follows: 5 μL 10× buffer, 67 μL H_2O; 8 μL 25 mM *$MgCl_2$*, 8 μL dNTPs (5 mM each), 2 μL each primer (50 pmol), 1 μL *Taq* polymerase (5 U/μL), 2 μL of ditags per reaction.
4. Estimation for PCR cycles number made on agarose gel (between 18 and 26 cycles). Load 10 μL of PCR products to the gel. Here, we chose 26 cycles.
5. Then, run 12 PCRs with same parameters including 2 μL of ditags. Pool all the PCR products obtaining a total volume of 1.2 ml. Load 10 μL of PCR products on a 2% agarose gel.
6. Divide PCR products into two parts of 600 μL. Purify and precipitate as described in **Section 3.3.5.** Wash with 1 ml 75% ethanol; air dry and dissolve in 80 μL loTE and 54 μL BB.

3.3.7. Purifying the 100 bp Ditags on Acrylamide Gel

1. Prepare a 12% acrylamide gel (1.5 mm thickness): 23.7 ml H_2O, 10.5 ml acrylamide 19:1, 0.7 ml TAE 50×, 30 μL TEMED, and 350 μL APS 10%.
2. Load the whole of the 107 μL ditags in the six wells of the gel. Run the gel for about 3 h at 160 V (42–45 mA) until BB is about 12 cm distant from the wells.
3. Reveal with 10 μL SYBR Green® I in 100 ml TAE 1×, stain 20 min in obscurity. Visualize on UV box and cut the six bands at 110 bp (**Fig. 20.1**).

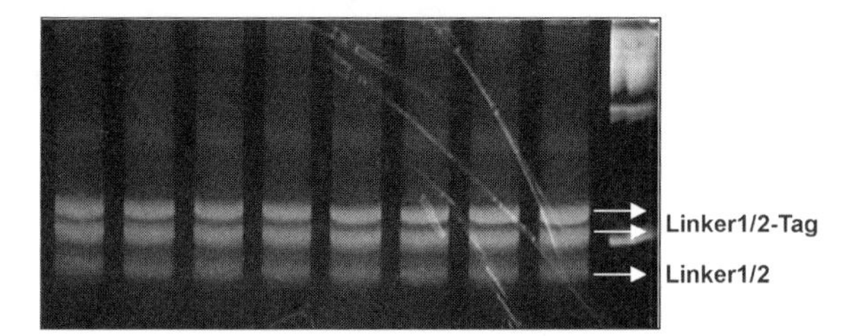

Fig. 20.1. Purifying the 110-bp ditags on acrylamide gel.

4. Pierce the bottom of two 0.5-ml tubes with a 18-gauge needle, put three bands per tube. Put each tube into a 2-ml tube and add a mix of: 474 μL loTE, 25 μL ammonium acetate, 1 μL glycogen. Spin for 5 min at 10,000 rpm.
5. Add 250 μL of the same mix at the bottom of the 2-ml tube. Incubate at 37°C for 1 h.
6. Use four Costar® Spin-X® centrifuge tube filters and add 1 μL glycogen to the bottom of the tubes. Load about 250 μL of the ditags on top of each filter. Spin for 10 min at 15,000 rpm.
7. Pool all ditags in a 2-ml tube (final volume is about 1 ml). Purify and ethanol precipitate as described. Wash twice with 1 ml ethanol 75%, vacuum dry and suspend in 300 μL loTE.

3.3.8. Amplification of Ditags by Second PCR "200 PCRs" Using Biotinylated Primers

1. Set up two series each of 100 PCRs including the purified ditags with the same parameters as described in **Section 3.3.7**. Instead of unmodified primers 1 and 2 use 5′-biotinylated primers.
2. After PCR, pool all PCR products (total volume: about 10 ml).
3. Purify with PCI as described, precipitate with isopropanol. Wash once with 1 vol ethanol 75%. Vacuum dry and suspend in 160 μL loTE.

3.3.9. Ditag Isolation by Digesting 110-bp Ditags with MboI

1. Prepare the following mix: 18.5 μL 10× NEB3 buffer, 1.8 μL 1% BSA, 9 μL *Mbo*I.
2. Add to 159 μL ditags. Save 1 μL of the ditags (110 bp DNA fragment) for check.
3. Incubate 7.5 h at 37°C.
4. Check for *Mbo*I digestion on 3% agarose gel: analyze 1 μL of the uncut fragment and 3 μL of the digestion product showed that most of the 80% the 110 bp fragment has been digested.
5. A second acrylamide electrophoresis is carried out with 12% acrylamide gel allowing to purify fragments from 22 to 28 bp (ditags). Purify, precipitate, and wash as described in **Section 3.3.8.**
6. For each series of 100 PCRs suspend in a 50 μL 1× W&B buffer and pool.

7. Wash 100 μL of streptavidin beads with 100 μL of 1× W&B on the MPC and add 100 μL of the ditags to 100 μL of the beads.
8. Wash once with 100 μL and then with 200 μL of 1× W&B buffer, pool the supernatants (about 500 μL) and add 2 μL of glycogen.
9. Purify by PCI, precipitate with isopropanol, wash twice with 1 ml ethanol 75%, vacuum dry, and dissolve in 6 μL H_2O.

3.3.10. Concatemer Synthesis and Separation on Acrylamide Gel

1. Prepare the mix for the ligation: 1 μL 10× ligase buffer, 1 μL 10 mM ATP, 1 μL T4 DNA ligase, 1 μL H_2O.
2. Add the 6 μL ditags and incubate at 16°C over-night.
3. Prepare a 8% acrylamide gel (1 mm thickness): 31 ml H_2O, 8 ml acrylamide 37.5:1, 800 μL 50× TAE, 35 μL TEMED, and 400 μL 10% APS.
4. Incubate the ligation reaction for 5 min at 55°C then for 5 min on ice.
5. Add 4 μL BB and load the total volume into one well of the gel. Run the gel for about 2.5 h at 160 V (42–45 mA) until BB is about 12 cm distant from the wells.
6. Reveal with SYBR Green® I and cut the fractions >1000 bp.
7. Purify with pierced tubes and Costar® Spin-X® Centrifuge Tube Filters as described in **Section 3.3.7**.
8. Purify with PCI, precipitate with isopropanol, dissolve in 6 μL H_2O, and generate the first library with concatemers of 600–2000 bp.
9. Use 2 μL for ligation to the vector and store the remaining 4 μL at 4°C.

3.3.11. Cloning and Screening

Concatemers can be cloned and sequenced in different vectors with a *BamHI* site using classical method. For the study we used pBluescript II (NEB, Beverly, USA) vector and Electromax DH10B. PCR screening was performed using M13 sense and antisense primers. Length of fragments were controlled and only those whose size is higher than 500 bp (or 12 tags) are used for sequencing (*see* **Note 5**).

3.3.12. DNA Sequencing and SAGE Data Analysis

1. Sequencing reactions were performed on DNA minipreps by using Big Dye terminator sequencing chemistry and run on 377-XL Applied Biosystems automated sequencers (Genome Express, Meylan, France).
2. C+tag (Skuld-tech) was used for automatic tag detection and counting. This program provided criteria for assessing the quality of the SAGE libraries (length distribution of ditags, frequency of replicate ditags, detection of linkers). Ditags with less than 20 bp were discarded and repeated ditags were not taken into account for calculation of tag numbers (*see* **Note 6**).

3. For tag-to-gene mapping, we downloaded Hs.seq.uniq and Hs.data files from the UniGene FTP site at NCBI (ftp://ncbi.nlm.nih.gov/repository/UniGene). A table (Preditag) was constructed by extracting virtual tags from the representative sequences associated with each UniGene cluster in Hs.seq.uniq files, then by parsing attributes associated to each cluster in Hs.data files as previously described *(3)*. Microsoft Access functions were used for tag-to-gene assignment and subsequent data management. A query using tag sequence as the primary key allowed us to match experimental sequences, virtual sequences (Preditag), and selected annotations, thus generating a table of results. SAGE reticulocytes library was edited manually to complete Unigene data.

4. Notes

1. The majority of the common cells contain approximately 5 μg of total RNA per 10^6 cells. This account is sufficient to realize the SAGE method. Concerning reticulocyte, we estimated that its RNA content is lower. For this reason, we chose to purify a higher number of reticulocytes in order to obtain at least an equivalent quantity in total RNA. Here, we purified 2.25 $\times 10^9$ reticulocytes.
2. Purity and integrity of RNA are critical elements for gene expression profiling. A check of quality of total RNA in our experiments was performed using agarose gel electrophoresis on test samples. The proportion of the ribosomal bands (28S/18S) is conventionally viewed as the primary indicator of integrity.
3. Several methodological improvements have been recently published concerning SAGE. The method has benefit from progress in DNA sequencing technology increasing the rate of acquisition and reducing the cost of analysis. We recently report data of a modified protocol and the successful high-throughput sequencing, preserving the main advantages of the digital method in terms of measurement of expression levels *(8)*. In contrast to the original SAGE library procedure based on concatemer preparation, this application could potentially allow one individual to prepare and sequence several SAGE libraries in a few days. As described, building a SAGE library requires several days of skillful work, (synthesis of cDNA, ditags purification and PCR amplification, ligation of concatemers inserted in cloning vectors, and sequenced). The last steps (from **Section 3.3.8**) are minimized in terms of time-scale, because ditags are processed directly as "molecular clones". The need to concatenate SAGE tags is inherent to the use of the classical Sanger method, which is economically

realistic only if several hundred nucleotides are sequenced in each run. Since the high-throughput sequencing (or massive sequencing) technologies allow hundreds of thousands individual DNA fragments to be processed at once by parallel sequencing, there is no need to concatenate them. The new strategy consists to simplify all the steps by preparing ditags and directly sequenced them after adaptator ligation as for experimental conditions designed for processing randomly fragmented genomic DNA fragments *(8)*.

4. A high rate of rRNA may be observed due to the low abundance of mRNA. Purified mRNA isolated with oligo(dT) magnetic beads are devoided of rRNA from commun cell types. Starting from purified reticulocytes, we found a significant amount of nuclear-encoded rRNAs corresponding to 5% of the tags. The tags were manually annoted and extracted from the analysis.
5. Purification and concatenation of SAGE ditags are critical for optimal performance. This is critical to obtain long concatemers to facilitate the sequencing step.
6. The major problems encountered in this study are related to the specificity of the cell type. Reticulocytes are cells with low mRNA levels comprising a high rate of hemoglobin (HBB) transcripts. A significant number of repeated ditags: The C+tag program provides criteria for assessing the quality of the SAGE library, such as ditags (ditags are the combination of two tags into a unit), frequency of repeated ditags, and detection of linkers. For the definition of these terms see *http://www.embl-heidelberg.de/info/sage*. The rate of contamination by linker sequences was usually less than 0.5%. For this particular reticulocyte library, repeated ditags (mainly from the *HBB* gene) represented 40% of the total ditag population. A high frequency reveals an artifactual loss of complexity during the library construction or, as expected in this cell, a low complexity of the original mRNA population observed in highly specialized tissues. In the aim to determine the most abundantly expressed genes and without the goal to quantitative comparison; such a library could be sequenced and explored. With the same protocol, the amount of repeated ditags was less than 1.5% in libraries constructed in our laboratory.

Acknowledgments

We acknowledge the support of the Languedoc-Roussillon Génopole® for this work.

References

1. Velculescu, V. E., Zhang, L., Vogelstein, B., et al. (1995) Serial analysis of gene expression. *Science* 270, 484–487.
2. Virlon, B., Cheval, L., Buhler, J. M., et al. (1999) Serial microanalysis of renal transcriptomes. *Proc Natl Acad Sci USA* 96, 15286–15291.
3. Bonafoux, B., Lejeune, M., Piquemal, D., et al. (2004) Analysis of remnant reticulocyte mRNA reveals new genes and antisense transcripts expressed in the human erythroid lineage. *Haematologica* 89, 1434–1438.
4. Piquemal, D., Commes, T., Manchon, L., et al. (2002) Transcriptome analysis of monocytic leukemia cell differentiation. *Genomics* 80, 361–371.
5. Quere, R., Manchon, L., Lejeune, M., et al. (2004) Mining SAGE data allows large-scale, sensitive screening of antisense transcript expression. *Nucleic Acids Res* 32, e163.
6. Beutler, E., Gelbart, T. (1986) The mechanism of removal of leukocytes by cellulose columns. *Blood Cells* 12, 57–64.
7. Chomczynski, P., Sacchi, N. (1987) Single-step method of RNA isolation by acid guanidinium thiocyanate-phenol-chloroform extraction. *Anal Biochem* 162, 156–159.
8. Quéré, R., Manchon, L., Pierrat, F., et al. (2007) Rapid and accurate pyrosequencing of serial analysis of gene expression ditags. Application Note N°4 Roche diagnostics.

Chapter 21

Real-Time PCR Analysis for Blood Cell Lineage Specific Markers

Louise Edvardsson and Tor Olofsson

Abstract

We here describe the methods for the isolation of distinct hematopoietic subpopulations, as defined by their immune phenotype by fluorescence-activated cell sorting, and how these cells can be analyzed even at a single-cell level for the gene expression of a number of transcription factors and other differentiation markers.

Key words: Hematopoiesis, transcription factors, cytokines, growth factor receptors, cell sorting, RT-PCR.

1. Introduction

Characterization of hematopoietic cell differentiation and maturation has evolved rapidly over the last decades. This has been driven by firstly, the development of a large number of monoclonal antibodies for detection of surface antigens applied in flow cytometric analysis, and secondly by the availability of a wide selection of specific primers and probes for analysis of gene expression patterns by real-time PCR (RT-PCR). The combination of fluorescence-activated cell sorting and quantitative RT-PCR now provides very powerful tools for characterization and analysis of even small subpopulations and small number of cells along the different hematopoietic maturation lineages.

Hematopoietic differentiation from stem cells (HSC) to mature blood cells occurs through a number of lineage choices and maturation steps governed by a set of transcription factors *(1–12)* and cytokines/chemokines and their respective receptors

Peter Bugert (ed.), *DNA and RNA Profiling in Human Blood: Methods and Protocols, vol. 496*

DOI 10.1007/978-1-59745-553-4_21 Springerprotocols.com

(13–17). The hematopoietic differentiation stages for isolation purposes are defined by their immune phenotype as analyzed by flow cytometry using a set of monoclonal antibodies, and there are several schemes established for isolation of stem cells, progenitor cells, and their progeny *(18–24)*.

Flow cytometry and cell sorting is the method of choice for the isolation of phenotypically defined populations of bone marrow and blood cells for functional assays and gene expression analysis. Freshly isolated cells are directly sorted into lysing buffer, taken to reverse transcription, and the cDNA analyzed by quantitative RT-PCR for a number of transcription factors and other differentiation markers, such as cytokine receptors or proteins synthesized at a specific maturation stage. We have used these methods to characterize human hematopoietic cells along the myeloid lineages.

2. Materials

2.1. Cell Isolation

1. Bone marrow from healthy volunteers after informed consent.
2. IMDM/FBS: Iscove´s Modified Dulbecco medium (IMDM) supplemented with 5% FBS (Gibco; Invitrogen Inc., Carlsbad, CA, USA).
3. Heparin, 5,000 IU/ml.
4. Lymphoprep^{TM} (Fresenius Kabi, Norway).
5. Indirect CD34 microbead kit and CD19 microbeads, LS columns, and MACS separator stand with magnet (Miltenyi Biotec, Bergisch Gladbach, Germany).
6. PBS/BSA/EDTA: sterile PBS, pH 7.2 supplemented with 0.5% BSA (w/v) and 2 mM EDTA (Sigma, St Louis, MO, USA).
7. Cell strainer (40 μm) and 5 ml polystyrene round-bottomed tubes with cell-strainer cap (Falcon^{TM}; BD, Becton Dickinson, Franklin Lakes, NJ, USA).

2.2. Cell Sorting

1. Cell sorter, e.g., FACS Vantage of FACS Aria (BD) with ACDU (automated cell deposition unit).
2. Monoclonal antibodies; *see* **Table 21.1**.
3. Lysing solution for collection of sorted cells: add 5 μL NP-40 to 995 μL RNase free water. Vortex to make a homogeneous solution. This solution should be prepared immediately before cell sorting and kept on ice.
4. Collection tubes: 0.2 ml low profile thin-walled eight tube and flat cap strips (Low Profile Thermo-Strips; Abgene Ltd, Epsom, UK).
5. Strip holder: MicroAmp 96-well base (Applied Biosystems, Foster City, CA, USA) (*see* **Note 1**).

Table 21.1
Examples of transcription factors (TF), growth factor receptors (GFR), and other lineage markers useful for characterization of hematopoietic differentiation

Gene	Type of marker/protein name	MoAb	Assay ID[a]
TAL1	TF SCL		Hs 00268434_ml
SPI1	TF PU.1		Hs 00231368_ml
GATA1	TF GATA-1		Hs 00231112_ml
GATA2	TF GATA-2		Hs 00231119_ml
NFE2	TF NF-E2		Hs 00232351_ml
CEBPA	TF C/EBPα		Hs 00269972_sl
CEBPE	TF C/EBPε		Hs 00357657_ml
KIT	Stem cell factor receptor	CD117	Hs 00174029_ml
FLT3	Flt3	CD135	Hs00174690_ml
IL3RA	IL-3Rα	CD123	Hs 00174356_ml
MPL	TpoR	CD110	Hs 00268434_ml
CSF2RB	β_C-Chain	CD131	Hs 00166144_ml
CSF3R	G-CSF receptor	CD114	Hs 01114427_ml
CSF2RA	GM-CSF receptor	CD116	Hs 00538896_ml
EPOR	Erythropoietin receptor		Hs 00181092_ml
GPIIB	Integrin GPIIb	CD41	Hs 00166246_ml
ABO	ABO transferase		Hs 00220850_ml
GYPA	Glycophorin A	CD235a	Hs 00266777_ml
GYPC	Glycophorin C	CD236R	Hs 00242583_ml
KEL	Kell antigen	CD238	Hs 00220850_ml
HBB	β-Globin		b
PRTN3	Proteinase 3		Hs 00160521_ml
LTF	Lactoferrin		Hs 00158924_ml
	18S rRNA		Hs 99999901_sl
B2M	Human β2-microglobulin		4326319E

[a]Assays-on-Demand Targets and Endogenous Control Assays (Applied Biosystems); [b]Forward primer: 5'-CACCTTTGCCACACTGAGTGA-3'; reverse primer: 5'-GTGATGGGCCAGCACACA-3'; probe: 5'-FAM-TGAGAACTTCAGGCTCCT-MGB-3'; when appropriate the corresponding monoclonal antibody (MoAB) for GFRs are given; other surface markers for cell sorting include CD45, CD34, CD133, CD7, CD19, CD13, CD33, CD15, CD14, CD11b, CD71, CD45RA.

2.3. Reverse Transcription and Real-Time PCR

1. Nuclease free water.
2. Nonidet P40 substitute (NP-40; USB Corporation, Cleveland, OH, USA).
3. RT buffer, 10× (Applied Biosystems).
4. TE buffer: 10 mM Tris–HCl, pH 8; 1 mM EDTA (Applied Biosystems).
5. dNTPs, 5 mM each (Applied Biosystems).
6. Random hexamers, 50 μM (Applied Biosystems).
7. RNAse inhibitor, 20 U/μL (Applied Biosystems).
8. Sensiscript RT Kit (Qiagen, Hilden Germany).
9. TaqMan gene expression assays (Applied Biosystems); *see* **Table 21.1**.
10. TaqMan PreAmp master mix or gene expression master mix (Applied Biosystems).
11. TaqMan Universal PCR master mix (Applied Biosystems).

3. Methods

3.1. Cell Isolation and Cell Sorting

1. Bone marrow cells obtained by aspiration (10 ml or more) are immediately transferred to a 50 ml tube containing an equal volume of IMDM with heparin (100 IE/ml) and mixed thoroughly.
2. Isolate bone marrow mononuclear cells (MNC) on a Lymphoprep gradient; 20 ml Lymphoprep in a 50 ml tube is overlayered with 20 ml bone marrow cell suspension and centrifuged for 20 min at 2,000 rpm in a swing out tube holder.
3. Carefully remove the interphase containing the MNC by pipetting, transfer to another 50 ml tube and add cold sterile PBS/BSA/EDTA to 40 ml and centrifuge for 10 min at 1,300 rpm; this will leave most of the platelets in the supernatant.
4. Resuspend the cells in 20 ml cold PBS/BSA/EDTA buffer, filter through a cell strainer 40 μm and make a cell count; spin down the cells for 10 min at 1,000 rpm.
5. Resuspend the cells in a small volume PBS/BSA/EDTA buffer (<500 μL).
6. The cells are now ready for isolation of CD34+ cells (*see* **Note 2**)
7. Use the Indirect CD34 microbead kit for isolation of CD34+ cells and follow the manufacturer´s instructions, keep buffers and cells cold and work fast without interruptions (*see* **Note 3**and **4**).
8. After elution of the CD34+ cells in 5 ml PBS/BSA/EDTA buffer, count the cells and add 5 ml of IMDM/FBS; spin down the cells at 1,000 rpm for 5 min.

9. Resuspend the cells in cold IMDM/FBS to a concentration of 10^7 cells per ml.
10. The cells are now ready for labeling with fluorochrome-conjugated monoclonal antibodies (e.g., FITC, PE, PE-TexasRed, PerCP-Cy5.5, PE-Cy7, APC, APC-Cy7, Pacific Blue, or any of the Alexa fluorochromes available).
11. Add 5–20 μL (as recommended by the manufacturer) of each antibody to the CD34+ cells, mix well, and incubate in a refrigerator for 15 min.
12. Wash the cells twice in 4 ml cold IMDM/FBS and resuspend in 500 μL of the same buffer.
13. During antibody labeling of the cells, prepare the lysing buffer for collection of the sorted cells; for lysing buffer, *see* **Sections 3.2.2** and 2.2.3. For single-cell sorting add 10 μL per tube and for 500 cells add 42 μL per tube and keep the strips on ice until sorting. Do not use the first and the last tube of the eight tube strips (*see* **Note 5**).
14. Before starting to sort, filter the cells through a 5 ml polystyrene round-bottomed tubes with cell-strainer cap (*see* **Note 6**).
15. Use a 70 μm nozzle for sorting to keep the addition of sheath fluid to the sample as little as possible and sort in "single-cell mode" at less than 1,000 events/s (*see* **Note 7**).
16. Sort the defined subpopulations directly into the 0.2 ml low profile eight tube strips. Immediately after sorting is completed, which will take only seconds, firmly cover the strip with the caps, tap the strip gently with your finger tips, and centrifuge the strip placed in a MicroAmp 96-well base strip holder in a centrifuge equipped with a 96-well plate holder at 1,000 rpm for 60 s. Then place the strip on ice until lysing and reverse transcription.
17. An example of sorting myeloid progenitors and subpopulations thereof is shown in **Fig. 21.1**. For further reading on stem cell isolation and cell sorting, the reader is referred to **Chapter 3** in Hematopoietic Stem Cell Protocols, ed JM Walker, no. 63 in the series Methods in Molecular Medicine, Humana Press, and Refs. *(18–24)*.

3.2. Lysing and Reverse Transcription

1. The following mixtures should be prepared immediately before cell sorting and kept on ice.
2. a) To make a 10 μL single-cell collection mixture, add 0.5 μL RNase inhibitor to 9.5 μL water/NP-40. Mix well and add 10 μL to each tube and keep the strip on ice.
   b) To make a 42 μL 500-cell collection mixture, add 2 μL RNAse inhibitor to 40 μL water/NP-40. Add 42 μL to each tube and keep on ice.
3. Immediately after cell sorting and centrifugation of the strips proceed to lysing by placing the strips with tightly secured caps

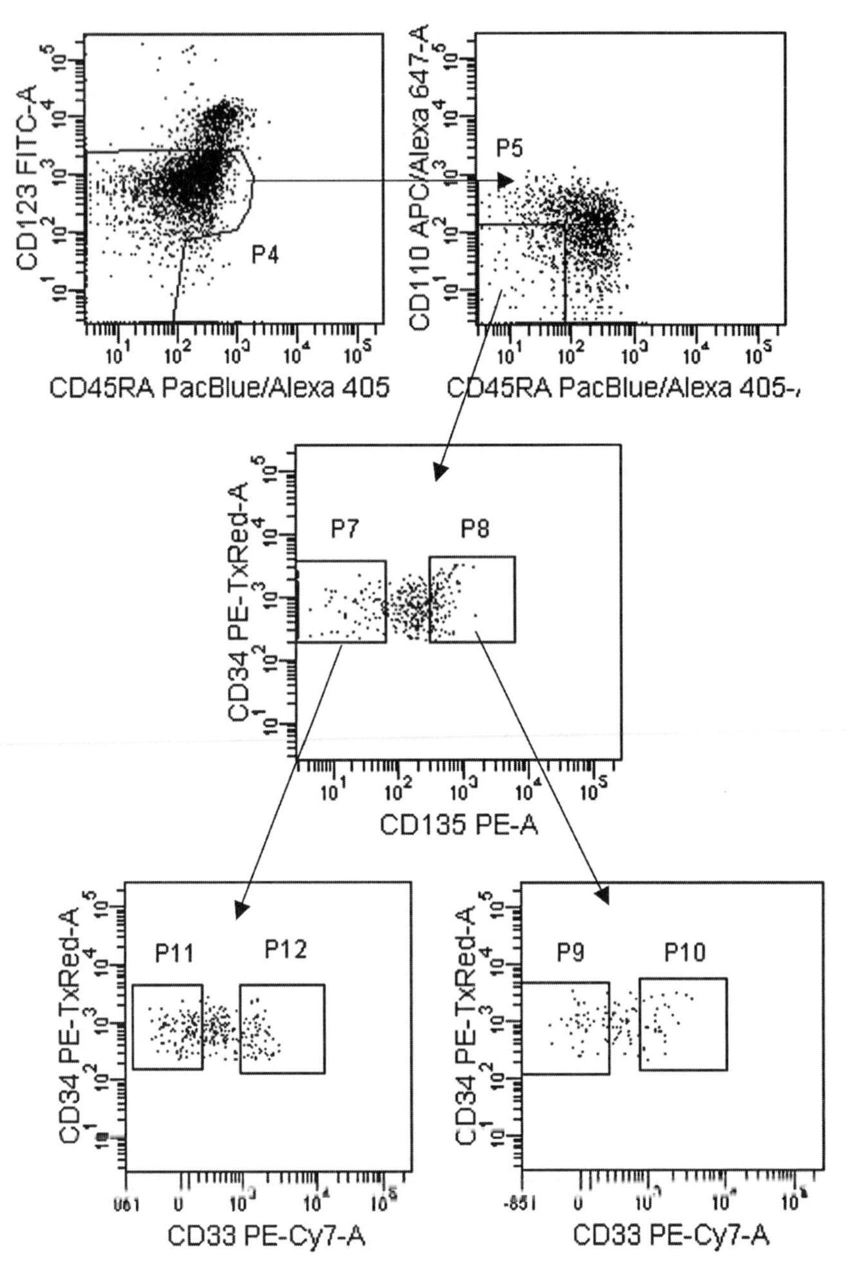

Fig. 21.1. Isolation of human common myeloid progenitor (CMP) subpopulations by cell sorting. CD34+/CD19- cells were labeled with CD123-FITC/CD135-PE/CD34-PE-TexRed/CD38-PerCP-Cy5.5/CD33-PE-Cy7/CD110-APC/CD45RA-PacBlue. P4 containing myeloid progenitors was displayed in the CD110/CD45RA plot where P5 contains the CMP; these were further separated in CD135+/− populations P7 and P8. P7 was further separated in CD33+/− cells in P11 and P12. P8 was separated in P9 and P10. Single-cell sorting was performed on P9, P10, P11 and P12 allowing a detailed analysis of the impact of CD33-expression of CD135+ and CD135- common myeloid progenitors. For example the results showed that CD135-/CD33+ cells expressed neutrophil markers (PR3+, G-CSFR+), whereas the CD135-/CD33-negative cells lacked expression of PR3 and G-CSFR, suggesting subtle pre-commitment steps of the cells within the CMP population.

in a BioRad iCycler and run the lysing program: 65°C, 1 min; 22°C, 3 min; 4°C hold (be sure not to use hot start).

4. Proceed with reverse transcription as soon as possible after lysing.
5. Prepare master mix for reverse transcription as follows:

   a) for single-cell (one reaction; multiply as needed):

      RT buffer (10 ×), 1.54 μL
      dNTPs (5 mM each), 1.54 μL
      Random hexamer (50 μM), 0.77 μL
      RNase free water (0.385 μL)
      RNase inhibitor (20 U/μL), 0.385 μL
      Sensiscript (0.77 μL).
      Mix thoroughly but do not vortex, centrifuge slightly, and add 5.4 μL of this master mix to each single-cell tube (10 μL preamplified cDNA), tap the tubes firmly with your finger tips and centrifuge.

   b) For 500-sorted cells (one reaction; multiply as needed):

      RT buffer (10 ×), 6.69 μL
      dNTPs (5 mM each), 6.69 μL
      Random hexamer (50 μM), 3.35 μL
      RNase free water (1.67 μL)
      RNase inhibitor (20 U/μL), 1.67 μL
      Sensiscript (3.35 μL)
      Mix thoroughly but do not vortex, centrifuge slightly, and add 23.5 μL of this master mix to each 500-cell tube (42 μL cDNA), tap the tubes firmly with your finger tips and centrifuge.

6. Place the strips in the BioRad iCycler and run the RT-reaction at 37°C, 60 min; 95°C, 5 min; 4°C hold (do not use hot start).
7. a) In the case of single-cells immediately proceed with preamplification.
   b) After reverse transcription, the resulting cDNA samples from 500 cells can be frozen at –80°C before RT-PCR analysis is performed. Each cDNA sample from 500 cells is sufficient for RT-PCR of four targets run in triplicates.

3.3. Preamplification of Single-Cell cDNA

1. Make a 1/100 pool of the TaqMan gene expression assays of interest; you can pool at least ten different assays, e.g., 2 μL of ten different assays added to 180 μL TE buffer; the assays contain both primers and probe
2. For each single-cell sample mix in a 0.2 ml PCR tube: 25 μL PreAmp master mix, 12.5 μL pooled assays, and 12.5 μL cDNA.

3. Run the following preamplification: 95°C, 10 min; 16 cycles with 95°C, 15 s and 60°C, 4 min; 4°C hold.
4. Transfer the tubes to ice as soon as possible after completion and add 200 μL TE buffer to each tube and transfer to 1 ml eppendorf tubes. This is the preamplified single-cell cDNA sample ready for RT-PCR analysis and is sufficient for RT-PCR of four targets run in triplicate.

3.4. Real-time PCR analysis

1. Thaw cDNA samples and perform gene expression assays on ice.
2. Make a mix of ten volumes of TaqMan Universal PCR master mix with one volume of the TaqMan gene expression assay (e.g., eight samples in triplicate and a negative control in duplicate will be analyzed for one target mix of 360 μL master mix with 36 μL gene expression assay).
3. a) For each sample take:

   Master mix (44 μL)+ assay
   cDNA (15.5 μL) from 500-cell samples or 20 μL from single-cell samples
   RNase free water to a total volume of 80 μL
   This will give you 3 × 25 μL triplicates.

   b) For the negative control:
   Master mix (32 μL)+ assay
   RNase free water (26 μL)
   This will give you 2 × 25 μL negative control.

4. a) In case of single-cell analysis, we use β2-microglobulin instead of 18S (*see* **Note 8**).
   b) In addition to the different specific targets to be analyzed, include a sample with 18S ribosomal RNA when analyzing 500 cells; the results from this will be used for normalization of target quantities.
5. Standard curves for each gene should be established by analyzing serial dilutions of cDNA.
6. As a calibrator, a cDNA from a pool of unseparated CD34+ cells or a mixture of CD34+ cell cDNA with cDNA from mononuclear bone marrow cells is recommended (*see* **Note 9**).
7. Use a high-quality optical 96-well PCR plate or strips and cautiously transfer 25 μL of samples (triplicates) and negative controls (duplicates) to the bottom of the wells; cover with an optical lid and centrifuge the plate or strips for 15 s at 1,000 rpm.
8. Put the plate/strips in the RT-PCR instrument (e.g., ABI Prism 7000 sequence detection system).
9. The PCR reaction is run as: 50°C, 2 min; 95°C, 10 min; 40 cycles with 95°C, 15 s and 60°C, 1 min.

10. Quantification of the relative levels of the specific mRNA of interest is achieved by the relative standard curve method (described in detail in Applied Biosystems User Bulletin No. 2: Relative Quantification of Gene Expression).

4. Notes

1. It is important that the strip holder allows free movement of the ACDU during cell sorting and we have found that MicroAmp 96-well Base works fine with the FACS Aria.
2. In case you are primarily interested in myeloid cells you can include a CD19+ cell depletion using MACS CD19 microbeads. This will remove all CD34+/CD19+ precursor B-cells that sometimes make up 25% of the CD34+ cells which could interfere with the isolation of pure myeloid progenitors.
3. The manufacturer gives the option of using the small MS column for separation of $<2 \times 10^9$ MNC, but we would recommend to always use the larger LS column.
4. You may save the CD34-negative cells passing through the column in case you would like to isolate more mature cells along the myeloid lineages.
5. We sometimes have had problems with leakage from the first and last tube of the strip during reverse transcription and therefore routinely avoid using these positions.
6. Do not forget this filtering step, it will save you a lot of time from cleaning a clogged nozzle.
7. We sort at high-speed settings with sheath pressure of 70 psi, which the stem cells and progenitor cells tolerated very well.
8. We usually first analyze for β2-microglobulin expression to certify that the sample actually contained a cell, before analyzing for more specific targets; this will save you negative results and money.
9. Making a pool of cDNA from CD34+ cells will give you the possibility to include the same calibrator over many separate experiments.

References

1. Attar, R. A., Scadden, D. T. (2004) Regulation of hematopoietic stem cell growth. *Leukemia* 18, 1760–1768.
2. Payne, K. J., Crooks, G. M. (2002) Human hematopoietic lineage commitment. *Immunol Rev* 187, 48–64.
3. Zhang, Y., Payne, K. J., Zhu, Y., et al. (2005) SCL expression at critical points in human hematopoietic lineage commitment. *Stem Cells* 23, 852–860.
4. Friedman, A. D. (2007) C/EBPα induces PU.1 and interacts with AP-1 and NF-κB

to regulate myeloid development. *Blood Cells, Mol Dis* 39, 340–343.
5. Ye, M., Graf, T. (2007) Early decisions in lymphoid development. *Curr Opin Immunol* 19, 123–128.
6. Kastner, P., Chan, S. (2008) PU.1: A crucial and versatile player in hematopoiesis and leukemia. *Int J Biochem Cell Biol* 40, 22–27.
7. Iwasaki, H., Somoza, C., Shigematsu, H., et al. (2005) Distinctive and indispensable roles of PU.1 in maintenance of hematopoietic stem cells and their differentiation. *Blood* 106, 1590–1600.
8. Ohnedo, K., Yamamoto, M. (2002) Roles of hematopoietic transcription factors GATA-1 and GATA-2 in the development of red blood cell lineage. *Acta Haematologica* 108, 237–245.
9. Shivdasani, R. A. (2001) Molecular and transcriptional regulation of megakaryocyte differentiation. *Stem Cells* 19, 397–407.
10. Matthias, P., Rolink, A. G. (2005) Transcriptional networks in developing and mature B cells. *Nat Rev Immunol* 6, 497–508.
11. Anderson, M. K. (2006) At the crossroads: diverse roles of early thymocyte transcriptional regulators. *Immunol Rev* 209, 191–211.
12. Zhu, J., Emerson, S. G. (2002) Hematopoietic cytokines, transcription factors and lineage commitment. *Oncogene* 21, 3295–3313.
13. Sitnicka, E., Buza-Vidas, N., Larsson, S., et al. (2003) Human CD34+ hematopoietic stem cells capable of multilineage engrafting NOD/SCID mice express flt3: distinct flt3 and c-kit expression and response patterns on mouse and candidate human hematopoietic stem cells. *Blood* 102, 881–886.
14. Youn, B. S., Mantel, C., Broxmeyer, H. E. (2000) Chemokines, chemokine receptors and hematopoiesis. *Immunol Rev* 177, 150–174.
15. Jelkmann, W. (2004) Molecular biology of erythropoeitin *Intern Med* 43, 649–659.
16. Kaushansky, K., Drachman, J. G. (2002) Molecular and cellular biology of thrombopoietin: the primary regulator of platelet production. *Oncogene* 21, 3359–3367.
17. Miranda, M. B., Johnson, D. E. (2007) Signal transduction pathways that contribute to myeloid differentiation. *Leukemia* 21, 1363–1377.
18. Weissman, I. L., Anderson, D. J., Gage, F. (2001) Stem and progenitor cells: origins, phenotypes, lineage commitments, and transdifferentiations. *Annu Rev Cell Dev Biol* 17, 387–403.
19. Manz, M. G., Miyamoto, T., Akashi, K., et al. (2002) prospective isolation of human clonogenic common myeloid progenitors. *Proc Natl Acad Sci USA* 99, 11872–11877.
20. Edvardsson, L., Dykes, J., Olsson, M. L., et al. (2004) Clonogenicity, gene expression and phenotype during neutrophil versus erythroid differentiation of cytokine-stimulated CD34+ human marrow cells in vitro. *Br J Haematol* 127, 451–463.
21. Edvardsson, L., Dykes, J., Olofsson, T. (2006) Isolation and characterization of human myeloid progenitor populations – TpoR as discriminator between common myeloid and megakaryocyte/erythroid progenitors. *Exp Hematol* 34, 599–609.
22. Adolfsson, J., Mansson, R., Buza-Vidas, N., et al. (2005) identification of Flt3+ lympho-myeloid stem cells lacking erythro-megakaryocyte potential: a revised road map for adult blood lineage commitment. *Cell* 91, 661–672.
23. Chen, L., Gao, Z., Zhu, J., et al. (2007) Identification of CD13+CD36+ cells as a common progenitor for erythroid and myeloid lineages in human bone marrow. *Exp Hematol* 35, 1047–1055.
24. Case, J., Mead, L. E., Bessler, W. K., et al. (2007) Human CD34+AC133+VEGFR-2+ cells are not endothelial progenitor cells but distinct, primitive hematopoietic progenitors. *Exp Hematol* 35, 1109–1118.

Chapter 22

Monitoring the Immune Response Using Real-Time PCR

Patrick Stordeur

Abstract

Induction of an immune response to a particular antigen is the basis of vaccination. This has been done for years to prevent infectious diseases, and has the potential for the treatment of cancer. The immune response is nowadays more precisely modulated rather than simply induced, like in case of immunotherapy of allergic diseases. Likewise, autoimmune diseases are associated with an inappropriate immune response, and many efforts are made for specifically inhibiting this unwanted response. A possible line of attack is the induction of an antigen-specific immune tolerance, which also has a use in the field of transplantation, where allogeneic responses are deleterious for the graft. In all of these fields of fundamental and clinical medicine, the modulation of immune response requires the assistance of laboratory tests, among which real-time PCR appears more and more helpful. This chapter describes a protocol to quantify immune-related mRNAs using reverse transcription-real-time PCR. The transcripts can be quantified in cultured cells or in cultured whole blood, after an incubation period in the presence of the antigen to which the immune response is analyzed. This is the typical approach to evaluate the efficacy of a vaccine. The transcripts can also be quantified directly in the biological sample, giving information about the in vivo immune status of the individual. The techniques to achieve these different methods are described, and are illustrated by the analysis of the response against the toxoid tetanus antigen.

Key words: Whole blood, peripheral blood mononuclear cells, immune monitoring, methods, messenger RNA, real-time PCR, cytokine, IL-2, IFN gamma, antigen, tetanus toxoid, transplantation, allergy, autoimmunity.

1. Introduction

The monitoring of the immune response is critical for the follow-up and treatment evaluation of numerous diseases. This is the case, for example, for the development of new vaccines including those directed against tumor cells, for the evaluation of new clinical transplantation protocols aiming to induce tolerance, and for the follow-up of immunotherapy of allergic and autoimmune

Peter Bugert (ed.), *DNA and RNA Profiling in Human Blood: Methods and Protocols, vol. 496*

DOI 10.1007/978-1-59745-553-4_22 Springerprotocols.com

diseases. Current techniques to monitor immune responses mainly include cytokine protein quantification and effector T cell frequency analysis. More recently, protein arrays provided a new tool particularly for the monitoring of the humoral response. Apart from the latter, these techniques are generally blood consuming and labor intensive, and results are obtained only after several days to weeks *(1–5)*. In contrast, real-time PCR can be performed in one or two days and requires small blood volumes. These advantages, added to the high sensitivity and precision of this technique, have paved the way to a new approach of monitoring immune responses *(6–8)*.

There are different ways to apply real-time PCR to immune monitoring.The high sensitivity of the technique can be taken advantage of to easily and rapidly detect immune-related mRNA transcripts in a limited number of cells where the corresponding protein could barely be measured. This approach has been used

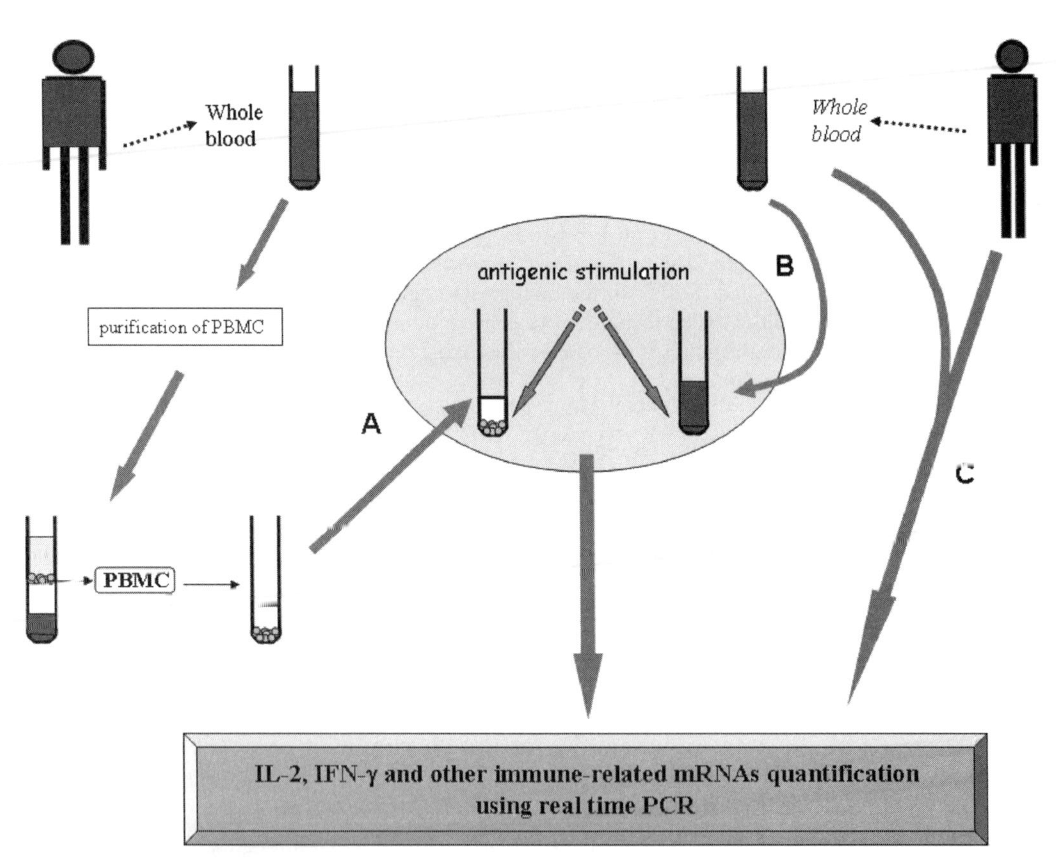

Fig. 22.1. Different ways to apply real-time PCR for immune monitoring. The antigen to which the immune response is analyzed is added to PBMC purified from peripheral blood (A) or directly to whole blood (B). This antigen can be any antigen like TT as described in this chapter. After several hours incubation, the culture is stopped and mRNAs are quantified by real-time PCR. A different approach consists in quantifying the transcripts directly in stabilized whole blood and other biological samples (C) (*see* **Note 8**).

to detect acute kidney rejection before clinical signs of renal failure appear. Indeed, low levels of different mRNAs involved in the immune response can be detected in urine in this clinical situation *(9)*. Real-time PCR can thus be performed in that way on different body fluids that contain a limited number of cells, or in which proteins like cytokines disappear rapidly contrary to their corresponding mRNA. Real-time PCR is also performed to detect the induction of different mRNAs in response to a particular antigen after in vitro incubation of blood samples. Two approaches are possible: the antigen to which the immune response is analyzed can be added to whole blood, or to peripheral blood mononuclear cells (PBMC) purified from blood. Use of whole blood presents several advantages: (1) it avoids time-consuming purification of PBMC, this step being able in addition to either non-specifically activate the cells or modulate cytokine mRNA levels; (2) it allows the use of smaller blood volumes; and (3) it is more representative of the in vivo situation than purified cells. However, most of the techniques used to monitor immune responses are still performed on PBMC. Therefore, the two alternatives are described here, using tetanus toxoid (TT) as antigen (**Fig. 22.1**).

2. Materials

2.1. Blood Samples

Blood should be collected in sodium heparinate. Avoid lithium heparinate. We observed heterogeneous and sometimes incomprehensible results when using lithium heparinate.

2.2. Cell Purification and Culture

1. AIM-V® medium (GIBCO, Invitrogen, Merelbeke, Belgium).
2. Hanks balanced salt solution (HBSS) ×10, without calcium and magnesium (GIBCO, Invitrogen).
3. PBS without calcium and magnesium (BioWhittakker™, Cambrex Bioscience, Verviers, Belgium).
4. Fetal bovine serum (FBS) (HyClone, Perbio Science, Erembodegem, Belgium).
5. PBS supplemented with 2% of FBS.
6. Lymphoprep™ (Axis-Shield, Oslo, Norway).
7. Trypan blue 0.4% w/v (StemCell technologies, Vancouver, Canada).
8. Tetanus toxoid (TT) (Statens Serum Institute, Denmark) (1 Lf/ml corresponds to 2.5 μg/ml).

2.3. mRNA Extraction

1. MagNA Pure™ instrument (Roche Applied Science, Vilvorde, Belgium).
2. PAXgene tubes (Qiagen Benelux, Venlo, The Netherlands).

3. MagNA Pure™ mRNA extraction kit "MagNa Pure™ LC mRNA Kit I" (Roche Applied Science).
4. TaqMan pre-developed assay reagent for human CD3 epsilon chain (Applied Biosystems, Foster City, CA).
   Facultative reagents:
5. TriPure™ isolation reagent (Roche Applied Science).
6. TRIzol® reagent (Invitrogen).
7. PAXgene blood RNA kit (Qiagen).

2.4. PCR

1. Lightcycler™ 2.0 instrument (Roche Applied Science).
2. Lightcycler™ RNA master hybridization probes kit (Roche Applied Science).
3. Oligonucleotides "OliGold®" (**Tables 22.1** and **22.2**) are from Eurogentec (Eurogentec, Seraing, Belgium).
4. Wizard SV gel clean-up system (Promega, Leiden, The Netherlands).
5. DNA from fish sperm, MB-grade (Roche Applied Science). Prepare a 10 mg/ml stock solution in TE buffer (pH 8.0). Keep at –20°C.

3. Methods

3.1. Isolation of Peripheral Blood Mononuclear Cells (PBMC)

1. Transfer 12 ml of heparinized blood in a 50 ml Falcon tube and make it up to 35 ml with HBSS ×1 (*see* **Note 1**).
2. Layer the diluted sample on top of the Lymphoprep™ or layer the Lymphoprep™ underneath the diluted sample. Be careful to minimize mixing of Lymphoprep™ and sample. Use 15 ml of Lymphoprep™.
3. Centrifuge for 20 min at 800–1,200 *g* (*see* **Note 2**) at room temperature, with the brake off.
4. Recover the PBMC fraction from the Lymphoprep™: plasma interface and transfer it in a 50 ml Falcon tube. Suspend the fraction to 50 ml with 1× HBSS.
5. Centrifuge at 800 *g* for 10 min. Discard the supernatant and repeat this washing step.
6. Suspend cell pellet by adding 5 ml of AIM-V to the cell pellet (centrifugation of suspended cells in AIM-V must be done without brake, to ensure complete sedimentation of all the cells).
7. Count the cells, evaluate their viability using Trypan blue (mix one volume of cell suspension with one volume of Trypan blue solution; dead cells appear light blue).
8. Adjust the volume with AIM-V to obtain 1 million cells per ml, for FACS analysis and for cell culture.

Table 22.1
Primers and probes for real-time and conventional PCR of immune-related mRNAs

mRNA Target	Oligonucleotide sequence (5′ → 3′)[a]	Product size(bp)	Oligonucleotide concentration (nM)[b]	Accession number[c]
IL-2	F273: CTCACCAGGATGCTCACATTTA R367: TCCAGAGGTTTGAGTTCTTCTTCT P304: Fam-TGCCCAAGAAGGCCACAGAACTG-BHQ1	95	F: 900 R: 900	X01586
IL-6	F288 : AACATGTGTGAAAGCAGCAAAG R397 : AGGCAAGTCTCCTCATTGAATC P326: Fam- CAACCTGAACCTTCCAAAGATGGCTG-BHQ1	73	F: 300 R: 300	NM_000600
IL-17A	F199: CATAACCGGAATACCAATACCAAT R302: GGATATCTCTCAGGGTCCTCATT P246: Fam- CAACCGATCCACCTCACCTTGGAAT-BHQ1	104	F: 300 R: 900	NM_002190
IFN-γ	F464: CTAATTATTCGGTAACTGACTTGA R538: ACAGTTCAGCCATCACTTGGA P491: Fam-TCCAACGCAAAGCAATACATGAAC-BHQ1	75	F: 600 R: 900	X13274
TGF-β	F1145 : GACTACTACGCCAAGGAGGTCA R1232 : TGCTGTGTGTACTCTGCTTGAAC P1169: Fam- CGCGTGCTAATGGTGGAAACCC-BHQ1	88	F: 900 R: 600	X02812

(continued)

Table 22.1 (continued)

mRNA Target	Oligonucleotide sequence (5′ → 3′)[a]	Product size(bp)	Oligonucleotide concentration (nM)[b]	Accession number[c]
FOXP3	F1155: GAGTTCCTCCACAACATGGACT R1319: ATGGTTTCTGAAGAAGGCAAAC P1190: Fam- CAACATGCGACCCCCTTTCACC-BHQ1	165	F: 600 R: 900	NM_014009
β-Actin	F976: GGATGCAGAAGGAGATCACTG R1065: CGATCCACACGGAGTACTTG P997: Fam-CCCTGGCACCCAGCACAATG-BHQ1	90[d]	F: 300 R: 300	X000351
RPLP0	F871: TGTCTGTGGAGACGGATTACAC R1027: TCTTCCTTGGCTTCAACCTTAG P935: Fam-ATCTGCCTTTGTGGCTGCTGCC-BHQ1	157	F: 900 R: 900	BC019014

For a full description, and other mRNA targets, *see* Refs. *(14, 15, 18)*

[a]F, R, and, P indicate forward and reverse primers and probes, respectively; numbers indicate the sequence position from Genebank accession numbers (see item[c]); for Taqman probes, the quencher BHQ1 is preferable to TAMRA (it has no fluorescence of its own, which enhances the signal-to-noise ratio)

[b]final concentration of forward (F) and reverse (R) primers

[c]Genebank accession number

[d]all primers were chosen to span intronic sequences so that genomic DNA amplification is not possible, excepted for β-actin for which a 112 bp larger fragment is obtained; this allows the detection of contaminating genomic DNA using this size difference on agarose gel

Table 22.2
Primers for conventional PCR of immune-related mRNAs

mRNA target	Oligonucleotide sequence (5′ → 3′)[a]	Product size(bp)	Annealing temperature (°C)
IL-2	F155: TGTCACAAACAGTGCACCTACT R672: AGTTACAATAGGTAGCAAACCATACA	518	58
IL-6	F217 : GAATTGACAAACAAATTCGGTACA R486 : CTCTGGCTTGTTCCTCACTACTC	270	58
IL-17A	F146: CAAATTCTGAGGACAAGAACTTCC R380: GAGTTCATGTGGTAGTCCACGTT	235	61
IFN-γ	F154: TTGGGTTCTCTTGGCTGTTA R632: AAATATTGCAGGCAGGACAA	479	58
TGF-β	F925 : GGGACTATCCACCTGCAAGA R1263 : CGGAGCTCTGATGTGTTGAA	339	60
FOXP3	F887: CCTCCAGAGAGAGATGGTACAGT R1505: CTCCTTTCCTTGATCTTGAGGT	619	59
β-Actin	F745: CCCTGGAGAAGAGCTACGA R1253: TAAAGCCATGCCAATCTCAT	509	58
RPLP0	F533: CCAGGCTTTAGGTATCACCACTAA R1071: AGTCAAAGAGACCAAATCCCATATC	539	61

Standard curves were generated from serial dilutions of PCR products prepared by conventional PCR, for which specific conditions were as follows: 35 cycles with 95°C for 20 s, indicated annealing temperature for 20 s and 72°C for 45 s; $MgCl_2$ final concentration is 1.5 mM

[a]F and R indicate forward and reverse primers, respectively; numbers indicate the sequence position from Genebank accession numbers (see Table 22.1)

9. Analyze 100,000 cells by flow cytometry. Use an antibody cocktail that gives at least the percent of total, CD4 and CD8 T cells, B cells, NK cells.

3.2. Antigenic Stimulation

1. Add TT to 10 μg/ml (equivalent to 4 Lf/ml) (*see* **Note 3**) to 200,000 PBMC (*see* **Note 4**). Because one works with cell suspensions at one million per ml, the total culture volume should be around 200 μL in AIM-V medium (*see* **Note 5**). Or:Add 10 μg/ml of TT to the volume of whole blood that contains 200,000 T cells (usually 200–300 μL in healthy subjects) (*see* **Note 6**).
2. Incubate for 16 h at 37°C in a 5% CO_2 atmosphere (*see* **Note 3**).
3. Stop the reaction by adding, for PBMC, 600 μL (i.e., 2.5 volume) of the reagent contained in the PAXgene tubes (*see* **Note 7**). Or: For whole blood, add 2.5 volumes (i.e., 250 μL per 100 μL of blood) of the reagent contained in the PAXgene tubes.
4. Incubate at least 1 h at room temperature. The lysate is stable at room temperature for up to 5 days, and for months at –80°C (*see* **Note 8**).

3.3. mRNA Extraction

Extracting mRNA using an automatic device such as the MagNA Pure™ (Roche Applied Science) provides reproducible amounts of high-quality mRNA. Moreover, the RT-PCR reaction mixtures containing all reagents, oligonucleotides and samples, are fully prepared directly in the PCR reaction vessel (in the case of the Lightcycler™, a capillary), by the MagNA Pure™ instrument. The sampling of RT-PCR components is fully automated, avoids manual sampling errors and thereby enhances the reproducibility and accuracy of the qPCR. In this approach, the RNA sampled by the device will be mRNA instead of total RNA. Reverse transcription and the PCR are successively performed in one step RT-qPCR reaction in the same reaction tube (or capillary). In the case of mRNA, the concentration is too low to be measurable at 260 nm. Hence, the same volume of mRNA (i.e., 5 μL) is always added to the PCR mixture, given the reproducibility ensured by the automation of the procedure. The technique described here is based on the use of the MagNa Pure™ LC mRNA Kit I (Roche Applied Science) on the MagNA Pure™ instrument. However, this expensive step in this method can be replaced by other conventional and less costly methods of RNA extraction. Different kits or reagents that provide high-quality RNA exist. A good choice is the kit provided by the manufacturer of the PAXgene tubes, the PAXgene blood RNA kit (Qiagen), but the use of reagents like TriPure™ (Roche Applied Science) and TRIzol® (Invitrogen) is also possible. If one of these alternatives is chosen, the concentration of total RNA should be measured on a spectrophotometer at 260 nm. Total RNA (100 ng) are then engaged in the PCR reaction.

1. Gently suspend the precipitate of the cell lysate of **Section 3.2.** and transfer 300 μL in a microtube.
2. Centrifuge for 5 min at 16,000 *g*.
3. Discard the supernatant.
4. Dissolve thoroughly the nucleic acid pellet in 300 μL of the lysis buffer contained in the MagNa Pure™ LC mRNA Kit I. The pellet is difficult to dissolve.
5. Vortexing for 45–60 s is often required. Some very small aggregates often remain undissolved (*see* **Note 9**).
6. Proceed with mRNA extraction following manufacturer's instructions.
7. Elute in 100 μL and use 5 μL by PCR reaction. If low number of PBMC (e.g., 50,000) or low blood volumes (e.g., 100 μL) are used, the elution volume can be reduced to 50 μL, and the volume added to the PCR kept at 5 μL.

3.4. RT-qPCR

Several assays are available to perform real-time PCR *(10–12)*. The choice of which one to use depends on the application of the technique (e.g., polymorphism detection and/or quantification). As far as mRNA quantification is concerned, three good choices are SYBR®Green, hybridization probes, and hydrolysis (Taqman) probes. The first one is the cheapest, but also the less specific and sensitive (SYBR®Green binds all double-stranded DNAs, including unspecific PCR products and primer dimers). The use of probes eliminates this problem. Both formats can be used, but the design of the primers and probes is somewhat easier with the Taqman chemistry (only three oligonucleotides in all (two primers and one probe) instead of four with hybridization probes (two primers and two probes)). The technique described in **Section 3.4.1** uses Taqman probes and the Lightcycler™ RNA Master Hybridization probes kit on the Lightcycler™ instrument (Roche Applied Science). For details on the design of oligonucleotides *see* **Note 10**.

3.4.1. Preparation and Execution of RT-qPCR Reaction

1. Prepare a RT-qPCR reaction mixture as follows (each volume can be multiplied by the number of samples (plus one per ten samples) to prepare the quantity needed for all samples of the experiment): RNA master hybridization probes (2.7× conc.) 7.5 μL; 1.3 μL 50 mM $Mn(OAc)_2$; 1, 2 or 3 μL of 6 pmoles/μL forward and reverse primers (final concentration 300, 600, or 900 nM, depending on the mRNA target; *see* **Note 11** and **Table 22.1**); 1 μL of 4 pmoles/μL TaqMan probe (final concentration 200 nM); 5 μL purified mRNA (or containing 100 ng of total RNA) or standard dilution; up to 20 μL H_2O.

2. Transfer the capillaries in the Lightcycler^{TM} and start the RT-PCR reaction.
3. The cycling conditions are as follows: after an incubation period of 20 min at 61°C to allow mRNA reverse transcription, and then an initial denaturation step at 95°C for 30 s, temperature cycling is initiated. Each cycle consists of 95°C for 0 s and 60°C for 20 s, the fluorescence being read at the end of this second step (F1/F2 channels when using FAM/TAMRA labeled probes, F1 channel when using FAM/BHQ1 labeled probes, no color compensation). Overall, 45 cycles are performed (*see* **Note 12**).

3.4.2. Generation of External Standards

mRNA levels can be expressed either in absolute copy numbers or in relative copy numbers normalized against a house-keeping gene (*see* **Note 13**). This is achieved by constructing, for each PCR run, a standard curve from serial dilutions of purified DNA. The latter consists in a PCR product that includes the quantified amplicon, and that is prepared by "classical" PCR from cDNA positive for the concerned target mRNA. These PCR products used as standards are purified from agarose gel by the "Wizard SV Gel Clean-up System" following manufacturer's instructions. The copy number of the standards is calculated from the DNA concentration measured by spectrophotometry *(13)*. Dilute from 10 to 10 in TE buffer (pH 8.0) supplemented with 10 μg/ml fish DNA, to get dilutions from 10^8 to 10^2 copies (*see* **Note 14**). Detailed information concerning these standards is given in **Table 22.2**.

4. Notes

1. Blood volume depends on the number of PBMC needed for the experiment. Generally, at least one million PBMC can be obtained from 1 ml of blood from a subject with normal WBC count.
2. To convert *g* to rpm, use the following formula: rpm = square root of RCF / (1.118 × 10^{-5} × radius), where RCF = relative centrifugal force (*g*), rpm = centrifuge speed in revolutions per minute, radius = radius of rotor in cm.
3. TT has been chosen to illustrate this protocol, the response to this antigen having been well studied. However, the technique can theoretically be applied to study the response to any antigen, what we did successfully for allergens, alloreactive cells, and autoantigens (**Fig. 22.2**) *(14–18)*. For each antigen, it is recommended to firstly perform kinetic experiments at a "supposed" working concentration. Depending

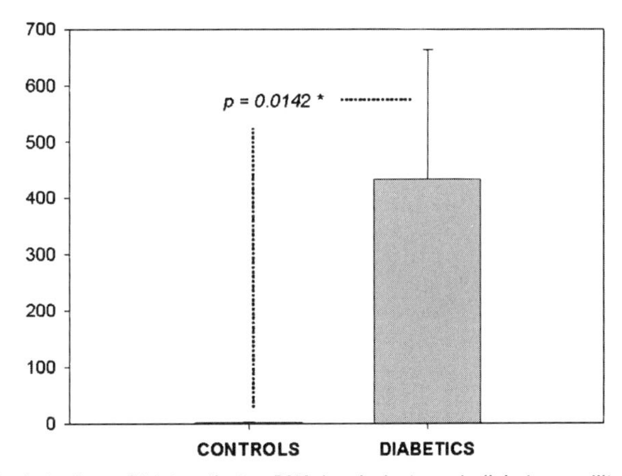

Fig. 22.2. Induction of higher IL-2 mRNA levels in type 1 diabetes mellitus patients compared to healthy subjects in response to GAD65 antigen. Type 1 diabetes mellitus is an autoimmune disease where T lymphocytes play a major role in the destruction of insulin-producing pancreatic β-cells. The isoform 65 of glutamic acid decarboxylase (GAD65) is one of the putative antigens recognized by autoreactive T cells. Whole blood from 10 diabetic patients and from 10 healthy subjects was incubated for 16 h in the presence of 10 μg/ml GAD65, and IL-2 mRNA was quantified as described. Y-axis: IL-2 mRNA copies per million of β-actin mRNA copies; mean + SEM are represented. Significantly higher levels of IL-2 mRNA were observed in diabetic patients compared to controls ($p = 0.0142$, Mann-Whitney test) (Laurent Crenier, personal communication). This suggests a possible use of this technique as a diagnosis tool in autoimmune diseases.

on the antigen and the mRNA studied, it could be necessary to test times spreading out from 1 h to 10 days. For example, maximum IL-2 mRNA levels are generally reached in 16 h when the whole antigen protein is added, while this maximum can be reached in 2 hours when antigenic peptides are used instead of the whole protein. Of course, these peptides have to be restricted to the major histocompatibility complex (MHC) antigens expressed on the cells used in the test. Likewise, some mRNAs are induced really more quickly than others. One example is, IL-4 mRNA is induced maximally in 1 h in whole blood in response to an allergen, whereas FOXP3 and IL-17 mRNAs generally require days to be induced in response to whole protein antigens, which implies the stimulation of PBMC. Indeed, the incubation time cannot exceed 20 h with whole blood. After this time, spontaneous lysis of red and white cells occurs. Once the time corresponding to the maximal induction level is determined, it is recommended to perform dose–response experiments to find the ideal working concentration of the antigen.

4. The number of PBMC needed to perform RT-PCR mainly depends on the yield of the technique used to isolate RNA. The technique described in this chapter, i.e., mRNA isolation on the MagNA Pure instrument allows a high-yield

RNA recovery. The minimum number of PBMC required in this case is 100,000 or even 50,000, but best results are obtained when using 200,000 PBMC. Manual RNA purification methods generally require larger samples. As a general rule, the ideal number of cells should be the first parameter to be tested as follows. Extract mRNA from different numbers of cells (e.g., from 100,000 to 1,000,000 by step of 100,000) and perform real-time RT-PCR for a house-keeping gene mRNA. Work with the number of cells that is in the middle of the curve which establishes a direct correlation between the starting number of cells and the copy number of mRNA (**Fig. 22.3**).

5. In the best case, the solution of antigen should be 20 × concentrated, in order to add 10 μL, a volume easy to take and which does not significantly dilute the PBMC or the blood.
6. Ideally, the volume of blood should be chosen to contain 100,000–200,000 responding cells. The best blood volume can also be determined as described in Note 4 for PBMC (see also **Fig. 22.3**).
7. This reagent induces complete cell lysis and, at the same time, nucleic acid precipitation. The nucleic acids can then be dissolved in guanidium/thiocyanate solutions similar to that described in 1987 by Chomczynski and Sacchi *(19)*, or to the guanidium/thiocyanate solution contained in the lysis buffer provided with the MagNA Pure LC kit for mRNA isolation (Roche Applied Science). The advantage of adding this reagent directly to the cell culture is that there is no need to discard the supernatant, reducing the potential loss of cells that can occur when working with such small volumes.
8. RNA can also be extracted directly from whole blood at the moment of blood collection using PAXgene tubes. The direct stabilization of mRNA by the reagent contained in the tube allows an accurate quantification of circulating mRNA levels without in vitro stimulation, making available an evaluation of peripheral immune status *(15)*. Alternatively, blood can be taken on sodium heparin and immediately added to the reagent of the tube, in order to take smaller blood volume like 200 μL. The analysis can also be performed from other body fluids such as urine and cerebrospinal fluid, and from biopsies (**Fig. 22.1**).
9. If RNA is extracted using TriPure or TRIzol, simply replace the lysis buffer of the MagNA Pure LC mRNA Kit I by the TriPure or TRIzol reagent, and proceed in the same way. Then follow the manufacturer's instructions as recommended for a "normal" cell lysate.
10. The primers and probes described here were designed with the Primer 3 software (http://www-genome.wi.mit.edu/cgi-bin/primer/primer3_www.cgi) *(20)*. The default param-

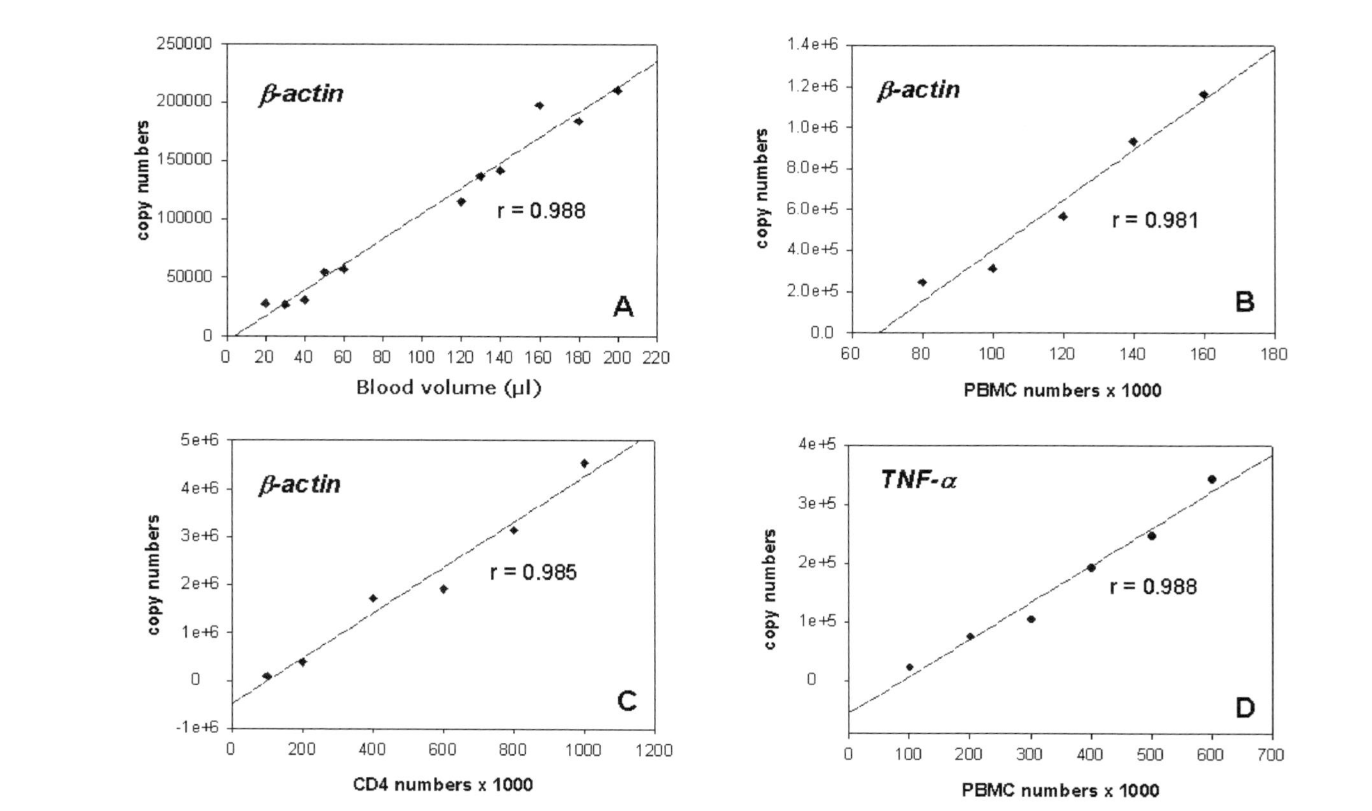

Fig. 22.3. Linear regression: mRNA copy number on starting blood volume or cell number. mRNA was extracted from various whole blood volumes (A) ranging from 20 to 200 μL, or various PBMC numbers (B and D) ranging from 80,000 to 600,000 or CD4 T cells numbers (C) ranging from 100,000 to 1,000,000. RT and real-time PCR for β-actin (A, B, C) or TNF-α (D) were performed. The Y-axis of each panel represents the raw copy numbers and the X-axis the blood volumes (A) or the cell numbers (B, C, D). The line is for linear regression.

eters of the program were applied except for the following: product size 90–150 bp, primer size 20–27 bp, primer T_m 59–62°C with a maximum T_m difference of 2.0°C, oligo T_m 69–72°C, product T_m 0–85°C, maximum self and 3′ self complementarily = 6.00, maximum poly X = 3, primer and Hyb Oligo penalty (penalty weights for primer pairs) = 2.0. In addition, the oligonucleotides were selected according to the following criteria (listed in order of importance): (1) intron spanning if possible; (2) no G at the probe 5′ end; (3) no more than two Gs or Cs within the five 3‘nucleotides for primers; (4) more Cs then Gs in the probe. The expected size of the real-time PCR product must be checked by agarose gel electrophoresis for every new mRNA target.

11. The same protocol is used for all target mRNAs, the only adjustment being the primer concentration. Using the standard at 10^5 copies or a corresponding cDNA, primer titration is performed at 300, 600, or 900 nM, the three concentrations being checked for each primer. The conditions specific for different mRNA targets are listed in **Table 22.1**.
12. To use this protocol on other instruments that do not use capillaries and air for heating/cooling, the following adjustments are needed: increase denaturation time to 10–20 s, and the combined hybridization/elongation step to 40–60 s. Keep the concentration of mRNA and oligonucleotides. Use the reagent mix provided by the manufacturer of the thermal cycler. Most of these reagents are formulated to be used at one “universal” concentration.
13. Best results are obtained with multiplex PCR, i.e., the co-amplification of the mRNA of the house-keeping gene together with the specific mRNA. This is not possible with SYBR®Green and therefore this requires the use of probes, which are labeled with different fluorochromes. The choice of these depends on the real time PCR instrument. The fluorochromes adapted to a particular instrument can be found on the website of several companies specialized in oligonucleotides synthesis, among which Biosearch Technologies (http://www.biosearchtech.com/hot/multiplexing.asp). Primer titration is performed as described in **Note 11**, separately for primers targeting the house-keeping gene mRNA and the mRNA of interest, and primers and probes of both mRNA targets are added together to the PCR mix described in **Section 3.4.1.** In some cases, the total volume can exceed 20 μL, which should not make any problem up to 3 excess μL.
14. In case of multiplex PCR, serial dilutions of both standards (i.e., standard for the mRNA of interest and standard for the mRNA of the house-keeping gene) are mixed, giving serial dilutions of 10^8–10^2 copies of both standards. The choice

of the house-keeping gene to be used as internal standard is critical, and nevertheless, never fully adequate *(12)*. One acceptable gene for whole blood and PBMC is the gene coding for human acidic ribosomal protein (HuPO) *(21)*. Moreover, when the aim of the method is the assessment of a T cell response, the most satisfactory house-keeping gene can be the epsilon chain of the CD3 complex. In this case, use per PCR reaction, 1 μL of the TaqMan pre-developed assay reagent for human CD3ε chain (a ready to use mix of probe and primers) from Applied Biosystem, and express the results in copies of the quantified mRNA per million of CD3ε mRNA copies. An alternative to avoid external standard curve is to express the results in $\Delta\Delta C_t$ *(10, 14)*.

References

1. Bouma, G. J., Schanz, U., Oudshoorn, M., et al. (1996) A cell-saving non-radioactive limiting dilution analysis-assay for the combined determination of helper and cytotoxic T lymphocyte precursor frequencies. *Bone Marrow Transplant* 17, 19–23.
2. van Besouw, N. M., Zuijderwijk, J. M., de Kuiper, P., et al. (2005) The granzyme B and interferon-gamma enzyme-linked immunospot assay as alternatives for cytotoxic T-lymphocyte precursor frequency after renal transplantation. *Transplantation* 79, 1062–1066.
3. Hernandez-Fuentes, M. P., Salama, A. (2006) In vitro assays for immune monitoring in transplantation. *Meth Mol Biol* 333, 269–290.
4. Keilholz, U., Weber, J., Finke, J.H., et al. (2002) Immunologic monitoring of cancer vaccine therapy: results of a workshop sponsored by the Society for Biological Therapy. *J Immunother* 25, 97–138.
5. Cekaite, L., Hovig, E., Sioud, M. (2007) Protein arrays: a versatile toolbox for target identification and monitoring of patient immune responses. *Meth Mol Biol* 360, 335–348.
6. Bustin, S. A., Benes, V., Nolan, T., et al. (2005) Quantitative real-time RT-PCR–a perspective. *J Mol Endocrinol* 34, 597–601.
7. Panelli, M. C., Wang, E., Monsurro, V. et al. (2002) The role of quantitative PCR for the immune monitoring of cancer patients. *Expert Opin Biol Ther* 5, 557–564.
8. Stordeur, P. (2005) Immune monitoring – Applications of real-time PCR, in *Encyclopedia of Medical Genomics and Proteomics* (Fuchs J. and Podda M., ed). Dekker encyclopedias, NJ, http://www.dekker.com/sdek/abstract~db=enc~content=a713596205
9. Li, B., Hartono, C., Ding, R., et al. (2001) Noninvasive diagnosis of renal-allograft rejection by measurement of messenger RNA for perforin and granzyme B in urine. *N Engl J Med* 344, 947–954.
10. Bustin, S. A. (2000) Absolute quantification of mRNA using real-time reverse transcription polymerase chain reaction assays. *J Mol Endocrinol* 25, 169–193.
11. Bustin, S. A. (2002) Quantification of mRNA using real-time reverse transcription PCR (RT-PCR): trends and problems. *J Mol Endocrinol* 29, 23–39.
12. Wong, M. L., Medrano, J. F. (2005) Real-time PCR for mRNA quantitation. *Biotechniques* 39, 75–85.
13. Overbergh, L., Valckx, D., Waer, M., et al. (1999) Quantification of murine cytokine mRNAs using real time quantitative reverse transcriptase PCR. *Cytokine* 11, 305–312.
14. Stordeur, P., Poulin, L. F., Craciun, L., et al. (2002) Cytokine mRNA quantification by real-time PCR. *J Immunol Meth* 259, 55–64. Erratum in (2002) *J Immunol Meth* 262: 229.
15. Stordeur, P., Zhou, L., Byl, B., et al. (2003) Immune monitoring in whole blood using real-time PCR. *J Immunol Meth* 276, 69–77.
16. Zhou, L., Toungouz, M., Donckier, V., et al. (2005) A rapid test to monitor alloreactive responses in whole blood using real-time polymerase chain reaction. *Transplantation* 80, 410–413.
17. Craciun, L., Stordeur, P., Troisi, R., et al. (2007) A rapid test of alloreactivity based on interleukin-2 mRNA expression might identify liver transplant recipients with donor-specific hyporesponsiveness. *Transplant Proc* 39, 2665–2667.

18. Ocmant, A., Michils, A., Schandene, L., et al. (2005) IL-4 and IL-13 mRNA real-time PCR quantification on whole blood to assess allergic response. *Cytokine* 31, 375–381.
19. Chomczynski, P., Sacchi, N. (1987) Single-step method of RNA isolation by acid guanidinium thiocyanate-phenol-chloroform extraction. *Anal Biochem* 162, 156–159.
20. Rozen, S., Skaletsky, H. (2000) WWW for general users and for biologist programmers, in *Bioinformatics Methods and Protocols: Methods in Molecular Biology* (Krawetz, S. and Misener, S. ed.), Humana, Totowa, NJ, pp. 365–386.
21. Dheda, K., Huggett, J. F., Bustin, S. A., et al. (2004) Validation of housekeeping genes for normalizing RNA expression in real-time PCR. *Biotechniques* 37, 112–119.

Index

Printed in the United States of America